30—

Image Processing and Pattern Recognition

Neural Network Systems Techniques and Applications

Edited by **Cornelius T. Leondes**

VOLUME 1. Algorithms and Architectures

VOLUME 2. Optimization Techniques

VOLUME 3. Implementation Techniques

VOLUME 4. Industrial and Manufacturing Systems

VOLUME 5. Image Processing and Pattern Recognition

VOLUME 6. Fuzzy Logic and Expert Systems Applications

VOLUME 7. Control and Dynamic Systems

Image Processing and Pattern Recognition

Edited by

Cornelius T. Leondes
Professor Emeritus
University of California
Los Angeles, California

VOLUME **5** OF

Neural Network Systems Techniques and Applications

ACADEMIC PRESS
San Diego London Boston New York Sydney Tokyo Toronto

This book is printed on acid-free paper. 

Copyright © 1998 by ACADEMIC PRESS

All Rights Reserved.
No part of this publication may be reproduced or transmitted in any form or by any means, electronic or mechanical, including photocopy, recording, or any information storage and retrieval system, without permission in writing from the publisher.

Academic Press
a division of Harcourt Brace & Company
525 B Street, Suite 1900, San Diego, California 92101-4495, USA
http://www.apnet.com

Academic Press Limited
24-28 Oval Road, London NW1 7DX, UK
http://www.hbuk.co.uk/ap/

Library of Congress Card Catalog Number: 97-80441

International Standard Book Number: 0-12-443865-2

PRINTED IN THE UNITED STATES OF AMERICA
97 98 99 00 01 02 ML 9 8 7 6 5 4 3 2 1

Contents

Comparison of Statistical and Neural Classifiers and Their Applications to Optical Character Recognition and Speech Classification

Ethem Alpaydın and Fikret Gürgen

Medical Imaging

Ying Sun and Reza Nekovei

Paper Currency Recognition

Fumiaki Takeda and Sigeru Omatu

Neural Network Classification Reliability: Problems and Applications

Luigi P. Cordella, Carlo Sansone, Francesco Tortorella, Mario Vento, and Claudio De Stefano

Parallel Analog Image Processing: Solving Regularization Problems with Architecture Inspired by the Vertebrate Retinal Circuit

Tetsuya Yagi, Haruo Kobayashi, and Takashi Matsumoto

Algorithmic Techniques and Their Applications

Rudy Setiono

Learning Algorithms and Applications of Principal Component Analysis

Liang-Hwa Chen and Shyang Chang

Learning Evaluation and Pruning Techniques

Leda Villalobos and Francis L. Merat

Contributors

Numbers in parentheses indicate the pages on which the authors' contributions begin.

Ethem Alpaydın (61), Department of Computer Engineering, Boğaziçi University, TR-80815 Istanbul, Turkey

Shyang Chang (321), Department of Electrical Engineering, National Tsing Hua University, Hsin Chu, Taiwan, Republic of China

Liang-Hwa Chen (321), Applied Research Laboratory, Telecommunication Laboratories, Chunghwa Telecom Co., Ltd., 12, Lane 551, Min-Tsu Road, Sec. 3, Yang-Mei, Taoyuan, Taiwan, Republic of China

Luigi P. Cordella (161), Dipartimento di Informatica e Sistemistica, Università degli Studi di Napoli "Federico II," Via Claudio, 21, I-80125 Napoli, Italy

Claudio De Stefano (161), Facoltà di Ingegneria di Benevento, Dipartimento di Ingegneria dell'Informazione ed Ingegneria Elettrica, Università degli Studi di Salerno, Piazza Roma, palazzo Bosco Lucarelli, I-82100 Benevento, Italy

Fikret Gürgen (61), Department of Computer Engineering, Boğaziçi University, TR-80815 Istanbul, Turkey

Haruo Kobayashi (201), Department of Electronic Engineering, Gumma University, 1-5-1 Tenjin-cho, Kiryu 376, Japan

Jorma Laaksonen (1), Laboratory of Computer and Information Science, Helsinki University of Technology, FIN-02150 Espoo, Finland

Jouko Lampinen (1), Laboratory of Computational Engineering, Helsinki University of Technology, FIN-02150 Espoo, Finland

Takashi Matsumoto (201), Department of Electrical, Electronics and Computer Engineering, Waseda University, Tokyo 169, Japan

Francis L. Merat (353), Electrical Engineering Department, Case Western Reserve University, Cleveland, Ohio 44106-7221

Reza Nekovei (89), Remote Sensing Laboratory, University of Rhode Island, Bay Campus, Narragansett, Rhode Island 02882

Erkki Oja (1), Laboratory of Computer and Information Science, Helsinki University of Technology, FIN-02150 Espoo, Finland

Sigeru Omatu (133), Department of Computer and Systems Sciences, College of Engineering, Osaka Prefecture University, Sakai, Osaka 593, Japan

Carlo Sansone (161), Dipartimento di Informatica e Sistemistica, Università degli Studi di Napoli "Federico II," Via Claudio, 21, I-80125 Napoli, Italy

Rudy Setiono (287), Department of Information Systems and Computer Science, National University of Singapore, Kent Ridge, Singapore 119260, Republic of Singapore

Ying Sun (89), Department of Electrical and Computer Engineering, University of Rhode Island, Kingston, Rhode Island 02881

Fumiaki Takeda (133), Technological Development Department, GLORY Ltd., 3-1, Shimoteno, 1-Chome, Himeji, Hyogo 670, Japan

Francesco Tortorella (161), Dipartimento di Informatica e Sistemistica, Università degli Studi di Napoli "Federico II," Via Claudio, 21, I-80125 Napoli, Italy

Mario Vento (161), Dipartimento di Informatica e Sistemistica, Università degli Studi di Napoli "Federico II," Via Claudio, 21, I-80125 Napoli, Italy

Leda Villalobos (353), Engineering School, University of Texas at El Paso, El Paso, Texas 79968-0521

Tetsuya Yagi (201), Kyushu Institute of Technology, 680-4 Kawazu, Iizuka-shi, Fukuoka Prefecture, 820 Japan

Preface

Inspired by the structure of the human brain, artificial neural networks have been widely applied to fields such as pattern recognition, optimization, coding, control, etc., because of their ability to solve cumbersome or intractable problems by learning directly from data. An artificial neural network usually consists of a large number of simple processing units, i.e., neurons, via mutual interconnection. It learns to solve problems by adequately adjusting the strength of the interconnections according to input data. Moreover, the neural network adapts easily to new environments by learning, and can deal with information that is noisy, inconsistent, vague, or probabilistic. These features have motivated extensive research and developments in artificial neural networks. This volume is probably the first rather comprehensive treatment devoted to the broad areas of algorithms and architectures for the realization of neural network systems. Techniques and diverse methods in numerous areas of this broad subject are presented. In addition, various major neural network structures for achieving effective systems are presented and illustrated by examples in all cases. Numerous other techniques and subjects related to this broadly significant area are treated.

The remarkable breadth and depth of the advances in neural network systems with their many substantive applications, both realized and yet to be realized, make it quite evident that adequate treatment of this broad area requires a number of distinctly titled but well-integrated volumes. This is the fifth of seven volumes on the subject of neural network systems and it is entitled *Image Processing and Pattern Recognition*. The entire set of seven volumes contains

Volume 1: *Algorithms and Architectures*
Volume 2: *Optimization Techniques*
Volume 3: *Implementation Techniques*
Volume 4: *Industrial and Manufacturing Systems*
Volume 5: *Image Processing and Pattern Recognition*
Volume 6: *Fuzzy Logic and Expert Systems Applications*
Volume 7: *Control and Dynamic Systems*

The first contribution to this volume is "Pattern Recognition," by Jouko Lampinen, Jorma Laaksonen, and Erkki Oja. Pattern recognition (PR) is the science and art of giving names to the natural objects in the real world. It is often considered part of *artificial intelligence*. However, the problem here is even more challenging because the observations are not in symbolic form and often contain much variability and noise. Another term for PR is *artificial perception*. Typical inputs to a PR system are images or sound signals, out of which the relevant objects have to be found and identified. The PR solution involves many stages such as making the measurements, preprocessing and segmentation, finding a suitable numerical representation for the objects we are interested in, and finally classifying them based on these representations. Presently, there are a growing number of applications for pattern recognition. A leading motive from the very start of the field has been to develop user-friendly and flexible user interfaces that understand speech and handwriting. Only recently have these goals become possible with the highly increased computing power of workstations. Document processing is emerging as a major application. In industrial problems as well as in biomedicine, automatic analysis of images and signals can be achieved with PR techniques. Remote sensing is routinely using automated recognition techniques, too. This contribution is a rather comprehensive presentation of the techniques and methods of neural network systems in pattern recognition. Several substantive examples are included. It is also worth noting as a valuable feature of this contribution that almost 200 references, which have been selectively culled from the literature, are included in the reference list.

The next contribution is "Comparison of Statistical and Neural Classifiers and Their Applications to Optical Character Recognition and Speech Classification," by Ethem Alpaydın and Fikret Gürgen. Improving person–machine communication leads to wider use of advanced information technologies. Toward this aim, character recognition and speech recognition are two applications whose automatization allows easier interaction with a computer. As they are the basic means of person-to-person communication, they are known by everyone and require no special training. Speech in particular is the most natural form of human communication and writing is the tool by which humanity has stored and transferred its knowledge for millennia. In a typical pattern recognition system, the first step is the acquisition of data. These raw data are preprocessed to suppress noise and normalize input. Features are those parts of the signal that carry information salient to its identity, and their extraction is an abstraction operation where the important information is extracted and the irrelevant is discarded. Classification is assignment of the input as an element of one of a set of predefined classes. The rules for classification are generally not known exactly and thus are estimated. A classifier is written as a parametric model whose parameters are computed using a given training sample to optimize particular error criterion. Approaches for classification differ in their assumptions about the model, in the way parameters are computed, or in the error criterion they optimize. This contribution treats what are probably the two principle approaches to classifiers as embodied by neural and statistical classifiers, and applies them to the major areas of optical character recognition and speech recognition. Illustrative examples are included as well as the literature for the two application categories.

The next contribution is "Medical Imaging," by Ying Sun and Reza Nekovei. The history of medical imaging began a century ago. The landmark discovery of X-rays by Wilhelm Conrad Röntgen in 1895 ushered in the development of noninvasive methods for visualization of internal organs. The birth of the digital computer in 1946 brought medical imaging into a new era of computer-assisted imagery. During the second half of the 20th century, medical imaging technologies have diversified and advanced at an accelerating rate. Today, clinical diagnostics rely heavily on the various medical imaging systems. In addition to conventional X-ray radiography, computer-assisted tomography and magnetic resonance imaging produce two-dimensional cross sections and three-dimensional imagery of the internal organs that drastically improve our capability to diagnose various diseases. X-ray angiography used in cardiac catheterization laboratories allows us to detect stenoses in the coronary arteries and guide treatment procedures such as balloon angioplasty and cardiac ablation. Ultrasonography has become a routine procedure for fetal examination. Two-fetal dimensional echocardiography combined with color Doppler flow imaging has emerged as a powerful and convenient tool for diagnosing heart valve abnormalities and for assessing cardiac functions. In the area of nuclear medicine, the scintillation gamma camera provides two-dimensional images of pharmaceuticals labeled by radioactive isotopes. Single photon emission computed tomography and positron emission tomography further allow for three-dimensional imaging of radioactive tracers. This contribution is a rather in-depth treatment of the important role neural network system techniques can play in the greatly significant area of medical imaging systems. Two major application areas are treated, i.e., detection of blood vessels in angiograms and image segmentation.

The next contribution is "Paper Currency Recognition," by Fumiaki Takeda and Sigeru Omatu. Three core techniques are presented. The first is the small size neurorecognition technique using masks. The second is the mask determination technique using the genetic algorithm. The third is the neurorecognition board technique using the digital signal processor.

Unification of these three techniques demonstrates that realization of neurorecognition machines capable of transacting various kinds of paper currency is feasible. The neurosystem technique enables acceleration in the commercialization of a new type of banking machine in a short period and in a few trials. Furthermore, this technique will be effective for various kinds of recognition applications owing to its high recognition ability, high speed transaction, short developing period, and reasonable cost. It can be presumed that it is so effective that it applies not only to paper currency and coins, but also to handwritten symbols such as electron systems or questionnaires.

The next contribution is "Neural Network Classification Reliability: Problems and Applications," by Luigi P. Cordella, Carlo Sansone, Francesco Tortorella, and Claudio De Stefano. Classification is a process according to which an entity is attributed to one of a finite set of classes or, in other words, it is recognized as belonging to a set of equal or similar entities, possibly identified by a name. In the framework of signal and image analysis, this process is generally considered part of a more complex process referred to as pattern recognition. In its simplest and still most commonly followed approach, a pattern recognition system is made of two distinct parts:

1. A description unit, whose input is the entity to be recognized, represented in a form depending on its nature, and whose output is generally a structured set of quantities, called features, which constitutes a description characterizing the input sample. A description unit implements a description scheme.
2. A classification unit, whose input is the output of the description unit and whose output is the assignment to a recognition class.

This contribution is a rather comprehensive treatment of pattern recognition in the classification problem by means of neural network systems. The techniques presented are illustrated by their application to two problem areas of major significance, i.e., handwritten character recognition and fault detection and isolation.

The next contribution is "Parallel Analog Image Processing: Solving Regularization Problems with Architecture Inspired by the Vertebrate Retinal Circuit," by Tetsuya Yagi, Haruo Kobayashi, and Takashi Matsumoto. Almost all digital image processors employ the same architecture for the sensor interface and data processing. A camera reads out the sensed image in a raster scan-out of pixels, and the pixels are serially digitized and stored in a frame buffer. The digital processor then reads the buffer serially or as blocks to smooth the noise in the acquired image, enhance the edges, and perhaps normalize it in other ways for pattern matching and object recognition. There have been several attempts in recent years to implement these functions in the analog domain, to attain low-power dissipation and compact hardware, or simply to construct an electrical model of these functions as they are found in biological systems. Analog implementations must have their performance evaluated in comparison with their digital counterparts, and systematic techniques for their design and implementation are evaluated therefrom. This contribution presents methods for the development of image processing parallel analog chips based on a class of parallel image processing algorithms. The architecture for these chips is motivated by physiological findings in lower vertebrates. The various aspects involved in this process are presented in an in-depth treatment, and illustrative examples are presented which clearly manifest the substantive effectiveness of the techniques presented.

The next contribution is "Algorithmic Techniques and Their Applications," by Rudy Setiono. Pattern recognition is an area where neural networks have been widely applied with much success. The network of choice for pattern recognition is a multilayered feedforward network trained by a variant of the gradient descent method known as the backpropagation learning algorithm. As more applications of these networks are found, the shortcomings of the backpropagation network become apparent. Two drawbacks often mentioned are the need to determine the architecture of a network before training can begin and the inefficiency of the backpropagation learning algorithm. Without proper guidelines on how to select an appropriate network for a particular problem, the architecture of the network is usually determined by trial-and-error adjustments of the number of hidden layers and/or hidden units. The backpropagation algorithm involves two parameters: the learning rate and the momentum rate. The values of these parameters have a significant effect on the efficiency of the learning process. However, there have been no clear guidelines for selecting their optimal values. Regardless of the values of the parameters, the backpropagation method is generally slow to converge and prone to get trapped at a local minimum of the error function. When designing a neural network system, the choice of a learning algorithm for training the network is crucial. As problems become more complex, larger networks are needed and the speed of training becomes critical. Instead of the gradient descent method, more sophisticated methods with faster convergence rate can be used to speed up network training. This contribution describes a variant of the quasi-Newtonian that can be used to reduce the network training time significantly. The substantively effective techniques presented in this contribution can be applied to a diverse array of significant problems, and several examples are included here. These are applications to the well-known spiral problem (described in this contribution), the multidisciplinary field of data mining in which it is desired to discover important patterns of interest that are hidden in databases, and the utilization of a neural network system as a means of distinguishing between benign and malignant samples in a breast cancer data set.

The next contribution is "Learning Algorithms and Applications of Principal Component Analysis," by Liang-Hwa Chen and Shyang Chang. The principal component analysis (PCA) learning network is one of a number of types of unsupervised learning networks. It is also a single layer neural network but the neurons are linear as described in this contribution. The learning is essentially based on the Hebb rule. It is utilized to perform PCA, i.e., to find the principle components embedded in the input data. PCA is one of the feature extraction methods, of which this contribution is a rather comprehensive treatment. Illustrative examples are included which demonstrate the substantive effectiveness of PCA (coupled with adaptive learning algorithms) to such problems as data compression, image coding, texture segmentation, and other significant applications.

The final contribution to this volume is "Learning Evaluation and Pruning Techniques," by Leda Villalobos and Francis L. Merat. In neural network system pruning, the process is initiated with a neural network system architecture that is larger than the minimum needed for learning. Such a neural network system architecture is then progressively reduced by pruning or weakening neurons and synaptic weights. This contribution is a rather comprehensive treatment of neural network system pruning techniques and their many significant applications. Not the least of the many applications noted in this contribution is that of evaluation and improvement of feature space in pattern recognition problems. Improving feature space quality has an unmeasurable value: a pattern recognition problem cannot be solved without good feature representation.

This volume on neural network systems techniques in image processing and pattern recognition systems clearly reveals the effectiveness and essential significance of the techniques available and, with further development, the essential role they will play in the future. The authors are all to be highly commended for their splendid contributions to this volume which will provide a significant and unique reference source for students, research workers, practitioners, computer scientists, and others on the international scene for years to come.

Cornelius T. Leondes

Pattern Recognition

Jouko Lampinen
Laboratory of Computational Engineering
Helsinki University of Technology
FIN-02150 Espoo, Finland

Jorma Laaksonen
Laboratory of Computer and Information Science
Helsinki University of Technology
FIN-02150 Espoo, Finland

Erkki Oja
Laboratory of Computer and Information Science
Helsinki University of Technology
FIN-02150 Espoo, Finland

I. INTRODUCTION

Pattern recognition (PR) is the science and art of giving names to the natural objects in the real world. It is often considered part of *artificial intelligence*. However, the problem here is even more challenging because the observations are not in symbolic form and often contain much variability and noise: another term for PR is *artificial perception*. Typical inputs to a PR system are images or sound signals, out of which the relevant objects have to be found and identified. The PR solution involves many stages such as making the measurements, preprocessing and segmentation, finding a suitable numerical representation for the objects we are interested in, and finally classifying them based on these representations.

Presently, there are a growing number of applications for pattern recognition. A leading motif from the very start of the field has been to develop user-friendly and flexible user interfaces, that would understand speech and handwriting. Only recently these goals have become possible with the highly increased computing power of workstations. Document processing is emerging as a major application. In industrial problems as well as in biomedicine, automatic analysis of images and

Image Processing and Pattern Recognition
Copyright © 1998 by Academic Press. All rights of reproduction in any form reserved.

signals can be achieved with PR techniques. Remote sensing is routinely using automated recognition techniques, too.

A central characteristic of the PR problem is that the number of different targets or objects that the system has to cope with is at least in principle unlimited, due to the variations caused, e.g., by viewing angles and illumination. Thus the problem cannot be solved by straightforward matching or data base searches. Still, the number of classes is finite and often relatively small. Each object has to be classified to one of the classes. The system is designed based on a sample of typical objects representing the different classes, and after this it must be able to classify also new, unknown objects with minimum error. This is often called *generalization*. A feasible design approach is to use some kind of model fitting or tuning based on the design set; traditionally, this has been called *learning*. Various adaptive and machine learning approaches have been popular in the PR system design problem.

Artificial neural networks (ANNs) are a class of flexible semiparametric models for which efficient learning algorithms have been developed over the years. They have been extensively used on PR problems. Even though realistic systems for such hard PR problems such as computer vision are hybrids of many methodologies including signal processing, classification, and relational matching, it seems that neural networks can be used to an advantage in certain subproblems, especially in feature extraction and classification. These are also problems amenable to statistical techniques, because the data representations are real vectors of measurements or feature values, and it is possible to collect training samples on which regression analysis or density estimation become feasible. Thus, in many cases neural techniques and statistical techniques are seen as alternatives. This approach has led on one hand to a fruitful analysis of existing neural networks, and on the other hand brought new viewpoints to current statistical methods, and sometimes produced a useful synthesis of the two fields. Recently, many benchmark and comparison studies have been published on neural and statistical classifiers [1–6]. One of the most extensive was the Statlog project [5] in which statistical methods, machine learning, and neural networks were compared using a large number of different data sets.

The purpose of the present review study is to discuss the ways in which neural networks can enter the PR problem and how they might be useful compared to other approaches. Comparisons are made both from an analytical and a practical point of view. Some illuminating examples are covered in detail. The contents of the subsequent sections are as follows: In Section II, we introduce the PR problem and show the general solution as a sequence of consequent, mutually optimized stages. The two stages in which neural networks seem to be the most useful are feature extraction and classification, and these will be covered in Sections III and IV. Then in Section V, applications will be explained, and Section VI presents some conclusions. An extensive publication list is given at the end of this chapter.

II. PATTERN RECOGNITION PROBLEM

This section presents an introduction to divergent aspects of pattern recognition. The operation of a pattern recognition system is presented as a series of consecutive processing stages. The functions of all these stages are elaborated, even though only few of them may actually be neural. The term pattern recognition can be defined in many ways, including the following [7]. **Pattern recognition** is an information-reduction process: the assignment of visual or logical patterns to classes based on the features of these patterns and their relationships.

The basic setting of pattern recognition is as follows. There is one unknown object presented as a set of signals or measurements in the input of a black box called a pattern recognition system. At the output of the system, there is a set of predefined classes. The purpose of the system is to assign the object to one of the classes. In a more general setting, there is more than one object to be recognized. In that case, the classification of the subsequent or nearby objects may or may not be interdependent. The list of classes may also contain a special reject class for the objects the system is unable to classify.

Depending on the measurements and the classes we are led to divergent areas of pattern recognition, including recognition of speech or speaker, detection of clinical malformations in medical images or time-signals, document analysis and recognition, etc. All these disciplines call for expertise in both the subject matter and the general theory and practice of pattern recognition. There exists an extensive amount of literature on both overall and specific questions of pattern recognition systems and applications. The classical textbook sources include, in the order of appearance, [8–17], some of which are actually revised versions of earlier editions. Recent developments—such as use of neural methods—are contained in such books as [18–23]. During the past thirty years, many valuable article collections have been edited in the field, including [24–28].

Technical systems are often considered as being comprised of consecutive blocks each performing its precisely defined task in the processing. The whole system can then be modeled in a bottom-up fashion as a block diagram. In the simplest case the flow of the data stream is one-directionally from left to right as shown in Fig. 1, presenting a general pattern recognition system. The diagram shown is naturally only one intuition of how to depict a view, and alternative structures can be seen, e.g., in [15, 17, 20, 29].

The following subsections shortly describe each of the stages with examples emanating principally from optical character recognition and speech recognition.

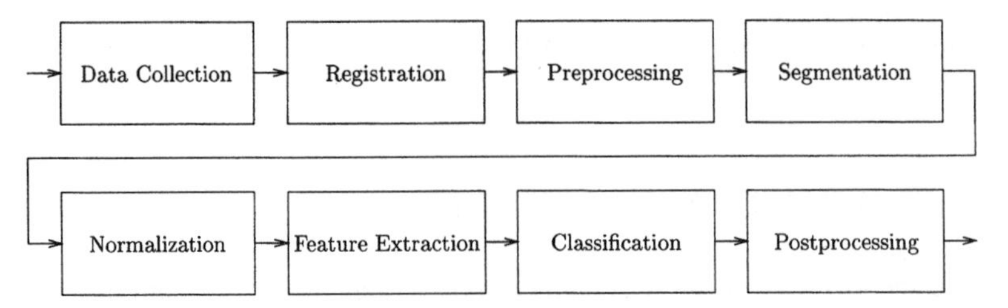

Figure 1 A block diagram of a generic pattern recognition system.

Some of the described stages may thus be obsolete or obscure in other types of pattern recognition systems.

A. DATA COLLECTION

The first stage in any pattern recognition system is data collection. Before a pattern vector is made up of a set of measurements, these measurements need to be performed using some technical equipment and converted to numerical form. In the case of image analysis or character recognition, such equipment includes video cameras and scanners; in the case of speech recognition, microphones, etc. The input data, whatever its form is, is sampled at fixed intervals in time or image metric domain and digitized to be presented with a preset number of bits per measurement. In any case, the data collection devices should record the objects with the highest fidelity available. Any additional noise will be disadvantageous to successful operation of the system. The data collection phase should also be designed in such a manner that the system will be robust to variations in operation of individual signal measurement devices.

The data collection stage possibly includes auxiliary storage for the collected data. The use of temporary storage is inevitable, if the recognition phase cannot be performed simultaneously with the data acquisition. More permanent data storage is needed for training material while a pattern recognition system is being constructed or tested. In some occasions, the amount of data storage needed may turn out to be a prohibitive factor in the development or use of an automated pattern recognition system. This discrepancy can be somewhat eased by compressing the stored data, but in the worst case, the fidelity of the data has to be sacrificed for the sake of storage shortage. This sacrifice is most often performed by reducing the spatial or temporal resolution of the data sampling or by presenting the measurements with a degraded accuracy using fewer bits per sample. Similar problems and solutions arise if the channel used in transferring the data is a bottleneck for the requirements of on-line processing.

B. REGISTRATION

In the registration of data, rudimentary model fitting is performed. The internal coordinates of the recognition system are somehow fixed to the actual data acquired. At least some *a priori* knowledge about the world surrounding the system is utilized in designing the registration stage. This external information mainly answers questions such as: How has the data been produced? Where or when does the sensible input begin and end? The registration process thus defines the framework in which the system operates so that it knows what to expect as valid input.

In speech recognition, the registration phase consists of ignoring epochs during which input is comprised of pure noise only and locating the beginnings and ends of utterances. In optical character recognition, the system must locate in the input image the area of interest. In the case of fill-in forms the area may be registered with some special printed marks, but in document analysis the system has to locate it automatically, based upon the overall layout of the page image.

C. PREPROCESSING

Real-world input data always contains some amount of noise and certain preprocessing is needed to reduce its effect. The term noise is to be understood broadly: anything that hinders a pattern recognition system in fulfilling its commission may be regarded as noise no matter how inherent this "noise" is in the nature of the data. Some desirable properties of the data may also be enhanced with preprocessing before the data is fed further in the recognition system.

Preprocessing is normally accomplished by some simple filtering method on the data. In the case of speech recognition, this may mean linear high-pass filtering aimed to remove the base frequency and to enhance the higher frequencies. In image recognition, the image may be median filtered to remove spurious point noise which might hamper the segmentation process. An advantageous preprocessing step for color images is decorrelation of the color components. Such a process transfers an image originally in the RGB (red-green-blue) coordinates linearly to the YIQ (luminosity-inphase-quadrature) system.

D. SEGMENTATION

The registered and preprocessed input data has to be split in subparts which make meaningful entities for classification. This stage of processing is called segmentation. It may either be a clearly separate process or tightly interwoven with previous or following processes. In either case, after the pattern recognition system has completed the processing of a totality of data, the resulting segmentation of the data to its subparts can be revealed. Depending on how the application has been realized, the segmentation block may either add the information regarding the segment boundaries to the data flow, or alternatively, copy all the segments in separate buffers and hand them over to the following stage one by one.

In speech recognition, a meaningful entity is most likely a single phoneme or a syllable containing a small but varying number of phonemes. In optical character recognition, the basic units for classification are single characters or some of the few composite glyphs such as fi and fl.

Some pattern recognition applications would be described better if, in Fig. 1, segmentation were placed after the classification stage. In such systems, the input data is partitioned with fixed-sized windows at fixed spatial or temporal intervals. The actual segmentation can take place only after the subparts have been labeled in the classification stage.

E. Normalization

A profound common characteristic of the environments where automated pattern recognition systems are used is the inherent variance of the objects to be recognized. Without this variance the pattern recognition problem would not exist at all. Instead, we would be concerned with deterministic algorithms such as those for sorting, searching, computer language compiling, Fourier transform, etc. The central question in pattern recognition, therefore, is how these variances can be accounted for. One possibility is to use feature extraction or classification algorithms which are *invariant* to variations in the outcomes of objects. For example, image features that are invariant to rotation are easy to define, but some types of natural variance will inevitably always evade the invariant feature extraction. Therefore, a separate normalization step is called for in almost all pattern recognition systems.

Normalization always causes as a side effect loss of degrees of freedom. This is reflected as dimension reduction in the intrinsic dimensionality of the data. If the normalization could be done ideally, only the dimensionality increase caused by the noise would be canceled out. This is unfortunately not true, but as will be explained in the following section, the dimensionality of the data has to be anyhow reduced. Insignificant loss in intrinsic dimensionality of the data during the otherwise beneficial normalization process is therefore not a serious problem.

For example, depending on individual habits, our handwriting is not straight upwards but somewhat slanted to left or right. Normalized characters can be achieved by estimating the slant and reverting it. In speech recognition, the loudness of speech can be normalized to a constant level by calculating the energy of an utterance and then scaling the waveform accordingly.

F. Feature Extraction

The meaning of the feature extraction phase is most conveniently defined referring to the purpose it serves [14]: feature extraction problem ... is that of extracting from the raw data the information which is most relevant for classification purposes, in the sense of minimizing the *within-class* pattern variability while enhancing the *between-class* pattern variability.

During the feature extraction process the dimensionality of data is reduced. This is almost always necessary, due to the technical limits in memory and computation time. A good feature extraction scheme should maintain and enhance those features of the input data which make distinct pattern classes separate from each other. At the same time, the system should be immune to variations produced both by the humans using it and the technical devices used in the data acquisition stage.

Besides savings in memory and time consumptions, there exists another important reason for proper dimensionality reduction in the feature extraction phase. It is due to the phenomenon known as the *curse of dimensionality* [30], that increasing the dimensionality of the feature space first enhances the classification accuracy but rapidly leads to sparseness of the training data and poor representation of the vector densities, thereby decreasing classification performance. This happens even though the amount of information present in data is enriched while its dimensionality is increased. The curse thus forces the system designer to balance between the amount of information preserved as the dimensionality of the data, and the amount of density information available as the number of training samples per unit cube in the feature vector space. A classical rule of thumb says that the number of training samples per class should be at least 5–10 times the dimensionality [31].

An issue connected to feature extraction is the *choice of metric*. The variances of individual features may vary orders of magnitude, which inevitably impairs the classifier. The situation can be eased by applying a suitable linear transform to the components of the feature vector.

In speech recognition, the features are most often based on first assuming momentary stability of the waveform. In that case spectral, cepstral, or linear prediction coefficients can be used as descriptive features. The diverse possibilities for feature extraction in recognition of handwritten characters include features calculated from the outline of the character, the distribution of mass and direction in the character area, etc. Neural networks provide some ways for dimensionality reduction and feature extraction. The connection of neural networks to feature extraction will be covered in depth in Section III.

G. Classification and Clustering

In addition to feature extraction, the most crucial step in the process of pattern recognition is classification. All the preceding stages should be designed and tuned aiming at success in the classification phase. The operation of the classification step can be simplified as being that of a transform of quantitative input data to qualitative output information. The output of the classifier may either be a discrete selection of one of the predefined classes, or a real-valued vector expressing the likelihood values for the assumptions that the pattern was originated from the corresponding class.

The primary division of the various classification algorithms used is that between *syntactic* and *statistical* methods. The statistical methods and neural networks are related in the sense that the same features can be used with both. Due to the centrality of classification methods to this text, they are not covered in this introductory section but analyzed in full depth in Section IV.

A topic closely related to classification is *clustering*. In clustering, either the existence of predefined pattern classes is not assumed, the actual number of classes is unknown, or the class memberships of the vectors are generally unknown. The task of the clustering process is therefore to group the feature vectors to clusters in which the resemblance of the patterns is stronger than between the clusters [32]. The processing blocks surrounding the classification stage in Fig. 1 are generally also applicable to clustering problems.

H. Postprocessing

In most pattern recognition systems, some data processing is performed also after the classification stage. These postprocessing subroutines, like the normalization processes, bring some *a priori* information about the surrounding world into the system. This additional expertise can be utilized in improving the overall classification accuracy. A complete postprocessing block may itself be a hybrid of successive or cooperative entities. In the context of this representation it however suffices to regard the postprocessor as an atomic operator.

The postprocessing phase is generally possible if the individual objects or segments make up meaningful entities such as bank account numbers, words, or sentences. The soundness or existence of these higher-level objects can be examined and if an error is indicated, further steps can be taken to correct the misclassification. The postprocessing phase thus resolves interdependencies between individual classifications. This is possible either by the operation of the postprocessing stage alone, or in cooperation with the segmentation and classification blocks as will be explained in the following section.

I. Loop-Backs between Stages

In Fig. 1, a block diagram of an idealized pattern recognition application was depicted. Such systems, in which the data flows exclusively from left to right, can hardly ever be optimal in the sense of recognition accuracy. By making the successive blocks interact, the overall performance of the system can be considerably enhanced. The system, of course, becomes much more complicated, but generally there is no other way to increase the classification accuracy.

Three possible routes for the backward links are drawn in Fig. 2 with dashed arrows and labeled (a), (b), and (c). The motivations behind these three configurations are:

(a) Information is fed back from postprocessing to classification. When the postprocessor detects an impossible or highly improbable combination of outputs from the classifier, it notifies the classifier. Either the postprocessor itself is able to correct the fault, or it asks the classifier for a new trial. In either case, the classifier ought to be able to revise its behavior and to not produce similar errors in the future. The classifier may also mediate this feedback information back to the segmentation block as will be explained below.

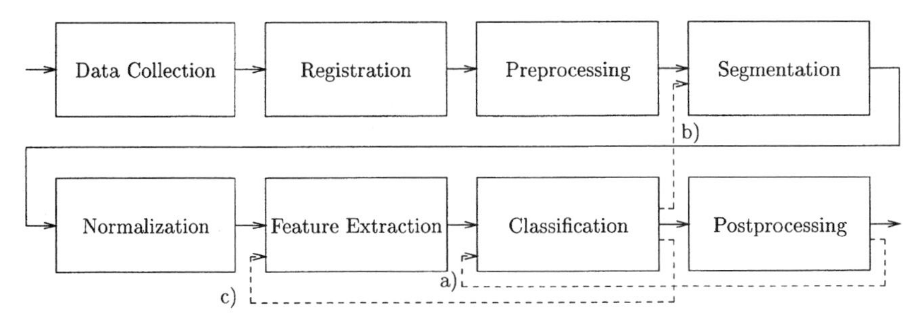

Figure 2 A block diagram of a pattern recognition system with some possible loop-back routes added.

(b) The classifier revises the segmentation phase. In this case, the classifier or the postprocessor has detected one or more successive patterns that are hard to classify. This might be an indication of malformed segmentation which should be located and corrected. This scheme can also be viewed as a segmentation algorithm probing the succeeding stages with tentative segments. It is then left for the classifier to select the most probable combination.

This view can also accommodate the possibility that segmentation is performed after classification. In this scheme, the data flows unmodified in its first pass through the segmentation block. When classification has taken place, the data is fed back to the segmenter and the actual segmentation is performed.

(c) The correctness of the classifications is used to revise the feature extractor. This kind of operation is mostly possible only during the training phase and generally necessitates the redesign of the classifier. This kind of scheme may be called *error-corrective feature extraction* [33].

J. Trainable Parts in a System

All the stages of a pattern recognition system contain parameters or variables which need to be given appropriate values. Some of these parameters are so delicate that they have to be selected by an expert of the application area and kept constant thereafter. Others may be tunable by trial and error or cross-validation processes in cooperation with an expert observing the overall performance of the system top-down. Profoundly more interesting are, however, parameters which the system is able to learn by itself from training with available data. Neural networks provide a whole new family of divergent formalisms for adaptive systems. Error-corrective neural training can be used in various parts of a pattern recognition system to improve the overall performance.

In most cases, the adaptive nature of the neural networks is only utilized during the training phase and the values of the free parameters are fixed at the end of it. A long-term goal, however, is to develop neural systems which retain their ability to adapt to slowly evolving changes in their operation environments. In such automata, the learning of the system would continue automatically and by itself endlessly. Evidently, the stability of such systems is more or less in doubt.

In many systems claimed to be neural, just a traditional classifier has been replaced by a neural solution. This is of course reasonable if it makes the system perform better. However, a more principled shift to bottom-up neural solution might be possible and called for. At least the normalization and feature extraction stages, together with classification, could be replaced with neural counterparts in many systems. Only then, the full potentiality of neural systems would be fulfilled.

III. NEURAL NETWORKS IN FEATURE EXTRACTION

A. FEATURE EXTRACTION PROBLEM

In real-world pattern recognition problems such as image analysis, the input dimensionality can be very high (of the order of hundreds) and the discriminant functions to be approximated are very nonlinear and complex. A classifier based on the measured objects (e.g., images) directly would require a large number of parameters in order to approximate and generalize well all over the input domain. Such a "black box" modeling approach is shown in Fig. 3. The central block could be a *supervised learning network*, such as the multilayer perceptron network, the radial basis function network, or the LVQ network. Together with their powerful training algorithms such as the error back-propagation, these networks provide highly efficient model-free methods to design nonlinear mappings or discriminant functions between inputs and outputs using a data base of training samples. Prominent examples are pattern recognition, optical character readers, industrial diagnostics, condition monitoring, modeling complex black box systems for control, and time series analysis and forecasting.

However, it is well known [34] that even neural networks cannot escape the parameter estimation problem, which means that the amount of training data must grow in proportion to the number of free parameters. Consequently, very large amounts of training data and training time are needed in highly complex and large-dimensional problems to form the input–output mappings [35]. Collecting the training samples would eventually be very expensive if not impossible. This seems to be a major limitation of the supervised learning paradigm. In conventional pattern recognition (see Section II), the answer is to divide the task in two parts: *feature extraction* which maps the original input patterns or images to a feature space of reduced dimensions and complexity, followed by *classification* in this space. This approach is shown in Fig. 4.

There is no well-developed theory for feature extraction; mostly features are very application oriented and often found by heuristic methods and interactive

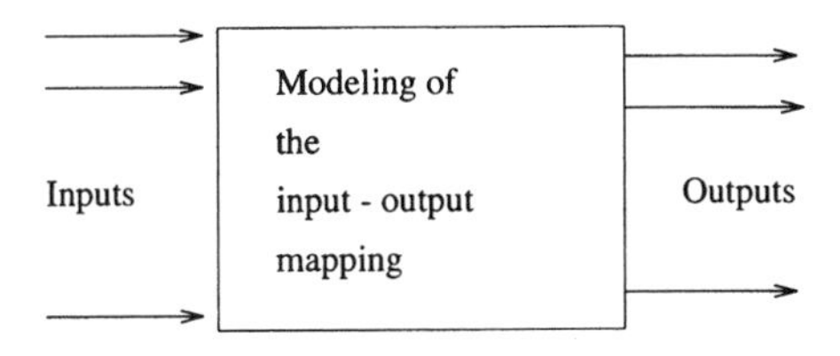

Figure 3 Black box modeling approach.

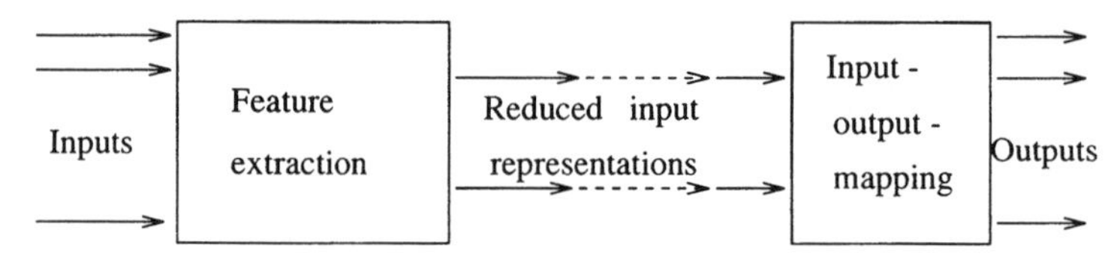

Figure 4 Feature extraction approach.

data analysis. It is not possible to give an overview of such interactive feature extraction methods; in any specific problem such as, e.g., character or speech recognition, there is an accumulated knowledge of the most feasible ways to extract the relevant information, and the reader is advised to look up review articles on the given application fields. Instead, some generic principles of neural-network-based feature extraction are reviewed here.

An important basic principle is that the feature extraction method should not depend on the class memberships of the objects, because by definition at the feature extraction stage these are not yet known. The same features are extracted from all the inputs, regardless of the target classes. It follows that if any learning methods are used for developing the feature extractors, they can be *unsupervised* in the sense that the target class for each object does not have to be known.

B. Two Classes of Unsupervised Neural Learning

Unsupervised learning algorithms are an important subclass of neural learning. The characteristic feature of unsupervised neural learning is that the training set only contains input samples. No desired outputs or target outputs are available at all. Basically, these algorithms fall into one of two categories [36]: first, extensions of the linear *transform coding* methods of statistics, especially principal component analysis, and second, learning *vector coding* methods that are based on competitive learning.

The first class of neural feature extraction and compression methods are motivated by standard statistical methods such as principal component analysis (PCA) or factor analysis (see, e.g., [37]), which give a reduced subset of linear combinations of the original input variables. Many of the neural models are based on the PCA neuron model introduced by one of the authors [38]. The additional advantage given by neural learning is that neural networks are nonlinear, and thus powerful nonlinear generalizations to linear compression can be obtained. Typically, the compressed representation thus obtained would be input to another neural net-

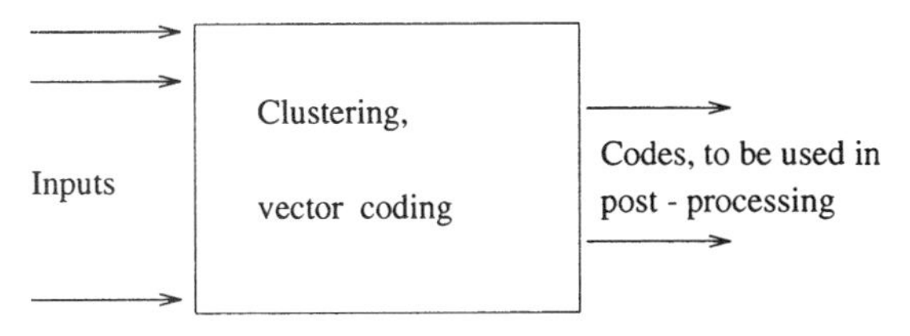

Figure 5 Clustering approach.

work working in the supervised mode, as shown in Fig. 4. These techniques will be covered in Sections III.C and III.D.

The second class of methods apply to cases when the entire problem to be solved is of the unsupervised nature: there are no target labels or values available at all. The results of the unsupervised neural network are used as such, as shown in Fig. 5. A typical application is *clustering* and *data compression*. It is of interest to find out what kind of typical clusters there are among the input measurement vectors. A competitive learning neural network gives an efficient solution to this problem. Section III.E reviews the best-known competitive learning network, the self-organizing map (SOM) introduced by Kohonen [39], and its use in massive data clustering.

This chapter is a review of the essential principles and theory underlying the two models of unsupervised learning, with some central references cited. It is not possible here to give even a rudimentary list of applications of these techniques. Instead, two large collections of references available on the Internet are cited: [40] and [41]. Together they give well over two thousand references to the use of unsupervised learning and feature extraction in neural networks.

C. Unsupervised Back-Propagation

In this section, it is shown that a powerful generalization of the linear principal component analysis method is given by a multilayer perceptron network that works in the *auto-associative* mode. To show this analogy, let us define some notation first. The overall input–output mapping formed by the network is $f\colon \mathbb{R}^d \to \mathbb{R}^d$, the input vector is $\mathbf{x} \in \mathbb{R}^d$, the output vector is $\mathbf{y} \in \mathbb{R}^d$, and there is a training sample $(\mathbf{x}_1, \ldots, \mathbf{x}_n)$ of inputs available. Let us require that the output is $\mathbf{x}$, too, i.e., $\mathbf{y} = f(\mathbf{x}) = \mathbf{x}$ for all $\mathbf{x}$. This mode of operation is called auto-associative, since the network is associating the inputs with themselves. Note that in back-propagation learning, the same training samples $\mathbf{x}_i$ are then used both as inputs and as desired outputs. Therefore, this is unsupervised learning.

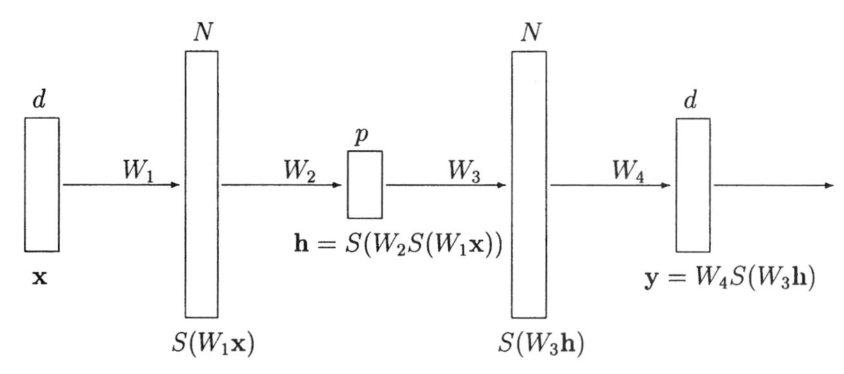

Figure 6 A five-layer network with linear output layer and three nonlinear hidden layers. The boxes denote layers; the number of units in each layer is given above the box, and the output vector of the layer is given under the box. The arrows give the transformations between the layers. The W_i are the weight matrices including the offsets, and S is the nonlinear neuron activation function.

To avoid the trivial solution, let us impose a constraint: the network has three or more layers, with the input and output layers having d units but one of the intermediate or hidden layers having a smaller number $p < d$ units [42–44]. This constraint means that the network has a *bottleneck layer*, giving the network the hourglass shape shown by the five-layer network in Fig. 6. Denoting the output vector of the bottleneck hidden layer by $\mathbf{h} \in \mathbb{R}^p$, the total mapping f from $\mathbf{x}$ to $\mathbf{y}$ breaks down to two parts: $\mathbf{h} = f_1(\mathbf{x}) = S(W_2 S(W_1 \mathbf{x}))$, $\mathbf{y} = f_2(\mathbf{h}) = W_4 S(W_3 \mathbf{h}))$. S is here a nonlinear scalar function, eventually a sigmoidal activation function. The expression $S(W^T \mathbf{x})$ is to be understood as a vector that is obtained from $W^T \mathbf{x}$ by applying the function S to each element of this vector separately.

In this network, the equality $f(\mathbf{x}) = \mathbf{x}$ cannot hold for all $\mathbf{x}$. Instead, we require that f must minimize the squared training set error

$$J_s(f) = \sum_{i=1}^{n} \|\mathbf{x}_i - f(\mathbf{x}_i)\|^2. \tag{1}$$

This is the standard cost function of MLPs and is minimized by back-propagation learning. It is a finite-sample estimate of

$$J_e(f) = E\{\|\mathbf{x} - f(\mathbf{x})\|^2\}. \tag{2}$$

Substituting the forms of the functions from Fig. 6 gives

$$J_e(f) = E\{\|\mathbf{x} - W_4 S(W_3 S(W_2 S(W_1 \mathbf{x})))\|^2\}. \tag{3}$$

It is now possible to interpret the function f_1 from the input vector $\mathbf{x}$ to the central hidden layer output vector $\mathbf{h}$ as the *feature extraction function*: the outputs $\mathbf{h}$ of the hidden layer can be interpreted as features. The data compression rate is adjusted by choosing the dimensions of $\mathbf{x}$ and $\mathbf{h}$, respectively d and p, and the faithfulness of the representation is measured by how well the original input $\mathbf{x}$ can be retrieved from the feature vector $\mathbf{h}$, i.e., by the criterion (1) or (3). If the criterion gets a low value, then obviously the network has been able to capture the information in $\mathbf{x}$ in a nonlinear function of reduced dimensionality. In theory, if such a compression is possible, then the multilayer nonlinear net of Fig. 6 can approximate it to an arbitrary accuracy, because of the well-known approximation properties of MLPs [45, 46]. The extra hidden layers of N units each are essential, because actually the two functions f_1 and f_2 must be represented and both require a hidden layer.

To operate this network after learning, a new input $\mathbf{x}$ is transformed to the compressed representation $\mathbf{h}$, which is then input to another postprocessing network according to Fig. 4. So, in most cases the last hidden layer and the output layer of the five-layer network are only used in learning and then discarded. A notable exception is data compression: then the first part of the net is used for compression and the second part is needed for decompression.

The network will now be shown to be a nonlinear generalization of principal components. The problem of PCA is to find a *linear* mapping W from the input vector $\mathbf{x} \in \mathbb{R}^d$ to the lower-dimensional feature vector $\mathbf{h} \in \mathbb{R}^p$ such that the information loss is minimal. The linear mapping can be represented by a matrix W: $\mathbf{h} = W^T\mathbf{x}$. There are several equivalent criteria for PCA [47], one of them being to minimize

$$J_{\text{PCA}}(W) = E\{\|\mathbf{x} - W\mathbf{h}\|^2\} = E\{\|\mathbf{x} - WW^T\mathbf{x}\|^2\}. \tag{4}$$

This means that a good approximation of $\mathbf{x}$ is obtained by applying the same matrix W to $\mathbf{h}$. The solution of Eq. (4) is that W will have orthonormal columns that span the same subspace as the p dominant eigenvectors of the data covariance matrix.

Comparing Eqs. (3) and (4) shows that the five-layer MLP network is indeed a nonlinear extension in the sense that the feature vector $\mathbf{h} = W^T\mathbf{x}$ of PCA is replaced by $\mathbf{h} = S(W_2S(W_1\mathbf{x}))$ in the MLP, and the reconstruction of $\mathbf{x}$, $W\mathbf{h}$, is replaced by $\mathbf{y} = W_4S(W_3\mathbf{h})$.

This is potentially a very powerful technique of nonlinear feature extraction and compression. Some demonstrations of the feature extraction ability of the five-layer MLP were given by [42] where a helix was faithfully mapped by the network, and by [48] who showed that image compression with lower error than PCA is possible using a large five-layer MLP network.

D. Nonlinear Principal Component Analysis

A problem in using the five-layer perceptron network is that it can be very large in practical applications. For example, when the inputs are 8×8 digital image windows and the hidden layers have moderate numbers of elements, the number of free weights will be in the thousands. Even for a 64-16-8-16-64 architecture, the number of weights is 2408. The training set size must be comparable, and the training times are very long.

A relevant question is whether similar improvements over the linear PCA technique could be obtained with a smaller network. A key property of a neural network is then its *nonlinearity*: a least-mean-square criterion involving nonlinear functions of input $\mathbf{x}$ means a deviation from the second-order statistics to higher orders which may have much more power in representing the relevant information. In general statistics, there is presently a strong trend to explore nonlinear methods, and neural networks are an ideal tool.

Starting from a simple linear neuron model proposed by the author in [38], that was able to learn the first principal component of the inputs using a constrained Hebbian learning rule, several linear and nonlinear extensions have been suggested over the years; for overviews, see [49] and [50]. The simplest extension of the linear PCA criterion (4) to a nonlinear one is

$$J_{\mathrm{nonl}}(W) = E\{\|\mathbf{x} - WS(W^T\mathbf{x})\|^2\}, \tag{5}$$

where S is again a nonlinear scalar function, eventually a sigmoidal activation function.

It was first shown in [51] that an associated learning rule minimizing Eq. (5) is

$$W_{k+1} = W_k + \gamma_k\big[\mathbf{x}_k S(\mathbf{x}_k^T W_k) - W_k S(W_k^T \mathbf{x}_k) S(\mathbf{x}_k^T W_k)\big]. \tag{6}$$

In learning, a set of training vectors $\{\mathbf{x}_k\}$ are input to the algorithm and W_k is updated at each step. The parameter γ_k is the usual learning rate of neural learning algorithms. After several epochs with the training set, the weight matrix W_k will converge to a "nonlinear PCA" weight matrix.

It has been shown recently by [52] and [53] that the nonlinear network is able to learn the separate components of inputs in the case when the input is an unknown weighted sum of independent source signals, a task that is not possible for the linear PCA technique. The neurons develop into feature detectors of the individual input components. However, to achieve this with the learning rule, a preliminary preprocessing is necessary that whitens or spheres the input vectors $\mathbf{x}_k$ in such a way that after sphering $E\{\mathbf{x}_k\mathbf{x}_k^T\} = I$. In signal processing, terms such as *independent component analysis* (ICA) or *blind source separation* (BSS) are used for this technique; some classical references are [54–56].

The algorithm (6) has an implementation in a one-layer network of nonlinear neurons with activation function S, that are learning by the constrained Hebbian learning principle. The first term in the update expression (6), $\mathbf{x}_k S(\mathbf{x}_k^T W_k)$, when taken element by element, is the product of the input to a neuron and the output of that neuron. The second term is a constraint term, forcing the weights to remain bounded. Preceded by a linear PCA neural layer that takes care of the input vector sphering, a two-layer ICA network is obtained [53]. Several applications of the ICA network in feature extraction have been reported in [57].

E. Data Clustering and Compression by the Self-Organizing Map

One of the best-known neural networks in the unsupervised category is the self-organizing map (SOM) introduced by Kohonen [39]. It belongs to the class of *vector coding* algorithms. In vector coding, the problem is to place a fixed number of vectors, called codewords, into the input space which is usually a high-dimensional real space $\mathbb{R}^d$. The input space is represented by a training set $(\mathbf{x}_1, \ldots, \mathbf{x}_n) \in \mathbb{R}^d$. For example, the inputs can be grayscale windows from a digital image, measurements from a machine or a chemical process, or financial data describing a company or a customer. The dimension d is determined by the problem and can be large.

Each codeword will correspond to and represent a part of the input space: the set of those points in the space which are closer in distance to that codeword than to any other codeword. Each such set is convex and its boundary consists of intersecting hyperplanes. This produces a so-called Voronoi tessellation into the space. The overall criterion in vector coding is to place the codewords in such a way that the average distances from the codewords to those input points belonging to their own Voronoi set are minimized. This is achieved by learning algorithms that are entirely data-driven and unsupervised.

Coding facilitates data compression and makes possible postprocessing using the discrete signal codes. Typically, the codewords are found to correspond to relevant clusters among the input training data, e.g., typical clusters of microfeatures in an image [35], and they can be efficiently used to cluster new inputs.

One way to understand the SOM [39, 58, 59] is to consider it as a neural network implementation of this basic idea: each codeword is the weight vector of a neural unit. However, there is an essential extra feature in the SOM. The neurons are arranged to a one-, two-, or multidimensional *lattice* such that each neuron has a set of neighbors; see Fig. 7. The goal of learning is not only to find the most representative code vectors for the input training set in the mean-square sense, but at the same time to realize a *topological mapping* from the input space to the grid of neurons. Mathematically, this can be defined as follows.

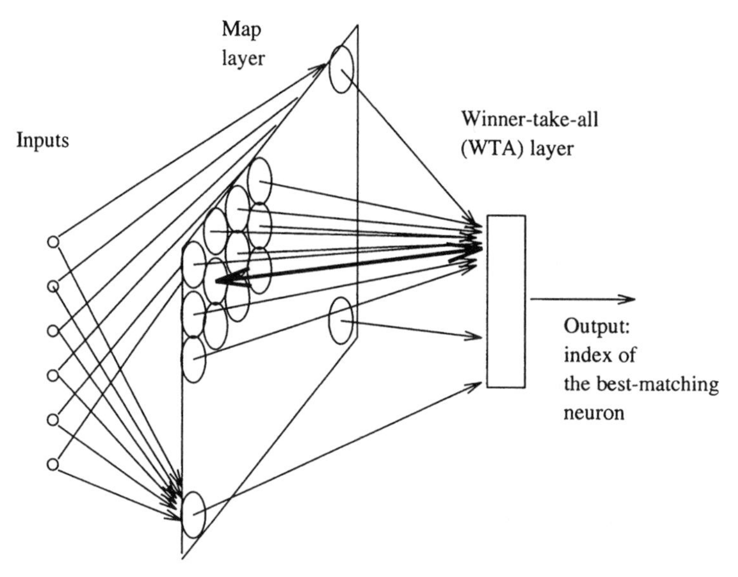

Figure 7 The SOM network. Each neuron in the map layer receives the same inputs. The best matching neuron (BMU) can be found by a Winner-take-all (WTA) layer which outputs its index. In learning, the BMU and its neighbors receive a learning signal from the WTA (only the signal to the BMU is shown by the thick arrow), telling them to update their weights.

For any data point $\mathbf{x}$ in the input space, one or several of the codewords are closest to it. Assume that $\mathbf{m}_i$ is the closest among all ℓ codewords:

$$\|\mathbf{x} - \mathbf{m}_i\| = \min\|\mathbf{x} - \mathbf{m}_j\|, \qquad j = 1, \ldots, \ell. \tag{7}$$

To make the correspondence unique, assume that the codeword with the smallest index is chosen if several codewords happen to be at exactly the minimum distance from $\mathbf{x}$. The unit i having the weight vector $\mathbf{m}_i$ is then called the *best-matching unit* (BMU) for vector $\mathbf{x}$, and index $i = i(\mathbf{x})$ can be considered as the *output* of the map. Note that for fixed $\mathbf{x}$, Eq. (7) defines the index i of the BMU, and for fixed i, it defines the Voronoi set of unit i as the set of points $\mathbf{x}$ that satisfy Eq. (7). By the above relation, the input space is mapped to the discrete set of neurons.

By a topological mapping the following property is meant: if an arbitrary point $\mathbf{x}$ is mapped to unit i, then all points in neighborhoods of $\mathbf{x}$ are mapped either to i itself or to one of the units in the neighborhood of i in the lattice. This implies that if i and j are two neighboring units on the lattice, then their Voronoi sets in the input space have a common boundary. Whether the topological property can hold for all units, however, depends on the dimensionalities of the input space and the neuron lattice: because no topological maps between two spaces of different dimensions can exist in the strict mathematical sense, a two-dimensional neural layer can only follow locally two dimensions of the multidimensional input space. Usually the input space has a much higher dimension, but the data cloud $(\mathbf{x}_1, \ldots, \mathbf{x}_n)$ used in training may be roughly concentrated on a lower-dimensional manifold that the map is able to follow at least approximately [60].

The fact that the mapping has a topological property has the advantage that it is more *error-tolerant*: a perturbation of the input $\mathbf{x}$ may cause the output $i(\mathbf{x})$ (the index of the BMU) to jump from the original unit to one of its neighbors, but usually not to an arbitrary position on the lattice, as would be the case if no neighborhood relation existed among the neurons. In a layered neural system in which the next layer "reads" the feature map but does not know the original inputs, such a property is essential to guarantee stable behavior.

The SOM network is shown as a feedforward network in Fig. 7. The role of the output "winner-take-all" layer is to compare the outputs from the map layer (equivalently, the distances $\|\mathbf{x} - \mathbf{m}_i\|$) and give out the index of the BMU. The SOM can be described without specifying the activation functions of the neurons; an equivalent network is obtained if the activation function is a radial basis function, hence the output of a neuron is a monotonically decreasing function of $\|\mathbf{x} - \mathbf{m}_i\|$.

The well-known Kohonen algorithm for self-organization of the code vectors is as follows [58]:

1. Choose initial values randomly for the weight vectors $\mathbf{m}_i$ of the units i.
2. Repeat Steps 3, 4 until the algorithm has converged.
3. Draw a sample $\mathbf{x}$ from the probability distribution of the input samples and find the best-matching unit i according to Eq. (7).
4. Adjust the weight vectors of all units by

$$\mathbf{m}_j := \mathbf{m}_j + \gamma h_r(\mathbf{x} - \mathbf{m}_j), \tag{8}$$

   where γ is a gain factor and h_r is a function of the distance $r = \|i - j\|$ of units i and j measured along the lattice.

(In the original version [39], the neighborhood function h_r was equal to 1 for a certain neighborhood of i, and 0 elsewhere. The neighborhood and the gain γ should slowly decrease in time.)

The convergence and the mathematical properties of this algorithm have been considered by several authors, e.g., [59] and [61].

The role of the SOM in feature extraction is to construct optimal codewords in abstract feature spaces. Individual feature values can then be replaced by these codes, which results in data compression. Furthermore, hierarchical systems can be built in which the outputs from the maps are again used as inputs to subsequent layers. The topological property of the feature maps is then essential for low-error performance [62].

In data clustering, the weight vectors of the SOM neurons develop into code vectors under unsupervised learning in which a representative training set of input vectors are used. The learning is slow, but it is an "off-line" operation. After the map has been formed, it can be used as such to code input vectors having similar statistical properties with the training vectors. Note that due to the unsupervised learning, the algorithm cannot give any semantic meanings to each unit, but this must be done by the user. The two-dimensional map is also a powerful tool for data visualization: e.g., a color code can be used in which each unit has its own characteristic color.

The unsupervised feature extraction scheme is especially suitable for general scene analysis in computer vision, since it is fairly inexpensive to collect large amounts of image data to be used in unsupervised training, as long as the images need no manual analysis and classification. One example is cloud classification from satellite images [63] in which even human experts have difficulties in giving class labels to cloud patches as seen by a weather satellite. The map can be used to cluster the patches, and after learning a human expert can go over the map and interpret what each unit is detecting.

A data base of well over two thousand applications of SOM is given by [40]. A recent review of the use of the SOM for various engineering tasks, including pattern recognition and robotics, is given by [64].

IV. CLASSIFICATION METHODS: STATISTICAL AND NEURAL

Numerous taxonomies for classification methods in pattern recognition have been presented. None has been so clearly more advantageous than the others that it would have gained uncontested status. The most profound dichotomy, however, is quite undisputed and goes between *statistical* and *syntactic* classifiers. The domain of this text is limited to the former, whereas the latter—also known as *linguistic* or *structural* approaches—is treated in many textbooks including [12, 15, 21].

The statistical alias *decision-theoretic* methods can further be divided in many ways depending on the properties one wants to emphasize. Opposing *parametric* and *nonparametric* methods is one often-used dichotomy. In parametric methods, a specific functional form is assumed for the feature vector densities, whereas nonparametric methods refer directly to the available exemplary data. Somewhere between these extremes, there are *semiparametric* methods which try to achieve the best of both worlds using a restricted number of adaptable parameters depending on the inherent complexity of the data [22].

One commonly stated (e.g., [20]) division goes between *neural* and classical *statistical* methods. It is useful only if one wants to regard these two approaches as totally disjoint competing alternatives. On the opposite extreme, neural methods have been seen only as iterative ways to arrive at the classical results of the traditional statistical methods (e.g., [23]). Better off, both methods can be described using common terms as was done by [65] and summarized in this text.

Neural methods may additionally be characterized by their learning process: *supervised* learning algorithms require all the exemplary data to be classified before the training phase begins, whereas *unsupervised* algorithms may utilize unlabeled data as well. Due to the general nature of classification, primarily only supervised methods are applicable to it. For clustering, data mining, and neural feature extraction, the unsupervised methods can be beneficial as well; see Section III.

If the pattern recognition problem is examined not from the viewpoint of mathematical theory but from the perspective of a user of a hypothetical system, a totally different series of dichotomies is obtained. Figure 8 represents one such taxonomy [66].

In the following sections, a set of classification algorithms are described and a taxonomy presented according to the structure of Table I. The methods are thus primarily grouped by belonging to either density estimators, regression methods, or others. The parametric or nonparametric nature of each method is discussed in the text. In Section IV.F, the neural characteristics of the various classification methods are addressed. In Table I, the algorithms regarded as neural are printed in italics.

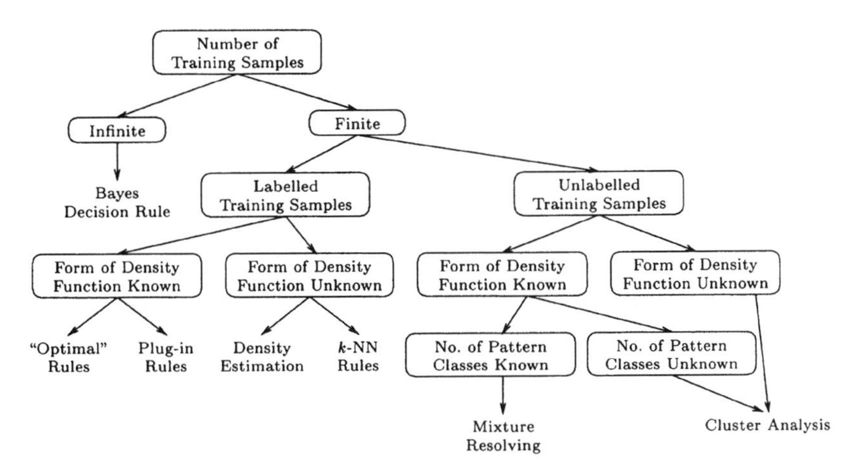

Figure 8 Dichotomies in the design of a statistical pattern recognition system, adapted from [66].

Table I

Taxonomy of Classification Algorithms Reviewed in the Text

	Density estimators	Regression methods	Others
Parametric	QDA LDA RDA		
Semiparametric	RKDA	MLP RBF	CLAFIC ALSM
Nonparametric	KDA PNN	MARS LLR	k-NN LVQ L-k-NN

A. Mathematical Preliminaries

In order to place the neural network classifiers in the context of statistical decision estimation, and to describe their functionality, we have to first define some mathematical concepts. A central mathematical notation in the theory of classifiers is the *classification function* $g\colon \mathbb{R}^d \mapsto \{1, \ldots, c\}$. For each real-valued d-dimensional *input feature vector* $\mathbf{x}$ to be classified, the value of $g(\mathbf{x})$ is thus an integer in the range $1, \ldots, c$, c being the number of classes. The classes are indexed with j when appropriate. The training set used in designing a classifier consists of n vectors $\mathbf{x}_i$, $i = 1, \ldots, n$, of which n_j vectors $\mathbf{x}_{ij}$, $i = 1, \ldots, n_j$, belong to class j.

The ordered pair $(\mathbf{x}, j)$ is stochastically speaking one realization of $(\mathbf{X}, J)$, an ordered pair of a random vector variable $\mathbf{X}$ and a discrete-valued random variable J. By assuming the realizations $(\mathbf{x}_i, j_i)$ to be stochastically independent and identically distributed, many considerations are simplified notably, although taking advantage of context-dependent information might certainly be beneficial in many applications.

The *a priori* probability of class j is denoted by P_j, its probability density function by $f_j(\mathbf{x})$, and that of the pooled data with all the classes combined by $f(\mathbf{x}) = \sum_{j=1}^{c} P_j f_j(\mathbf{x})$. Naturally, the priors have to meet the condition $\sum_{j=1}^{c} P_j = 1$. Using this notation, the Bayes classifier that minimizes the nonweighted misclassification error [14] is defined by

$$g_{\mathrm{BAYES}}(\mathbf{x}) = \operatorname*{argmax}_{j=1,\ldots,c} P_j f_j(\mathbf{x}). \tag{9}$$

We may alternatively consider the *a posteriori* probability $q_j(\mathbf{x}) = P(J = j \mid \mathbf{X} = \mathbf{x})$ of class j given $\mathbf{x}$ and use the rule

$$g_{\mathrm{BAYES}}(\mathbf{x}) = \operatorname*{argmax}_{j=1,\ldots,c} q_j(\mathbf{x}). \tag{10}$$

The rules of Eqs. (9) and (10) are equivalent since

$$q_j(\mathbf{x}) = P(J = j \mid \mathbf{X} = \mathbf{x}) = \frac{P_j f_j(\mathbf{x})}{f(\mathbf{x})}. \tag{11}$$

However, in practice the classifiers Eq. (9) and Eq. (10) have to be estimated from training data $(\mathbf{x}_1, j_1), \ldots, (\mathbf{x}_n, j_n)$ of pattern vectors with known classes, and then two distinct approaches emerge. The use of rule Eq. (9) requires explicit estimation of the class-conditional probability density functions f_j. For Eq. (10), some regression technique can be used to estimate the posterior probabilities q_j directly without separate consideration of the class-conditional densities.

The probability of a vector $\mathbf{x}$ to be misclassified is notated $\epsilon(\mathbf{x})$. Using the Bayes rule it is $\epsilon_{\text{BAYES}}(\mathbf{x}) = 1 - \max_{j=1,\ldots,c} q_j(\mathbf{x})$. The *overall misclassification rate* (ϵ) of the Bayes classifier is thus

$$\epsilon_{\text{BAYES}} = 1 - \int_{\mathbf{x} \in \mathbb{R}^d} f_{g_{\text{BAYES}}(\mathbf{x})}(\mathbf{x})\, dx. \tag{12}$$

B. Density Estimation Methods

In the density estimation approach one needs estimates for both the prior probabilities P_j and the class-conditional densities f_j in Eq. (9). The former estimation task is quite straightforward and the difficult and underdetermined part is to estimate the class-conditional densities. A classical *parametric* approach is to model the class-conditional densities as multivariate Gaussians. Depending on whether unequal or equal class covariances are assumed, the logarithm of $P_j f_j(\mathbf{x})$ is then either a quadratic or linear function of $\mathbf{x}$, giving rise to *quadratic discriminant analysis* (QDA) and *linear discriminant analysis* (LDA). A recent development is *regularized discriminant analysis* (RDA) which interpolates between LDA and QDA.

The success of these methods heavily depends on the validity of the normality assumption. If the class-conditional densities truly are normal, near-Bayesian classification error level can be achieved. On the other hand, if the densities are neither unimodal nor continuous, disastrous performance may follow. However, the critical areas for the classification accuracy are those where the distributions of the classes overlap. If the normality assumption holds there, the classification accuracy may be good even though the overall performance of the density estimation would be poor.

In *nonparametric* density estimation no fixed parametrically defined form for the estimated density is assumed. Kernel or Parzen estimates as well as k-nearest neighbor methods with large k are examples of popular nonparametric density estimation methods. They give rise to *kernel discriminant analysis* (KDA) and k-nearest neighbor (k-NN) classification rules (see Section IV.D.1).

In another approach the densities are estimated as finite mixtures of some standard probability densities by using the *expectation-maximization* (EM) algorithm or some other method [67–71]. Such an approach can be viewed as an economized KDA or an instance of the *radial basis function* (RBF) approach [22]. The self-organizing reduced kernel density estimator estimates densities in the spirit of radial basis functions, and the corresponding classification method is here referred to as *reduced kernel discriminant analysis* (RKDA).

1. Discriminant Analysis Methods

Quadratic discriminant analysis (QDA) [72] is based on the assumption that pattern vectors from class j are normally distributed with mean vector $\boldsymbol{\mu}_j$ and covariance matrix $\boldsymbol{\Sigma}_j$. Following the density estimation approach then leads to the rule

$$g_{\mathrm{QDA}}(\mathbf{x}) = \operatorname*{argmax}_{j=1,\dots,c}\Big[\log \widehat{P}_j - \tfrac{1}{2}\log\det\widehat{\boldsymbol{\Sigma}}_j - \tfrac{1}{2}(\mathbf{x}-\widehat{\boldsymbol{\mu}}_j)^T\widehat{\boldsymbol{\Sigma}}_j^{-1}(\mathbf{x}-\widehat{\boldsymbol{\mu}}_j)\Big]. \quad (13)$$

Here $\widehat{\boldsymbol{\mu}}_j$ and $\widehat{\boldsymbol{\Sigma}}_j$ denote the sample mean and the sample covariance estimates of the corresponding theoretical quantities.

If one assumes that the classes are normally distributed with different mean vectors but with a common covariance matrix $\boldsymbol{\Sigma}$, then the previous formula simplifies to the *linear discriminant analysis* (LDA) [72] rule

$$g_{\mathrm{LDA}}(\mathbf{x}) = \operatorname*{argmax}_{j=1,\dots,c}\Big[\log \widehat{P}_j + \widehat{\boldsymbol{\mu}}_j^T\widehat{\boldsymbol{\Sigma}}^{-1}(\mathbf{x}-\tfrac{1}{2}\widehat{\boldsymbol{\mu}}_j)\Big], \quad (14)$$

where a natural estimate for $\boldsymbol{\Sigma}$ is the pooled covariance matrix estimate

$$\widehat{\boldsymbol{\Sigma}} = \sum_{j=1}^{c}\widehat{P}_j\widehat{\boldsymbol{\Sigma}}_j.$$

Regularized discriminant analysis (RDA) [73] is a compromise between LDA and QDA. The decision rule is otherwise the same as Eq. (13) but instead of $\widehat{\boldsymbol{\Sigma}}_j$ one uses regularized covariance estimates $\widehat{\boldsymbol{\Sigma}}_j(\lambda,\gamma)$ with two regularizing parameters. Parameter λ controls the shrinkage of the class-conditional covariance estimates toward the pooled estimate and γ controls the shrinkage toward a multiple of the identity matrix. Let us denote by $\mathbf{K}_j$ the matrix

$$\sum_{i=1}^{n_j}(\mathbf{x}_{ij}-\widehat{\boldsymbol{\mu}}_j)(\mathbf{x}_{ij}-\widehat{\boldsymbol{\mu}}_j)^T, \quad \text{and let} \quad \mathbf{K} = \sum_{j=1}^{c}\mathbf{K}_j.$$

Then

$$\widehat{\boldsymbol{\Sigma}}_j(\lambda, \gamma) = (1 - \lambda)\widehat{\boldsymbol{\Sigma}}_j(\lambda) + \frac{\gamma}{d}\mathrm{tr}\big(\widehat{\boldsymbol{\Sigma}}_j(\lambda)\big)I, \tag{15}$$

where

$$\widehat{\boldsymbol{\Sigma}}_j(\lambda) = \frac{(1 - \lambda)\mathbf{K}_j + \lambda\mathbf{K}}{(1 - \lambda)n_j + \lambda n}. \tag{16}$$

One obtains QDA when $\lambda = 0,\ \gamma = 0$, and LDA when $\lambda = 1,\ \gamma = 0$, provided one uses the estimates $\widehat{\boldsymbol{\Sigma}}_j = \mathbf{K}_j/n_j$ and $\widehat{P}_j = n_j/n$.

2. Kernel Discriminant Analysis and Probabilistic Neural Network

In *kernel discriminant analysis* (KDA) [74, 75] one forms kernel estimates $\hat{f}_j$ of the class-conditional densities and then applies rule Eq. (9). The estimate of the class-conditional density of class j is

$$\hat{f}_j(\mathbf{x}) = \frac{1}{n_j}\sum_{i=1}^{n_j} K_{h_j}(\mathbf{x} - \mathbf{x}_{ij}), \qquad j = 1, \ldots, c, \tag{17}$$

where, given a fixed probability density function $K(\cdot)$ called the kernel, $h_j > 0$ is the smoothing parameter of class j, and K_h denotes the scaled kernel $K_h(\mathbf{x}) = h^{-d}K(\mathbf{x}/h)$. This scaling ensures that K_h and hence also each $\hat{f}_j$ is a probability density. A popular choice is the symmetric Gaussian kernel $K(\mathbf{x}) = (2\pi)^{-d/2}\exp(-\|\mathbf{x}\|^2/2)$. The choice of suitable values for the smoothing parameters is crucial and several approaches have been proposed in the literature; see, e.g., [72, 76–78].

The selection of the smoothing parameters can be based on cross-validated error count. In the first method, KDA1, all the smoothing parameters h_j are fixed to be equal to a parameter h. Optimal value for h is then selected using cross-validation (see Section IV.G) as the value which minimizes the cross-validated error count. In the second method, KDA2, the smoothing parameters are allowed to vary separately starting from a common value selected in KDA1.

In the second method the nonsmoothness of the object function is troublesome. Instead of minimizing the error count directly, it is advantageous to minimize a smoothed version of it. In a smoothing method described in [79], the class-conditional posterior probability estimates $\hat{q}_j(\mathbf{x})$ corresponding to the current smoothing parameters are used to define the functions u_j

$$u_j(\mathbf{x}) = \exp\big(\gamma\hat{q}_j(\mathbf{x})\big)\Big/\sum_{k=1}^{c}\exp\big(\gamma\hat{q}_k(\mathbf{x})\big), \tag{18}$$

where $\gamma > 0$ is a parameter. Then the smoothed error count is given by $n - \sum_{i=1}^{n} u_{j_i}(\mathbf{x}_i)$. As $\gamma \to \infty$, this converges towards the true error count. Since the smoothed error count is a differentiable function of the smoothing parameters, one can use a gradient-based minimization method for the optimization.

The *probabilistic neural network* (PNN) [80] is the neural network counterpart of KDA. Basically, all training vectors are stored and used as a set of Gaussian densities. In practice, only a subset of the kernels are actually evaluated when the probability values are calculated.

3. Reduced Kernel Density Analysis and Radial Basis Functions

The standard kernel density estimate suffers from the curse of dimensionality: as the dimension d of data increases, the size of a sample $\mathbf{x}_1, \ldots, \mathbf{x}_n$ required for an accurate estimate of an unknown density f grows quickly. On the other hand, even if there are enough data for accurate density estimation, the application at hand may limit the complexity of the classifier one can use in practice. A kernel estimate with a large number of terms may be computationally too expensive to use. One solution is to *reduce* the estimate, that is, to use fewer kernels but to place them at optimal locations. One can also introduce kernel-dependent weights and smoothing parameters. Various reduction approaches have been described in [81–85]. Some of these methods are essentially the same as the *radial basis function* (RBF) [22] approach of classification.

The self-organizing reduced kernel density estimate [86] has the form

$$\hat{f}(\mathbf{x}) = \sum_{k=1}^{\ell} w_k K_{h_k}(\mathbf{x} - \mathbf{m}_k), \tag{19}$$

where $\mathbf{m}_1, \ldots, \mathbf{m}_\ell$ are the reference vectors of a self-organizing map [59], $w_1, \ldots, w_\ell$ are nonnegative weights with $\sum_{k=1}^{\ell} w_k = 1$, and h_k is a smoothing parameter associated with the kth kernel. In order to achieve substantial reduction one takes $\ell \ll n$. The kernel locations $\mathbf{m}_k$ are obtained by training the self-organizing map using the whole available sample $\mathbf{x}_1, \ldots, \mathbf{x}_n$ from f. The weights w_k are computed iteratively and they reflect the number of training data in the Voronoi regions of the corresponding reference vectors. The smoothing parameters are optimized via stochastic gradient descent that attempts to minimize a Monte Carlo estimate of the integrated squared error $\int(\hat{f} - f)^2$. Simulations have shown that when the underlying density f is multimodal, the use of the feature map algorithm gives better density estimates than k-means clustering, the approach proposed in [87]. *Reduced kernel discriminant analysis* (RKDA) constitutes using estimates Eq. (19) for the class-conditional densities in the classifier Eq. (9). A drawback of RKDA in pattern classification applications is that the smoothing parameters of the class-conditional density estimates used in the approximate Bayes classifier are optimized from the point of view of integrated squared error and not discrimination performance which is the true focus of interest.

C. Regression Methods

In the second approach to classification the class posterior probabilities $q_j = P(J = j \mid \mathbf{X} = \mathbf{x})$ are directly estimated using some regression technique. *Parametric* methods include linear and logistic regression. Examples of *nonparametric* methodologies are projection pursuit [88, 89], additive models [90], multivariate adaptive regression splines (MARS), *local linear regression* (LLR), and the Nadaraya–Watson kernel regression estimator [78, 91], which is also called the *general regression neural network* [92]. Neural network approaches include multilayer perceptrons and *radial basis function* (RBF) expansions [22, 36].

One can use "one-of-c" coding to define the response $\mathbf{y}_i$ to pattern $\mathbf{x}_i$ to be the unit vector $[0, \ldots, 0, 1, 0, \ldots, 0]^T \in \mathbb{R}^c$ with 1 in the j_ith place. In the *least-squares* approach one then tries to minimize

$$\frac{1}{n}\sum_{i=1}^{n}\sum_{j=1}^{c}\left(\mathbf{y}_i^{(j)} - \mathbf{r}^{(j)}(\mathbf{x}_i)\right)^2 = \min_{\mathbf{r}\in\mathcal{R}}! \tag{20}$$

over a family $\mathcal{R}$ of $\mathbb{R}^c$-valued functions $\mathbf{r}$, where we denote the jth component of a vector $\mathbf{z}$ by $\mathbf{z}^{(j)}$. The corresponding mathematical expectation is minimized by the vector of class posterior probabilities, $\mathbf{q} = [q_1, \ldots, q_c]^T$. Of course, this ideal solution may or may not belong to the family $\mathcal{R}$, and besides, sampling variation will anyhow prevent us from estimating $\mathbf{q}$ exactly even when it does belong to $\mathcal{R}$ [93, 94].

The least-squares fitting criterion Eq. (20) can be thought to rise from using the maximum likelihood principle to estimate a regression model where errors are distributed normally. The *logistic* approach [72, Chap. 8] uses binomially distributed error, clearly the statistically correct model. One natural multivariate logistic regression approach is to model the posterior probabilities as the *softmax* [95] of the components of $\mathbf{r}$,

$$P(J = j \mid \mathbf{X} = \mathbf{x}) = q_j(\mathbf{x}) = \frac{\exp(\mathbf{r}^{(j)}(\mathbf{x}))}{\sum_{k=1}^{c}\exp(\mathbf{r}^{(k)}(\mathbf{x}))}. \tag{21}$$

Note that this also satisfies the natural condition $\sum_{k=1}^{c} q_k = 1$. A suitable fitting criterion is to maximize the conditional log-likelihood of $\mathbf{y}_1, \ldots, \mathbf{y}_n$ given that $\mathbf{X}_1 = \mathbf{x}_1, \ldots, \mathbf{X}_n = \mathbf{x}_n$. In the case of two classes this approach is equivalent to the use of the cross-entropy fitting criterion [22].

A very natural approach would be a regression technique that uses the error rate as the fitting criterion to be minimized [96]. *Classification and regression trees* (CART) are an example of a nonparametric technique that estimates the posterior probabilities directly but uses neither the least-squares nor the logistic regression approach [97].

1. Multilayer Perceptron

In the standard *multilayer perceptron* (MLP), there are d inputs, ℓ hidden units, and c output units, all the feedforward connections between adjacent layers are included, and the logistic activation function is used in the hidden and output layers [22, 36]. Such a network has $(d+1)\ell + (\ell+1)c$ adaptable weights, which are determined by minimizing the sum of squared errors criterion Eq. (20).

Using the notation of Section IV.C, one can use the vector $\tilde{\mathbf{y}}_i = 0.1 + 0.8\mathbf{y}_i$ as the desired output for input $\mathbf{x}_i$, i.e., the vectors $\mathbf{y}_i$ are scaled to better fit within the range of the logistic function. Then the scaled outputs $1.25(\mathbf{r}^{(j)}(\mathbf{x}) - 0.1)$ of the optimized network can be regarded as estimating the posterior probabilities $P(J = j \mid \mathbf{X} = \mathbf{x})$. A good heuristic is to start the local optimizations from many random initial points and to keep the weights yielding the minimum value for the sum of squared errors to prevent the network from converging to a shallow local minimum. It is advisable to scale the random initial weights so that the inputs to the logistic activation functions are of the order unity [22, Chap. 7.4].

In weight decay regularization [22, Chap. 9.2], one introduces a penalty for weights having a large absolute value in order to encourage smooth network mappings. When training MLPs with weight decay (MLP+WD), one minimizes the criterion

$$\frac{1}{n}\left[\sum_{i=1}^{n}\sum_{j=1}^{c}\left(\tilde{\mathbf{y}}_i^{(j)} - \mathbf{r}^{(j)}(\mathbf{x}_i, \mathbf{w})\right)^2 + \lambda \sum_{w \in W} w^2\right]. \tag{22}$$

Here $\mathbf{w}$ comprises all the weights and biases of the network, W is the set of weights between adjacent layers excluding the biases, and λ is the weight decay parameter. The network inputs and the outputs of the hidden units should be roughly comparable before the weight decay penalty in the form given above makes sense. It may be necessary to rescale the inputs in order to achieve this.

2. Local Linear Regression

Local linear regression (LLR) [78, 98] is a nonparametric regression method which has its roots in classical methods proposed for the smoothing of time series data; see [99]. Such estimators have received more attention recently; see, e.g., [100]. The particular version described below is also called LOESS [98, 99].

Local linear regression models the regression function in the neighborhood of each point $\mathbf{x}$ by means of a linear function $\mathbf{z} \mapsto \mathbf{a} + \mathbf{B}(\mathbf{z} - \mathbf{x})$. Given training data $(\mathbf{x}_1, \mathbf{y}_1), \ldots, (\mathbf{x}_n, \mathbf{y}_n)$, the fit at point $\mathbf{x}$ is calculated as follows. First one solves the weighted linear least-squares problem

$$\sum_{i=1}^{n} \|\mathbf{y}_i - \mathbf{a} - \mathbf{B}(\mathbf{x}_i - \mathbf{x})\|^2 w\big(\|\mathbf{x}_i - \mathbf{x}\|/h(\mathbf{x})\big) = \min_{\mathbf{a},\mathbf{B}}! \tag{23}$$

and then the fit at $\mathbf{x}$ is given by the coefficient $\mathbf{a}$. A reasonable choice for the function w is the tricube weight function [98], $w(u) = \max((1 - |u|^3)^3, 0)$. The local bandwidth $h(\mathbf{x})$ is controlled by a neighborhood size parameter $0 < \alpha \leq 1$: one takes k equal to αn rounded to the nearest integer and then takes $h(\mathbf{x})$ equal to the distance to the kth closest neighbor of $\mathbf{x}$ among the vectors $\mathbf{x}_1, \ldots, \mathbf{x}_n$. If the components of $\mathbf{x}$ are measured in different scales, then it is advisable to select the metric for the nearest neighbor calculation carefully. At a given $\mathbf{x}$, the weighted linear least-squares problem can be reduced to inverting a $(d+1) \times (d+1)$ matrix, where d is the dimensionality of $\mathbf{x}$; see, e.g., [78, Chap. 5].

3. Tree Classifier, Multivariate Adaptive Regression Splines, and Flexible Discriminant Analysis

The introduction of tree-based models in statistics dates back to [101] although their current popularity is largely due to the seminal book [97]. For Euclidean pattern vectors $\mathbf{x} = [x_1, \ldots, x_d]^T$, a classification tree is a binary tree where at each node the decision to branch either to left or right is based on a test of the form $x_i \geq \lambda$. The cutoff values λ are chosen to optimize a suitable fitting criterion. The tree growing algorithm recursively splits the pattern space $\mathbb{R}^d$ into hyperrectangles while trying to form maximally pure nodes, that is, subdivision rectangles that ideally contain training vectors from one class only. Stopping criteria are used to keep the trees reasonably sized, although the commonly employed strategy is to first grow a large tree that overfits the data and then use a separate pruning stage to improve its generalization performance. A terminal node is labeled according to the class with the largest number of training vectors in the associated hyperrectangle. The tree classifier therefore uses the Bayes rule with the class posterior probabilities estimated by locally constant functions. The particular tree classifier described here is available as a part of the S-Plus statistical software package [102–104]. This implementation uses a likelihood function to select the optimal splits [105]. Pruning is performed by the minimal cost-complexity method. The cost of a subtree T is taken to be

$$R_\alpha(T) = \epsilon(T) + \alpha \cdot \text{size}(T), \tag{24}$$

where $\epsilon(T)$ is an estimate of the classification error of T, size of T is measured by the number of its terminal nodes, and $\alpha > 0$ is a cost parameter. An overfitted tree is pruned by giving α increasingly large values and selecting nested subtrees that minimize R_α.

MARS [106] is a regression method that shares features with tree-based modeling. MARS estimates an unknown function r using an expansion

$$\hat{r}(\mathbf{x}) = a_0 + \sum_{k=1}^{M} a_k B_k(\mathbf{x}), \tag{25}$$

where the functions B_k are multivariate splines. The algorithm is a two-stage procedure, beginning with a forward stepwise phase which adds basis functions to the model in a deliberate attempt to overfit the data. The second stage of the algorithm is standard linear regression backward subset selection. The maximum order of variable interactions (products of variables) allowed in the functions B_k, as well as the maximum value of M allowed in the forward stage, are parameters that need to be tuned experimentally. Backward model selection uses the generalized cross-validation criterion introduced in [107].

The original MARS algorithm fits only scalar-valued functions and is therefore not well suited to discrimination tasks with more than two classes. A recent proposal called *flexible discriminant analysis* (FDA) [108] with its publicly available S-Plus implementation in the StatLib program library contains vector-valued MARS as one of its ingredients. However, FDA is not limited to just MARS as it allows the use of other regression techniques as its building blocks as well. In FDA, one can first train c separate MARS models $\mathbf{r}^{(j)}$ with equal basis function sets but different coefficients a_k to map training vectors $\mathbf{x}_i$ to the corresponding unit vectors $\mathbf{y}_i$. Then a linear map A is constructed to map the regression function output space $\mathbb{R}^c$ onto a lower-dimensional feature space $\mathbb{R}^\ell$ in a manner that optimally facilitates prototype classification based on the transformed class means $A(\mathbf{r}(\widehat{\mu}_j))$ and a weighted Euclidean distance function.

D. Prototype Classifiers

One distinct branch of classifiers appearing under the title **others** in Table I are prototype classifiers LVQ, k-NN, and L-k-NN. They share in common the principle that they keep copies of training samples in memory, and the classification decision $g(\mathbf{x})$ is based on the distances between the memorized prototypes and the input vector $\mathbf{x}$. Either the training vectors are retained as such or some sort of a training phase is utilized to extract properties of a multitude of training vectors to each of the memorized prototypes. In either case, the prototype classifiers are typical representatives of the *nonparametric* classification methods.

1. k-Nearest Neighbor Classifiers

In a *k-nearest neighbor* (k-NN) classifier each class is represented by a set of prototype vectors [27]. The k closest neighbors of an input pattern vector are found among all the prototypes and the class label is decided by the majority voting rule. A possible tie of two or more classes can be broken, e.g., by decreasing k by one and revoting.

In classical pattern recognition, the nonparametric k-NN classification method has been very popular since the first publication by Fix and Hodges [109] and an important limiting accuracy proof by Cover and Hart [110]. The k-NN rule should even now be regarded as a sort of a baseline classifier, against which other statistical and neural classifiers should be compared [111]. Its advantage is that no time is needed in training the classifier, and the corresponding disadvantage is that huge amounts of memory and time are needed during the classification phase. An important improvement in memory consumption—while still keeping the classification accuracy moderate—may be achieved using some *editing* method [112]. An algorithm known as *multiedit* [14] removes spurious vectors from the training set. Another algorithm known as *condensing* [113] adds new vectors to the classifier when it is unable to classify the pattern correctly. In both methods, a vector set originally used as a k-NN classifier is converted to a smaller edited set to be used as a 1-NN classifier.

2. Learning Vector Quantization

The *learning vector quantizer* (LVQ) algorithm [59] produces a set of prototype or *codebook* pattern vectors $\mathbf{m}_i$ that can be used in a 1-NN classifier. Training consists of moving a fixed number ℓ of codebook vectors iteratively toward or away from the training samples $\mathbf{x}_i$. The variations of the LVQ algorithm differ in the way the codebook vectors are updated. The LVQ learning process can be interpreted either as an iterative movement of the decision boundaries between neighboring classes, or as a way to generate a set of codebook vectors whose density reflects the shape of the function s defined as

$$s(\mathbf{x}) = P_j f_j(\mathbf{x}) - \max_{k \neq j} P_k f_k(\mathbf{x}), \tag{26}$$

where $j = g_{\mathrm{BAYES}}(\mathbf{x})$. Note that the zero set of s consists of the Bayes optimal decision boundaries.

3. Learning k-NN Classifier

Besides editing rules, iterative learning algorithms can be applied to k-NN classifiers [114]. The learning rules of the *learning k-NN* (L-k-NN) resemble those of LVQ but at the same time the classifier still utilizes the improved classification accuracy provided by the majority voting rule. The performance of the standard k-NN classifier depends on the quality and size of the training set, and the performance of the classifier decreases if the available computing resources limit the number of training vectors one can use. In such a case, the learning k-NN rule is better able to utilize the available data by using the whole training set to optimize the classification based on a smaller set of prototype vectors.

For the training of the k-NN classifier, three slightly different training schemes have been presented. As in the LVQ, the learning k-NN rules use a fixed number of *code vectors* $\mathbf{m}_{ij}$ with predetermined class labels j for classification. Once the code vectors have been tuned by moving them to such positions in the input space that give a minimal error rate, the decision rule for an unknown input vector is based on the majority label among its k closest code vectors.

The objective of all the learning rules is to make the correct classification of the training samples more probable. This goal is achieved by incrementally moving some of the vectors in the neighborhood of a training input vector toward the training sample and some away from it. For all the rules, the modifications to the code vectors $\mathbf{m}_i$ are made according to the LVQ rule:

$$\mathbf{m}_i(t+1) = \mathbf{m}_i(t) \pm \alpha(t)\big(\mathbf{x}(t) - \mathbf{m}_i(t)\big), \tag{27}$$

where $\mathbf{x}(t)$ is the training sample at the step t. With a positive sign of $\alpha(t)$, the movement of the code vector is directed toward the training sample, and with negative sign away from it. The learning rate $\alpha(t)$ should decrease slowly in order to make the algorithm convergent; in practice it may be sufficient to use a small constant value.

E. Subspace Classifiers

The motivation for the subspace classifiers originates from compression and optimal reconstruction of multidimensional data. The use of linear subspaces as class models is based on the assumption that the data within each class approximately lie on a lower-dimensional subspace of the pattern space $\mathbb{R}^d$. A vector from an unknown class can then be classified according to its shortest distance from the class subspaces.

The sample mean $\widehat{\boldsymbol{\mu}}$ of the whole training set is first subtracted from the pattern vectors. For each class j, the correlation matrix $\widehat{\mathbf{R}}_j$ is estimated and its first few eigenvectors $\mathbf{u}_{1j}, \ldots, \mathbf{u}_{\ell_j j}$ are used as columns of a basis matrix $\mathbf{U}_j$. The classification rule of the *class-featuring information compression* (CLAFIC) algorithm [115] can then be expressed as

$$g_{\mathrm{CLAFIC}}(\mathbf{x}) = \operatorname*{argmax}_{j=1,\ldots,c} \big\| \mathbf{U}_j^T \mathbf{x} \big\|^2. \tag{28}$$

The *averaged learning subspace method* (ALSM) introduced by one of the current authors [47] is an iterative learning version of CLAFIC, in which the unnormalized sample class correlation matrices $\widehat{\mathbf{S}}_j(0) = \sum_{i=1}^{n_j} \mathbf{x}_{ij}\mathbf{x}_{ij}^T$ are slightly modified according to the correctness of the classifications,

$$\widehat{\mathbf{S}}_j(k+1) = \widehat{\mathbf{S}}_j(k) + \alpha \sum_{i \in A_j} \mathbf{x}_i \mathbf{x}_i^T - \beta \sum_{i \in B_j} \mathbf{x}_i \mathbf{x}_i^T. \tag{29}$$

Here $\mathbf{x}_{1j}, \ldots, \mathbf{x}_{n_j j}$ is the training sample from class j, α and β are small positive constants, A_j is the set of indices i for which $\mathbf{x}_i$ comes from class j but is classified erroneously to a different class, and B_j consists of those indices for which $\mathbf{x}_i$ is classified to j although it actually originates from a different class. The basis matrices $\mathbf{U}_j$ are recalculated after each training epoch as the dominant eigenvectors of the modified $\widehat{\mathbf{S}}_j$. The subspace dimensions ℓ_j need to be somehow fixed. One effective iterative search algorithm and a novel weighting solution have been recently presented [116].

F. Special Properties of Neural Methods

In the previous discussion we characterized some popular classification techniques in terms of the mathematical principles they are based on. In this general view many neural networks can be seen as representatives of certain larger families of statistical techniques. However, this abstract point of view fails to identify some key features of neural networks that characterize them as a distinct methodology.

From the very beginning of neural network research [117–119] the goal was to demonstrate problem-solving without explicit programming. The neurons and networks were supposed to learn from examples and store this knowledge in a distributed way among the connection weights.

The original methodology was exactly opposite to the goal-driven or top-down design of statistical classifiers in terms of explicit error functions. In neural networks, the approach has been bottom-up: starting from a very simple linear neuron that computes a weighted sum of its inputs, adding a saturating smooth nonlinearity, and constructing layers of similar parallel units, it turned out that "intelligent" behavior such as speech synthesis [120] emerged by simple learning rules. The computational aspect has always been central. At least in principle, everything that the neural network does should be accomplished by a large number of simple local computations using the available input and output signals, as in real neurons, but unlike heavy numerical algorithms involving such operations as matrix inversions. Perhaps the best example of a clean-cut neural network classifier is the LeNet system [4, 121] for handwritten digit recognition (see Section V.B.1).

Such a computational model supports well the implementation in regular VLSI circuits.

In the current neural network research, these original views are clearly becoming vague as some of the most fundamental neural networks such as the one-hidden-layer MLP or RBF networks have been shown to have very close connections to statistical techniques. The goal remains, however, of building much more complex artificial neural systems for demanding tasks such as speech recognition [122] or computer vision [35], in which it is difficult or eventually impossible to state the exact optimization criteria for all the consequent processing stages.

Figure 9 is an attempt to assess the neural characteristics of some of the classification methods discussed earlier. The horizontal axis measures the flexibility of a classifier architecture in the sense of the richness of the discriminant function family encompassed by a particular method. High flexibility of architecture is a property often associated with neural networks. In some cases (MLP, RBF, CART, MARS) the flexibility can also include algorithmic model selection during learning.

In the vertical dimension, the various classifiers are categorized on the basis of how they are designed from a training sample. Training is considered nonneural

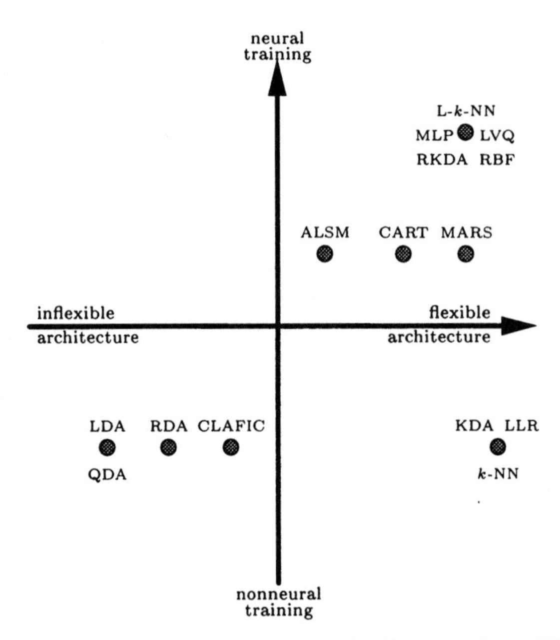

Figure 9 Neural characteristics of some classifiers according to [65].

if the training vectors are used as such in classification (e.g., k-NN, KDA), or if some statistics are first estimated in batch mode and the discriminant functions are computed from them (e.g., QDA, CLAFIC). Neural learning is characterized by simple local computations in a number of real or virtual processing elements. Neural learning algorithms are typically of the error correction type; for some such algorithms, not even an explicit cost function exists. Typically, the training set is used several times (epochs) in an on-line mode. Note, however, that for some neural networks (MLP, RBF) the current implementations in fact often employ sophisticated optimization techniques which would justify moving them downwards in our map to the lower half plane.

In this schematic representation, the classical LDA and QDA methods are seen as least neural with the RDA and CLAFIC possessing at least some degree of flexibility in their architecture. The architecture of KDA, k-NN, and LLR is extremely flexible. Compared to CLAFIC, the ALSM method allows for both incremental learning and flexibility of architecture in the form of subspace dimensions that can change during learning. In this view, neural classifiers are well exemplified in particular by such methods as MLP, RBF, RKDA, LVQ, and learning k-NN (L-k-NN), but also to some degree by ALSM, CART, and MARS.

G. Cross-Validation in Classifier Design

In order to get reliable estimates of classifier performance, the available data should first be divided into two separate parts: the training sample and the testing sample. The whole process of classifier design should then be based strictly on the training sample only. In addition to parameter estimation, the design of some classifiers involves the choice of various tuning parameters and model or architecture selection. To utilize the training sample efficiently, cross-validation [123] (or "rotation," cf. [14, Chap. 10.6.4]) can be used. In v-fold cross-validation, the training sample is first divided into v disjoint subsets. One subset at a time is then put aside; a classifier is designed based on the union of the remaining $v-1$ subsets and then tested for the subset left out. Cross-validation approximates the design of a classifier using all the training data and then testing it on an independent set of data, which enables defining a reasonable object function to be optimized in classifier design. For example, for a fixed classifier, the dimension of the pattern vector can be selected so that it minimizes the cross-validated error count. After optimization, one can obtain an unbiased estimate of the performance of the optimized classifier by means of the separate testing sample. Notice that the performance estimates might become biased if the testing sample were in any way used during the training of the classifier.

H. Rejection

Other criteria than minimum classification error can be important in practice, including use of class-dependent misclassification costs and Neyman–Pearson-style classification [11, 124]. The use of a *reject class* can help reduce the misclassification rate ϵ in tasks where exceptional handling (e.g., by a human expert) of particularly ambiguous cases is feasible. The decision to reject a pattern $\mathbf{x}$ and to handle it separately can be based on its probability to be misclassified, which for the Bayes rule is $\epsilon(\mathbf{x}) = 1 - \max_{j=1,\ldots,c} q_j(\mathbf{x})$. The highest misclassification probability occurs when the posterior probabilities $q_j(\mathbf{x})$ are equal and then $\epsilon(\mathbf{x}) = 1 - 1/c$. One can therefore select a rejection threshold $0 \leq \theta \leq 1 - 1/c$ and reject $\mathbf{x}$ if

$$\epsilon(\mathbf{x}) > \theta. \tag{30}$$

The notation $g(\mathbf{x})$ used for the classification function can be extended to include the rejection case by denoting with $g(\mathbf{x}) = 0$ all the rejected vectors $\mathbf{x}$. When the overall rejection rate of a classifier is denoted by ρ, the rejection-error balance can be depicted as a curve in the $\rho\epsilon$ plane, parameterized with the θ value. In recognition of handwritten digits, the rejection-error curve is found to be generally linear in the $\rho \log \epsilon$ plane [125]. This phenomenon can also be observed in Fig. 18.

I. Committees

In practice, one is usually able to classify a pattern using more than one classifier. It is then quite possible that combining the opinions of several parallel systems results in improved classification performance. Such hybrid classifiers, classifier ensembles, or *committees*, have been studied intensively in recent years [126].

Besides improved classification performance, there are other reasons to use a committee classifier. The pattern vectors may be composed of components that originate from very diverse domains. Some may be statistical quantities such as moments and others discrete structural descriptors such as numbers of endpoints, loops, and so on. There may not be an obvious way to concatenate the various components into a single pattern vector suitable for any single classifier type. In some other situations, the computational burden can be reduced either during training or in the recognition phase if the classification is performed in several stages.

Various methods exist for forming a committee of classifiers even when their output information is of different types. In the simplest case, a classifier only outputs its decision about the class of an input pattern, but sometimes some measure of the certainty of the decision is also provided. The classifier may propose a set of classes in the order of decreasing certainty, or a measure of decision certainty may be given for all the classes. Various ways to combine classifiers with such types of output information are analyzed in [127–130].

The simplest decision rule is to use a majority rule among the classifiers in the committee, possibly ignoring the opinion of some of the classifiers [131]. Two or more classifiers using different sets of features may be combined to implement rejection of ambiguous patterns [132–135]. A genetic algorithm can be applied in searching for optimal weights to combine the classifier outputs [136]. Theoretically more advanced methods may be derived from the EM algorithm [128, 129, 137–139] or from the Dempster-Shafer theory of evidence [127, 140].

The outputs of several regression-type classifiers may be combined linearly [141] or nonlinearly [142] to reduce the variance of the posterior probability estimates. A more general case is the reduction of variance in continuous function estimation: a set of MLPs can be combined into a committee classifier with reduced output variance and thus smaller expected classification error [143–146]. A separate confidence function may also be incorporated in each of the MLPs [147].

Given a fixed feature extraction method, one can either use a common training set to design a number of different types of classifiers [148] or, alternatively, use different training sets to design several versions of one type of classifier [149–153].

J. On Comparing Classifiers

Some classification accuracies attained using the classification algorithms described in the previous sections will be presented later in this text in Section V.B.4. Such comparisons need, however, to be considered with utmost caution.

During the last years, a large number of papers have been published in which various neural and other classification algorithms have been described and analyzed. The results of such experiments cannot generally be compared due to the use of different raw data material, preprocessing, and testing policies. In [154] the methods employed in experimental evaluations concerning neural algorithms in two major neural networks journals in 1993 and 1994 were analyzed. The bare conclusion was that the quality of the quantitative results—if presented at all—was poor. For example, the famous NETtalk experiments [120] were in [155] replicated and compared to the performance of a k-NN classifier. The conclusion was that the original results were hard to reproduce and the regenerated MLP results were outperformed by the k-NN classifier.

Some larger evaluations or *benchmarking* studies have also been published in which a set of classification algorithms have been tried to be assessed in a fair and impartial setting. Some of the latest in this category include [2, 5, 6, 156, 157]. The profound philosophical questions involved in comparisons are addressed in [158]. In [159] the distribution-free bounds for the difference between the achieved and achievable error levels are calculated for a set of classification algorithms in the cases of both finite and infinite training sets.

V. NEURAL NETWORK APPLICATIONS IN PATTERN RECOGNITION

A. Application Areas of Neural Networks

Neural computing has proved to be a useful solution technique in many application areas that are difficult to tackle using conventional computing. In a recent ESPRIT research project Siena [160], a large number of commercial neural network applications developed in Europe were reviewed. Figure 10 shows the distribution of the cases by application type. About 9% of the cases in the study were clear pattern recognition applications. To solve some part of the whole task, pattern recognition was applied in a much larger number of the applications; many prediction and identification problems contain similar recognition and classification stages as used in pattern recognition applications.

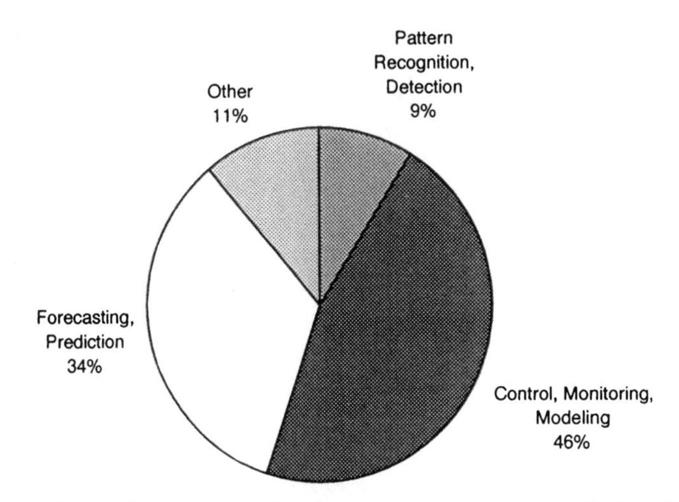

Figure 10 Distribution of categories of commercial neural network applications [160].

As neural networks provide rather general techniques for modeling and recognition, they have found applications in many diverse engineering fields. Table II presents some neural network application areas together with some typical applications compiled from case listing in [160]. Note that pattern recognition is needed in three of the five categories in the table: recognition, classification, and visual processing.

Table II

Neural Network Application Areas and Case Applications [160]

Application type	Case applications
Recognition and identification	Oil exploration
	Fiber optic image transmission
	Automated data entry
	Number plate recognition
	Fingerprint analysis
Assessment and classification	Credit risk management
	Medical diagnosis
	Bridge construction analysis
	Fruit grading
	TV picture quality control
	Industrial nondestructive testing
	Tyre quality control
	Improving hospital treatment and reducing expenses
	Property valuation
	Product pricing sensitivity analysis
	Route scheduling optimization
	Quality control in livestock carcasses
Monitoring and control	Machine health monitoring
	Dynamic process modeling
	Chemical synthesis
	Chemical manufacture
	Bioprocess control
Forecasting and prediction	Stock market prediction
	Classifying psychiatric care
	Holiday preference prediction
	Traffic jam reduction
	Survey analysis
	TV audience prediction
	Future business demand forecasting
Sensory and visual	Automated industrial inspection
	Railway track visual maintenance inspection
	Mail sorting

In the early days of neural computing, the first applications were in pattern recognition, but since then neural computing has spread to many other fields of computing. Consequently, large engineering fields, modeling and control, together with prediction and forecasting, made up two-thirds of the cases in the study.

Still, the relative impact of neural network techniques is perhaps largest in the area of pattern recognition. In some application types, such as optical character recognition, neural networks have already become a standard choice in commercial products. The main reasons for the success of neural network methods in such problems are outlined in the previous chapters—by carefully designed preprocessing and feature extraction the main difficulties in the applications are in

Table III
Examples of Neural Pattern Recognition Applications

Application	Neural network solution
Problem Domain	
Identification and verification	
Face recognition	Classification of small images by MLP tree [161]
	Dynamic link matching, Gabor-jet features [162]
Face identification	ZN-Face[1)] system [163], based on [162]
Paper currency recognition	Geometric features, MLP classifier [164]
Signature verification	Wavelet decomposition, MLP classifier [165]
	Fisher's discriminant analysis enhanced with NN[1)] [166]
Ultrasonic weld testing	Manually selected features, MLP classifier[1)] [167]
Wood defect recognition	Self-organizing features[1)], see Section V.B.3
Medical applications	
Blood vessel detection	Convolution filter bank, MLP classifier [168]
Contour finding in MRI	MLP detection of contour pixels [169]
Aerial imaging and reconnaissance	
Radar target classification	Spectral features, MLP classifier [170]
Automatic target recognition	Biological vision modeling [171]
	See [172] for survey on ATR
Character recognition	
Numeric handprint recognition	LeNet architecture[1)] [121], see Section V.B.1
	Zernike moment features, MLP classifier [173]
	Geometric and moment features, MLP classifier [174]
Handwritten form processing	Selected features, MLP classifier[1)] [167]
On-line recognition	Dynamic stroke features, RBF classification [175]
Speech processing	
"Phonetic typewriter"	Cepstral feature classification by SOM [122]
Speech recognition	Phoneme classification by time delay neural network [176]

[1)]Commercial products are marked by [1)] in the table.

determining the nonlinear class boundaries, which is a very suitable problem for neural network classifiers.

In Table III we have collected recent neural network applications in pattern recognition. Typical architecture of neural pattern recognition algorithms follows that shown in Fig.1. In most of the applications listed in Table III, conventional features, such as moment invariants or spectral features, are computed from the segmented objects and neural networks are used for the final classification.

Then the value of using neural networks in the application depends on the goodness of the classifier. Although any classifier cannot solve the actual recognition problem if the selected features do not separate the target classes adequately, the choice of the most efficient classifier can give the few extra percent in recognition rate to make the solution sufficient in practice. The advantages of neural classifiers compared to other statistical methods were reviewed in Section IV.F.

In the next section we review some more integral neural network pattern recognition systems, in which the feature extraction is integrated to the learning procedure.

B. Examples of Neural Pattern Recognition Systems

In this section we review some pattern recognition systems, in which neural network techniques have a central role in the solution, including the lower levels of the system. As the vast majority of neural network solutions in pattern recognition are based on carefully engineered preprocessing and feature extraction, and neural network classifier, the most difficult parts of the recognition problem, such as invariances, are thus solved by hand before they ever reach the network.

Moreover, the handcrafted feature presentations cannot produce similar invariances and tolerance to varying conditions that are observed in biological visual systems. A possible direction to develop more capable pattern recognition systems might be to include the feature extraction stage as part of the adaptive trained system.

In the pattern recognition systems considered here also a considerable amount of the lower parts of the recognition problem are solved by neural networks. In Table III, examples of such systems are, e.g., [121, 163, 171].

1. System Solution with Constrained MLP Architecture—LeNet

The basic elements of virtually all pattern recognition systems are preprocessing, feature extraction, and classification, as elaborated in previous sections. The methods and practices to design the feature extraction stage to be efficient with neural network classifiers were reviewed in Section III.A, including methods such as manual selection, and data reduction by, e.g., principal component analysis.

In theory it is possible to integrate the feature extraction and classification in one processing block and to use supervised learning to train the whole system. However, the dimensionality of the input patterns causes a serious challenge in this approach. In a typical visual pattern recognition application the input to the feature extraction stage is an image comprising thousands or even hundreds of thousands of pixels, and in the feature extraction stage this very high-dimensional space is mapped to the feature space of much reduced dimensionality. A system with the original (sub)image as the input would have far too many free parameters to generalize correctly, with any practical number of training samples.

LeNet Architecture

The solution proposed by LeCun *et al.* [121, 177] is based on constraining the network structure with prior knowledge about the recognition problem. The network architecture, named LeNet, is rather similar to the Neocognitron architecture (see Section V.B.2): the feature extraction is carried out by scanning the input image with neurons that have local receptive fields to produce convolutional feature maps (corresponding to S layers in the Neocognitron), followed by subsampling layer to reduce the dimensionality of the feature space and to bring in distortion tolerance to the recognition (corresponding to the C layers in the Neocognitron). Figure 11 shows the basic architecture of a LeNet with two layers of feature detectors.

In the Neocognitron the feature extracting neurons are trained with unsupervised competitive learning, while in the LeNet network back-propagation is used to train the whole network in a supervised manner. This has the considerable advantage that the features are matched to separate the target classes, while in unsupervised feature extraction the features are independent of the target classes. The trade-off is that a rather large number of training samples are needed and the training procedure may be computationally expensive.

Example of the LeNet Network

The following example of the architecture of the LeNet network was reported in [177]. The task was to recognize handwritten digits, that were segmented and transformed to fit in 16×16 pixel images in preprocessing. The network had four feature construction layers (named H1, H2, H3, and H4) and an output layer with ten units. Layers H1 and H3 corresponded to the feature map layers in Fig. 11, and H2 and H4 to the resolution reduction layers, respectively.

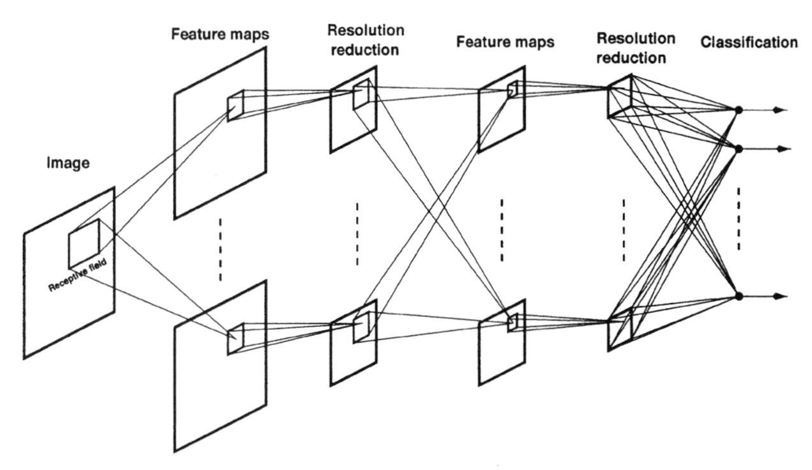

Figure 11 Schematic diagram illustrating the basic structure of many successful neural pattern recognition systems, such as the Neocognitron and LeNet. The main differences in the networks are in the training algorithm, the number of feature map layers, and the connection pattern of the classifier.

• The layer H1 contained four different feature detectors with 5 × 5 pixel receptive fields. Thus the output of the H1 layer contained four maps produced by scanning the input image with each of the feature detector neurons.

• The following layer H2 performed averaging and subsampling of the H1 feature maps: in the layer H2 there was a neuron connected with equal fixed weights to each nonoverlapping 2 × 2 area in the H1 feature map.

• Layer H3 constructed higher-order features from combinations of the primary features in H2 maps. The layer had 12 different feature detecting neurons, each neuron connected to one or two of the H2 maps by 5 × 5 receptive fields. In an earlier version of the system the H3 neurons were connected to all H2 maps, resulting in a large number of free parameters in this stage [121]. The reduced connection patterns were determined by pruning the network with the optimal brain damage technique [178].

• The layer H4 was identical to layer H2, averaging and subsampling the H3 feature maps. The output layer was fully connected to layer H4.

The network was trained on a large data base of manually labeled digits, and was able to produce state-of-the-art level recognition [177]. The example shows that it is possible to use back-propagation–based supervised learning techniques to solve large parts of the pattern recognition problem, by carefully constraining the network structure and weights according to prior knowledge about the task

A comparison of this architecture, including several variations in the number of feature maps, and other learning algorithms for handwritten digit recognition is presented in [179]. The report concentrates on methods where there is no separate handcrafted feature extraction stage, but the feature extraction is combined with classification and trained together.

2. Invariant Recognition with Neocognitron

One of the first pattern recognition systems based solely on neural network techniques was the Neocognitron paradigm, developed by Fukushima *et al.* [180]. The architecture of the network was originally inspired by Hubel and Wiesel's hierarchy model of the visual cortex [181]. According to the model, cells at the higher layers in the visual cortex have a tendency to respond selectively to more complicated features of the stimulus patterns and, at the same time, have larger receptive fields.

The basic structure of the Neocognitron is shown in Fig. 11. It consists of alternating feature detector and resolution reduction layers, called S and C layers, respectively. Each S layer contains several feature detector arrays called cell planes, shown as the small squares inside the layers in Fig. 11. All neurons in a cell plane have similar synaptic connections, so that functionally a cell plane corresponds to a spatial convolution, since the neurons are linear in weights. The S layers are trained by competitive learning, so that each plane will learn to be sensitive to a different pattern.

The C layers are essential to the distortion tolerance of the network. Each cell plane in the S layer is connected by fixed weights to a similar but smaller cell plane in the successive C layer. The weights of the C cells are chosen so that one active S layer cell in its receptive field will turn the C cell on. The purpose of the C layers is to allow positional variation to the features detected by the preceding S layer. The successive S layer is of the same size as the previous C layer, and the S cells are connected to all the C planes. Thus the next-level cell planes can detect any combinations of the previous level features. Finally the sizes of the cell planes decrease so that the last C plane contains only one cell, with receptive field covering the whole input plane.

In Fig. 12 the tolerance to small distortions is elucidated; the dashed circles show the areas where the key features distinguishing "A" must be found. The features may appear in any place inside the circles.

In the later versions of the Neocognitron [182] a selective attention mechanism is implemented to allow segmentation and recognition of overlapping patterns, as in cursive handwriting.

3. Self-Organizing Feature Construction System

In this section we review a neural pattern recognition system based on self-organizing feature construction. The system is described in more detail in [35, 183, 184].

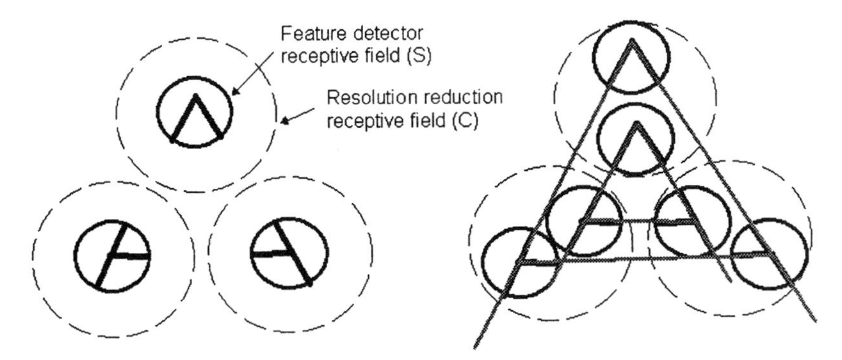

Figure 12 Illustration of the principle for recognizing deformed patterns by the Neocognitron.

The basic principle in the system is to define a set of generic local primary features, which are assumed to contain pertinent information of the objects, and then to use unsupervised learning techniques for building higher-order features from the primary features and reducing the number of degrees of freedom in the data. Then the final supervised classifiers can have a comparably small number of free parameters and thus require a small amount of preclassified training samples.

The feature extraction–classification system is composed of a pipelined block structure, where the number of neurons and connections decrease and the connections become more adaptive in higher layers. The major elements of the system are the following.

Primary features: The primary features should detect local, generic shape-related information from the image. A self-similar family of Gabor filters (see, e.g., [185]) is used for this task, since the Gabor filters have optimal combined resolution in spatial and frequency domains.

Self-organized features: To form complex features the Gabor filter outputs are clustered to natural, possibly nonconvex clusters by a multilayer self-organizing map.

Classifier: Only the classifier is trained in a supervised manner in the highly reduced feature space.

Figure 13 shows the principle of the self-organizing feature construction in face recognition [35]. At the lowest levels, two banks of eight Gabor filters were used. The two filter banks had different spatial resolution and eight orientations, as shown in Fig. 13. The primary feature was thus comprised of the two eight-dimensional vectors of the filter outputs.

The complex features were then produced by a two-layer self-organizing map. The first-level map contained 10×10 units, so that the eight-dimensional feature vectors of both resolutions were separately mapped through the 10×10

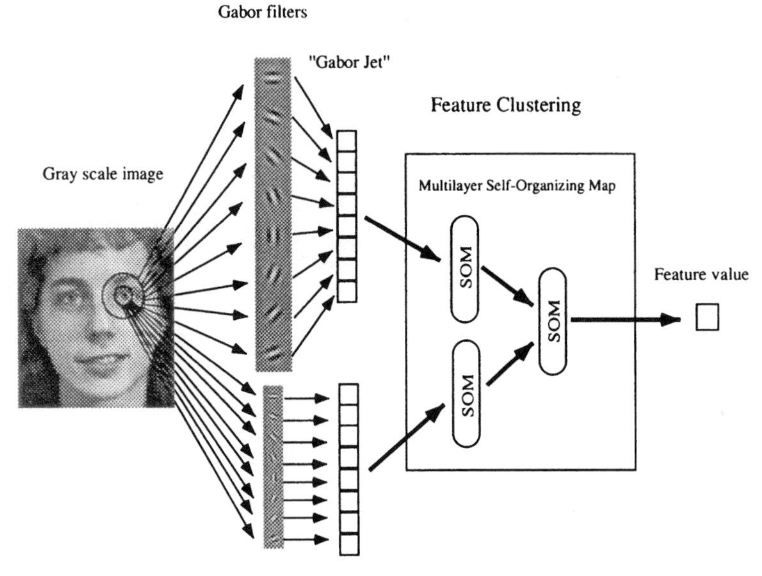

Figure 13 Schematic drawing of the feature extraction system. Left part: eight-dimensional Gabor vectors at two resolutions are extracted from every pixel location in the 128×128 digital image. Right part: the two-layer SOM produces a feature value $c(p)$ for each pixel location p.

map, to produce two two-dimensional vectors. These were stacked to form a four-dimensional input vector for the second-layer map, that had 100 units in a one-dimensional lattice. Thus the feature extraction stage maps a neighborhood of a pixel to a feature value, such that similar details are mapped to nearby features. A special virtue of the multilayer SOM is that the cluster shapes can be also nonconvex [186]. Figure 14 shows an example of feature mapping, where a face image is scanned with the feature detector and the resulting feature values are shown as gray scales.

It was shown in [186] and [35] that such feature images can be classified with very simple classifiers. Often it is sufficient to take feature histograms of the object regions, to form translation-invariant classification features.

The role of the classifier is more important in this feature construction system than with manually selected features, since the features are not directly related to the object classes. For any given class, many of the filters, and features, are irrelevant, and the classifier must be able to pick up the combination of the relevant features. Thus the pure Euclidean distance of the feature histograms cannot be used as the basis of the classification. The most suitable classifiers are then methods that are based on hyperplanes, such as subspace classifiers and multilayer perceptron, while the distance-based methods, such as nearest neighbor classifiers and radial basis function networks, might be less effective.

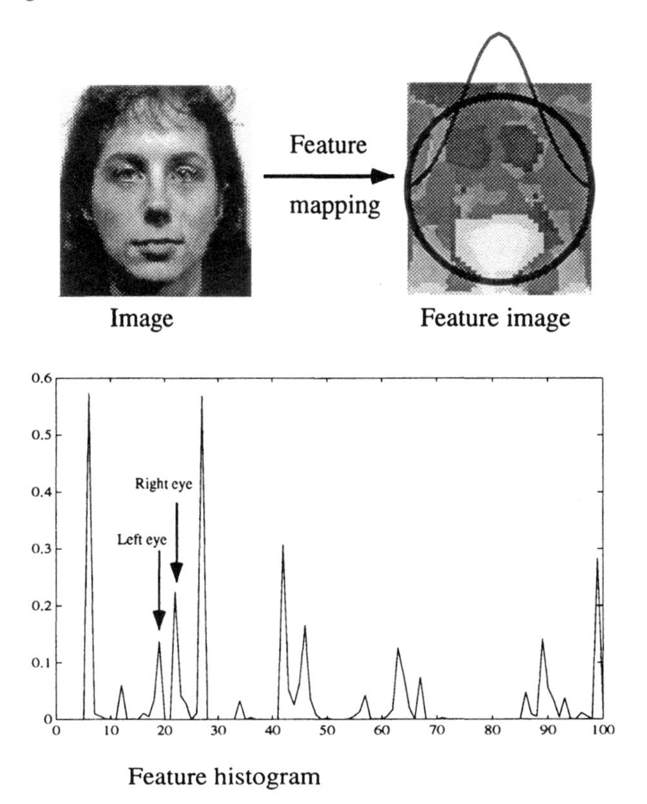

Figure 14 Upper part: an example image and the feature image. The image was a part of a 128×128 image. The 100 feature values are represented by gray levels in the feature image. The circle gives the approximate face area to be used in computing the feature histogram. Lower part: the Gaussian weighted feature histogram. The Gaussian weight function had width $R = 50$ and was centered as shown by the circle of radius R in the feature image.

Practical Example: Recognition of Wood Surface Defects

The proposed self-organizing feature construction method has been applied in some industrial pattern recognition problems, as described in [184] in detail. Here we give a short review on the recognition of wood surface defects.

As a natural material, wood has significant variation both within and between species, making it a difficult material for automatic grading. In principle, the inspection and quality classification of wood is straightforward: the quality class of each board depends on its defects and their distribution, as dictated by the quality standard. However, the definitions of the defects are based on their biological origin, appearance, or cause, so that the visual appearance of defects in the same class has substantial variation. The Finnish standards alone define 30 different de-

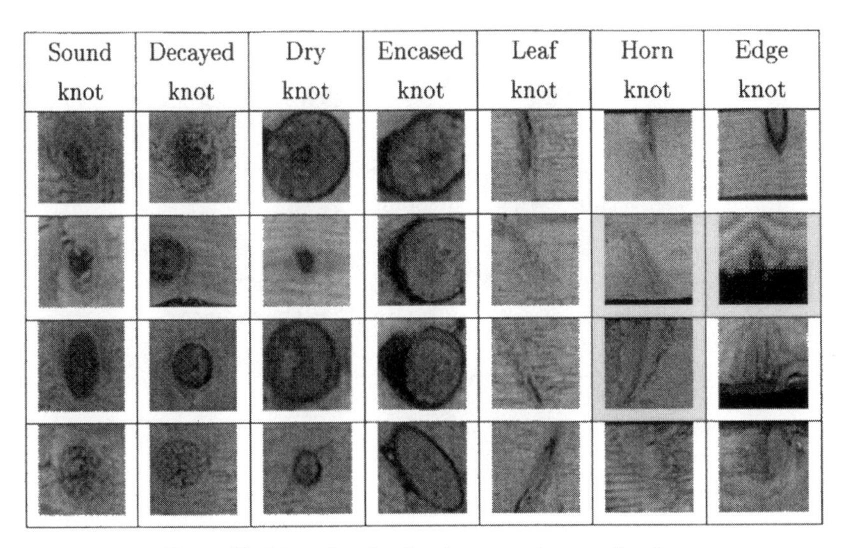

Figure 15 Examples of various knot types in spruce boards.

fect classes, such as sound, dry, encased, and decayed knots, resin pockets, splits, bark, wane, mould, etc., each with various degrees of seriousness.

Knots are the most common defect category and have a crucial role in sorting lumber. Figure 15 shows the most important knot classes on spruce boards.

Figure 16 shows a schematic of a wood surface defect recognition system, where the shape-related information is encoded by a self-organizing feature construction system into a "shape histogram," and the color histogram is collected by another multilayer SOM as an additional classification feature. A third type of information used as a classification feature, in addition to the shape and color feature histograms, was the energy of each Gabor filter over the whole image. It corresponds to a logarithmically sampled frequency spectrum of the image, and yields about 2% better recognition rates.

The image set used in the knot identification tests consisted of 438 spruce samples. The imaging was done at 0.5 mm $\times$ 0.5 mm resolution by a 3-CCD matrix camera with 8 bits/pixel quantization. Half of the samples (219) were used for training the classifier and the other half for evaluating the results.

Table IV shows the confusion matrix in the knot classification [184]. The recognition rate was about 85%, yielding about 90% correctness in the final grading of the boards, which is clearly better than the sustained performance of manual grading (about 70–80%). Based on these results, an industrial machine-vision–based system for automatic wood surface inspection has been developed, and is reported in [187]. The system is implemented on signal processors, so that it can

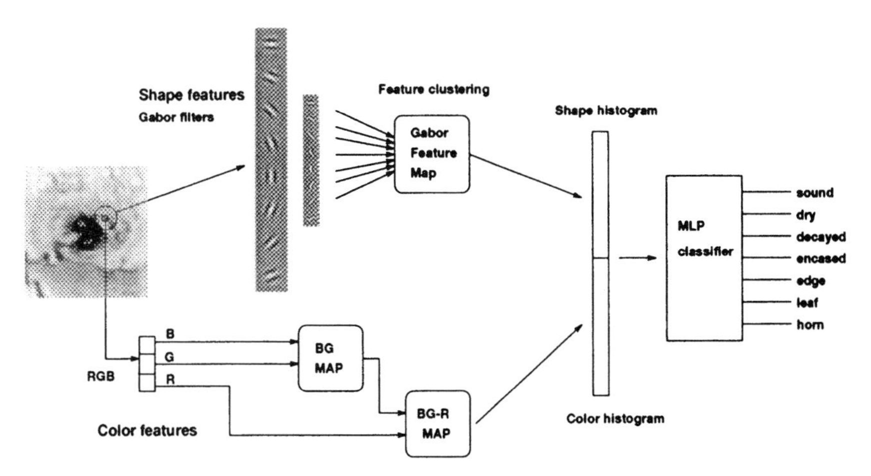

Figure 16 A schematic of the classification system combining shape-based and color-based information.

process more than one 2 × 2-m veneer sheet in a second, with imaging resolution of 1 mm, with about 20 defects on an average sheet.

4. Classification of Handwritten Digits

This section summarizes the results of a large comparison between various neural and classical statistical classification methods [157]. The data used in the experiments consisted of handwritten digits. Eight hundred ninety-four fill-

Table IV
Classification Results of Wood Surface Defects

	Dry	Encased	Decayed	Leaf	Edge	Horn	Sound	*N*	To other cl. %
Dry	26	1	0	1	0	0	4	32	19
Encased	1	10	0	0	0	0	2	13	23
Decayed	5	0	1	0	0	0	3	9	89
Leaf	0	0	1	24	0	0	3	28	14
Edge	0	0	0	0	34	2	0	36	6
Horn	0	0	0	0	6	10	0	16	37
Sound	4	0	0	0	0	0	81	85	4
N	36	11	2	25	40	12	93	219	
From other cl. %	28	9	50	4	15	17	13		15

in forms were digitized using an automatically fed flat binary scanner with the resolution of 300 × 300 dots per inch. The form was designed to allow simple segmentation of digits: each digit was written in a separate box so that for most cases there was no connecting, touching, or overlapping of the numerals. The size of each digit was normalized retaining the original aspect ratio to fit to a 32 × 32-pixel box. In the direction of the smaller size, the image was centered, and then the slant of writing was eliminated. The resulting image was finally concatenated to form a 1024-dimensional pattern vector having component values of ± 1 representing black and white pixels, respectively. The whole handwritten digit corpus of 17880 vectors was divided equally to form separate training and testing sets. The former was used in computing the Karhunen–Loève transform which was applied to both sets. The feature vectors so created were 64-dimensional, but each classification algorithm was allowed to select a smaller input vector dimensionality using training set cross-validation.

Figure 17 displays a sample of the digit images in the leftmost column. In the remaining columns, images reconstructed from an increasing number of features are shown. For the clarity of the visualization, the mean of the training set has been first subtracted from the digit images and then added back after the reconstruction.

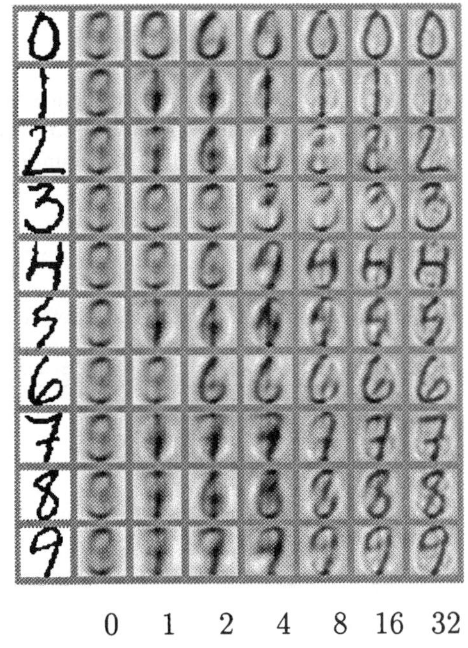

Figure 17 Some handwritten digits on the left and their reconstruction from varying number of features. The number of features used is shown below the images.

It can be noted how rapidly the reconstruction fidelity is increased due to the optimal information-preserving property of the Karhunen–Loève transform.

In the experiments, the maximum feature vector dimension was thus 64. Due to the effects of the curse of dimensionality, cross-validation indicated smaller input dimensionality to be optimal for some classifiers. Each classifier algorithm had its own set of cross-validated parameters. The cross-validation procedure was ten-fold: 8046 vectors were used in training a classifier and the remaining 894 vectors of the training set were used to evaluate the classification accuracy. This procedure was then repeated nine times until all the vectors in the training set had been used exactly nine times in training and once in evaluation. The cross-validated classification accuracy for the given set of parameter values was then calculated as the mean of the ten evaluations. By varying the parameter values, an optimal combination was found and it was used in creating the actual classifier using the whole training set. The final classification accuracy was calculated with that classifier and the original testing set. The classification error percentages are collected in Table V. Shown are testing set classification errors and, in parentheses, estimated standard deviation in ten independent trials for certain stochas-

Table V

Classification Accuracies for Handwritten Digit Data

Classifier	Error %		Cross-validated parameters
LDA	9.8		$d = 64$
QDA	3.7		$d = 47$
RDA	3.4		$d = 61,\ \gamma = 0.25,\ \lambda = 0$
KDA1	3.7		$d = 32,\ h = 3.0$
KDA2	3.5		$d = 36,\ h_1, \ldots, h_{10}$
RKDA	5.2	(.1)	$d = 32,\ \ell = 35$
MLP	5.4	(.3)	$d = 36,\ \ell = 40$
MLP+WD	3.5	(.1)	$[d = 36,\ \ell = 40],\ \lambda = 0.05$
LLR	2.8		$d = 36,\ \alpha = 0.1$
Tree classifier	16.8		$d = 16$, 849 terminal nodes
FDA/MARS	6.3		$d = 32$, 195 terms, second order
1-NN	4.2		$d = 64$
3-NN	3.8		$d = 38$
L-3-NN	3.6	(.1)	$[d = 38,\ \alpha = 0.1],\ \ell = 5750$, #epochs = 7
LVQ	4.0	(.1)	$[d = 38, \alpha(0) = 0.2,\ w = 0.5$, 10 epochs LVQ1], $\ell = 8000$, 1 epoch LVQ2
CLAFIC	4.3		$[d = 64],\ D = 29$
ALSM	3.1		$[d = 64,\ D = 29],\ \alpha = \beta = 3.1$, #epochs = 9
Committee	2.5		[LLR, ALSM, L-3-NN]

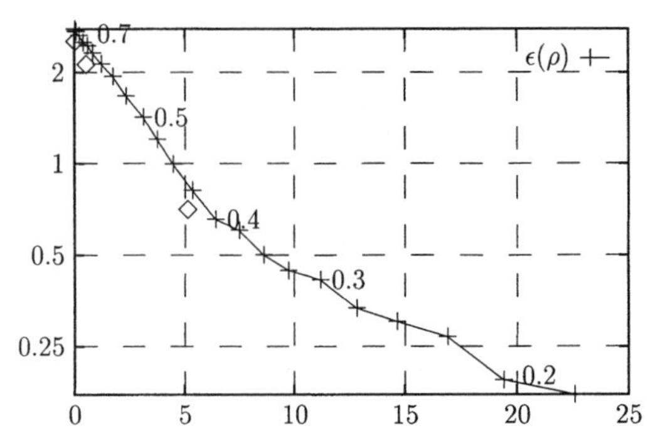

Figure 18 Error-reject curve for the LLR classifier. The rejection percentages are shown on the horizontal axis whereas the logarithmic vertical axis displays the remaining error percentages. The threshold parameter θ is given at selected points. The diamonds indicate the results obtained with the committee classifier using different voting strategies.

tic classifiers. The cross-validated parameters are given and parameters selected without cross-validation are shown in brackets.

Some evident conclusions can be drawn from the classification accuracies of Table V. First, the discriminant analysis methods, e.g., QDA, LDA, KDA, perform surprisingly well. This can be interpreted as an indirect indication that the distribution of the data closely resembles Gaussian in the Bayesian class border areas. Second, MLP performs surprisingly badly without the weight decay regularization modification. The tree classifier and MARS also disappoint. Third, the learning or adaptive algorithms such as ALSM and LVQ perform better than their nonadaptive counterparts such as CLAFIC and k-NN.

The committee classifier, the results of which are shown in the last line of Table V, was formed utilizing the majority voting principle from the LLR, ALSM, and L-3-NN classifiers. It can be seen that the committee quite clearly outperforms all the individual classifiers. Rejection option was also implemented. By using the LLR classifier and varying the rejection threshold θ of Eq. (30), the reject-error curve shown in Fig. 18 was obtained. The three diamonds in the figure display reject-error trade-off points obtained using the above described committee classifier with voting strategies allowing for rejection.

VI. SUMMARY

This chapter gave a review of neural network systems, techniques, and applications in pattern recognition (PR). Our point of view throughout the chapter is that, at the present state of the art, neural techniques are closely related with more conventional feature extraction and classification algorithms, which emanate from general statistical principles such as data compression, Bayesian classification, and regression. This helps in understanding the advantages and shortcomings of neural network models in pattern recognition tasks. Yet, we argue that neural networks have indeed brought new and valuable additions and insights to the PR theories, especially in their large flexible architectures and their emphasis on data-driven learning algorithms for massive training sets. It is no accident that the popularity of neural networks has coincided with the growing accessibility of computing power provided by the modern workstations.

We started the chapter by giving an overview of the problem and by introducing the general PR system, consisting of several consequent processing stages, neural or nonneural. We then concentrated on the two most important stages, feature extraction and classification. These are also the system components in which neural network techniques have been used most widely and to their best advantage. The most popular neural network approaches to these problems were given and contrasted with other existing solution methods.

Several concrete applications of neural networks on PR problems were then outlined partly as a literature survey, partly by summarizing the authors' own experiences in the field. Our original applications deal with face recognition, wood surface defect recognition, and handwritten digit recognition, in all of which neural networks have provided flexible and powerful PR methods. We hope that these case studies indicate that neural networks really work, but also that their use is not simple. As with any other engineering methodology, neural networks have to be carefully integrated into the total PR system in order to get out maximal performance.

REFERENCES

[1] B. D. Ripley. *J. Roy. Statist. Soc. Ser. B* 56:409–456, 1994.

[2] B. Cheng and D. Titterington. *Statist. Sci.* 9:2–54, 1994.

[3] Y. Idan, J.-M. Auger, N. Darbel, M. Sales, R. Chevallier, B. Dorizzi, and G. Cazuguel. In *Proceedings of the International Conference on Artificial Neural Networks* (I. Aleksander and J. Taylor, Eds.), Vol. 2, pp. 1607–1610. North-Holland, Brighton, England, 1992.

[4] L. Bottou, C. Cortes, J. S. Denker, H. Drucker, I. Guyon, L. D. Jackel, Y. LeCun, U. A. Müller, E. Säckinger, P. Y. Simard, and V. Vapnik. In *Proceedings of 12th International Conference on Pattern Recognition*, Vol. II, pp. 77–82. IEEE Computer Society Press, Jerusalem, 1994.

[5] D. Michie, D. J. Spiegelhalter, and C. C. Taylor (Eds.). *Machine Learning, Neural and Statistical Classification*. Ellis Horwood Limited, 1994.

[6] F. Blayo, Y. Cheneval, A. Guérin-Dugué, R. Chentouf, C. Aviles-Cruz, J. Madrenas, M. Moreno, and J. L. Voz. Deliverable R3-B4-P task B4: Benchmarks. Technical Report ESPRIT Basic Research Project Number 6891, 1995.

[7] Britannica Online. Encyclopædia Britannica on the Internet, 1996. Available at <http://www.eb.com/>.
[8] H. C. Andrews. *Introduction to Mathematical Techniques in Pattern Recognition.* John Wiley & Sons Inc., New York, 1972.
[9] R. O. Duda and P. E. Hart. *Pattern Recognition and Scene Analysis.* John Wiley & Sons Inc., New York, 1973.
[10] J. T. Tou and R. C. Gonzalez. *Pattern Recognition Principles.* Addison-Wesley, Reading, MA, 1974.
[11] T. Y. Young and T. W. Calvert. *Classification, Estimation and Pattern Recognition.* Elsevier Science Publishers, New York, 1974.
[12] R. Gonzalez and M. Thomason. *Syntactic Pattern Recognition.* Addison-Wesley, Reading, MA, 1978.
[13] J. Sklansky and G. N. Wassel. *Pattern Classifiers and Trainable Machine.* Springer-Verlag, Berlin/New York, 1981.
[14] P. A. Devijver and J. Kittler. *Pattern Recognition: A Statistical Approach.* Prentice-Hall International, London, 1982.
[15] K. S. Fu. *Syntactic pattern recognition and applications.* Prentice-Hall, Englewood Cliffs, NJ, 1982.
[16] K. Fukunaga. *Introduction to Statistical Pattern Recognition*, 2nd ed. Academic Press, New York, 1990.
[17] S.-T. Bow. *Pattern Recognition and Image Preprocessing.* Marcel Dekker, Inc., New York, 1992.
[18] Y.-H. Pao. *Adaptive Pattern Recognition and Neural Networks.* Addison-Wesley, Reading, MA, 1989.
[19] C. W. Therrien. *Decision, Estimation, and Classification.* John Wiley and Sons, New York, 1989.
[20] R. J. Schalkoff. *Pattern Recognition: Statistical, Structural and Neural Approaches.* John Wiley & Sons, New York, 1992.
[21] M. Pavel. *Fundamentals of Pattern Recognition*, 2nd ed. Marcel Dekker, New York, 1993.
[22] C. M. Bishop. *Neural Networks for Pattern Recognition.* Oxford University Press, London/New York, 1995.
[23] B. D. Ripley. *Pattern Recognition and Neural Networks.* Cambridge University Press, London/New York, 1996.
[24] J. Kittler, K. S. Fu, and L. F. Pau (Eds.). *Pattern Recognition Theory and Applications; Proceedings of the NATO Advanced Study Institute.* D. Reidel, Dordrecht, 1982.
[25] L. N. Kanal (Ed.). *Progress in Pattern Recognition 1.* North-Holland, Amsterdam, 1981.
[26] L. N. Kanal (Ed.). *Progress in Pattern Recognition 2.* North-Holland, Amsterdam, 1985.
[27] B. V. Dasarathy. *Nearest Neighbor Pattern Classification Techniques.* IEEE Computer Society Press, New York, 1991.
[28] C. H. Chen, L. F. Pau, and P. S. P. Wang. *Handbook of Pattern Recognition and Computer Vision.* World Scientific Publishing, Singapore, 1993.
[29] K. S. Fu and A. Rosenfeld. *Computer* 17:274–282, 1984.
[30] R. Bellman. *Adaptive Control Processes: A Guided Tour.* Princeton University Press, Princeton, NJ, 1961.
[31] A. K. Jain and B. Chandrasekaran. In *Handbook of Statistics* (P. R. Krishnaiah and L. N. Kanal, Eds.), Vol. 2, pp. 835–855. North-Holland, Amsterdam, 1982.
[32] J. Hartigan. *Clustering Algorithms.* John Wiley & Sons, New York, 1975.
[33] J. Laaksonen and E. Oja. In *Proceedings of the International Conference on Engineering Applications of Neural Networks*, Stockholm, Sweden, 1997.
[34] B. Widrow and M. Lehr. *Proc. IEEE* 78:1415–1442, 1990.

[35] J. Lampinen and E. Oja. *IEEE Trans. Neural Networks* 6:539–547, 1995.

[36] S. Haykin. *Neural Networks: A Comprehensive Foundation.* Macmillan College Publishing Company, Inc., New York, 1994.

[37] S. Wold. *Pattern Recog.* 8:127–139, 1976.

[38] E. Oja. *J. Math. Biol.* 15:267–273, 1982.

[39] T. Kohonen. *Biol. Cybernet.* 43:59–69, 1982.

[40] T. Kohonen *et al.* 2358 studies of the self-organizing map (SOM) and learning vector quantization (LVQ), 1996. Available at <http://www.cis.hut.fi/nnrc/>.

[41] E. Oja *et al.* A list of references related to PCA neural networks, 1996. Available at <http://www.cis.hut.fi/projects/pca/>.

[42] D. DeMers and G. Cottrell. In *Neural Information Processing Systems 5* (S. Hanson, J. Cowan, and L. Giles, Eds.), pp. 580–587. Morgan Kaufmann Publishers, San Francisco, CA, 1993.

[43] E. Oja. In *Proceedings of the International Conference on Artificial Neural Networks* (T. Kohonen, K. Mäkisara, O. Simula, and J. Kangas, Eds.), Vol. 1, pp. 737–745. North-Holland, Espoo, Finland, 1991.

[44] S. Usui, S. Nakauchi, and M. Nakano. In *Proceedings of the International Conference on Artificial Neural Networks* (T. Kohonen, K. Mäkisara, O. Simula, and J. Kangas, Eds.), pp. 867–872. North-Holland, Espoo, Finland, 1991.

[45] K. Funahashi. *Neural Networks* 2:183–192, 1989.

[46] K. Hornik, M. Stinchcombe, and H. White. *Neural Networks* 2:359–368, 1989.

[47] E. Oja. *Subspace Methods of Pattern Recognition.* Research Studies Press Ltd., Letchworth, England, 1983.

[48] Z. Wang and J. V. Hanson. In *Proceedings of the World Congress on Neural Networks*, pp. IV–605–608. Lawrence Erlbaum Associates, Hillsdale, NJ, 1993.

[49] E. Oja. *Neural Networks* 5:927–935, 1992.

[50] E. Oja and J. Karhunen. Nonlinear PCA: Algorithms and applications. Technical Report A18, Laboratory of Computer and Information Science, Helsinki University of Technology, 1993.

[51] J. Karhunen and J. Joutsensalo. *Neural Networks* 7:13–127, 1994.

[52] E. Oja. The nonlinear PCA learning rule and signal separation—mathematical analysis, Technical Report A26, Laboratory of Computer and Information Science, Helsinki University of Technology, 1995.

[53] J. Karhunen, E. Oja, L. Wang, R. Vigário, and J. Joutsensalo. *IEEE Trans. Neural Networks*, to appear.

[54] C. Jutten and J. Herault. *Signal Process.* 24:1–10, 1991.

[55] P. Comon. *Signal Process.* 36:287–314, 1994.

[56] J.-F. Cardoso and B. Laheld. *IEEE Trans. Signal Process.* 44, 1996.

[57] J. Karhunen, A. Hyvärinen, R. Vigário, J. Hurri, and E. Oja. In *Proceedings of the IEEE 1997 International Conference on Acoustics, Speech, and Signal Processing*, Munich, Germany, 1997.

[58] T. Kohonen. *Proc. IEEE* 78:1464–1480, 1990.

[59] T. Kohonen, *Self-Organizing Maps.* Springer-Verlag, Berlin/New York, 1995.

[60] T. Kohonen. *Self-Organization and Associative Memory*, 2nd ed. Springer-Verlag, Berlin, Heidelberg, New York, 1988.

[61] H. Ritter, T. Martinetz, and K. Schulten. *Neural Computation and Self-Organizing Maps: An Introduction.* Addison-Wesley, Reading, MA, 1992.

[62] J. Lampinen. Neural pattern recognition: Distortion tolerance by self-organizing maps. Ph.D. Thesis, Lappenranta University of Technology, Lappeenranta, Finland, 1992.

[63] A. Visa, K. Valkealahti, and O. Simula. In *Proceedings of the International Joint Conference on Neural Networks (IJCNN)*, pp. 1001–1006, Singapore, 1991.

[64] T. Kohonen, E. Oja, O. Simula, A. Visa, and J. Kangas. *Proc. IEEE* 84:1358–1384, 1996.

[65] L. Holmström, P. Koistinen, J. Laaksonen, and E. Oja. Comparison of neural and statistical classifiers—theory and practice. Technical Report A13, Rolf Nevanlinna Institute, Helsinki, 1996.
[66] A. K. Jain and J. Mao. In *Computational Intelligence Imitating Life* (J. M. Zurada, R. J. Marks II, and C. J. Robinson, Eds.). Chap. IV-1, pp. 194–212. IEEE Press, New York, 1994.
[67] R. A. Redner and H. F. Walker. *SIAM Rev.* 26, 1984.
[68] H. G. C. Tråvén. *IEEE Trans. Neural Networks* 2:366–377, 1991.
[69] C. E. Priebe and D. J. Marchette. *Pattern Recog.* 24:1197–1209, 1991.
[70] C. E. Priebe and D. J. Marchette. *Pattern Recog.* 26:771–785, 1993.
[71] T. Hastie and R. Tibshirani. *J. Roy. Statist. Soc.* (*Series B*) 58:155–176, 1996.
[72] G. J. McLachlan, *Discriminant Analysis and Statistical Pattern Recognition*. John Wiley & Sons, New York, 1992.
[73] J. H. Friedman. *J. Amer. Statist. Assoc.* 84:165–175, 1989.
[74] D. J. Hand. *Kernel Discriminant Analysis*. Research Studies Press, Chichester, 1982.
[75] B. W. Silverman and M. C. Jones. *Internat. Statist. Rev.* 57:233–247, 1989.
[76] B. W. Silverman. *Density Estimation for Statistics and Data Analysis*. Chapman & Hall, London/New York, 1986.
[77] D. W. Scott. *Multivariate Density Estimation: Theory, Practice, and Visualization*. John Wiley & Sons, New York, (1992.
[78] M. P. Wand and M. C. Jones. *Kernel Smoothing*. Chapman & Hall, London, New York, 1995.
[79] G. E. Tutz. *Biometrika* 73:405–411, 1986.
[80] D. F. Specht. *Neural Networks* 3:109–118, 1990.
[81] K. Fukunaga and J. M. Mantock. *IEEE Trans. Pattern Anal. Machine Intell.* PAMI-6:115–118, 1984.
[82] K. Fukunaga and R. R. Hayes. *IEEE Trans. Pattern Anal. Machine Intell.* 11:423–425, 1989.
[83] I. Grabec. *Biol. Cybernet.* 63:403–409, 1990.
[84] P. Smyth and J. Mellstrom. In *Advances in Neural Information Processing Systems 4* (J. Moody, S. Hanson, and R. Lippmann, Eds.), pp. 667–674. Morgan Kaufmann, San Mateo, CA, 1992.
[85] L. Wu and F. Fallside. *Computer Speech Lang.* 5:207–229, 1991.
[86] L. Holmström and A. Hämäläinen. In *Proceedings of the 1993 IEEE International Conference on Neural Networks*, Vol. 1, pp. 417–421, San Francisco, California, 1993.
[87] J. MacQueen. In *Proceedings of the Fifth Berkeley Symposium on Mathematical Statistics and Problems* (L. M. LeCam and J. Neyman, Eds.), pp. 281–297. U.C. Berkeley Press, Berkeley, CA, 1967.
[88] J. H. Friedman and W. Stuetzle. *J. Amer. Statist. Assoc.* 76:817–823, 1981.
[89] T. E. Flick, L. K. Jones, R. G. Priest, and C. Herman. *Pattern Recog.* 23:1367–1376, 1990.
[90] T. J. Hastie and R. J. Tibshirani. *Generalized Additive Models*. Chapman & Hall, London/New York, 1990.
[91] P. Koistinen and L. Holmström. In *Advances in Neural Information Processing Systems 4* (J. E. Moody, S. J. Hanson, and R. P. Lippman, Eds.), pp. 1033–1039. Morgan Kaufmann, San Mateo, CA, 1992.
[92] D. F. Specht. *IEEE Trans. Neural Networks* 2:568–576, 1991.
[93] H. White. *Neural Comput.* 1:425–464, 1989.
[94] M. D. Richard and R. P. Lippman. *Neural Comput.* 3:461–483, 1991.
[95] J. S. Bridle. In *Advances in Neural Information Processing Systems 2* (D. Touretzky, Ed.), pp. 211–217. Morgan Kaufmann, San Mateo, CA, 1990.
[96] W. H. Highleyman. *Proc. IRE* 50:1501–1514, 1962.
[97] L. Breiman, J. Friedman, R. Olshen, and C. Stone. *Classification and Regression Trees*. Chapman & Hall, London/New York, 1984.
[98] W. S. Cleveland and S. J. Devlin. *J. Amer. Statist. Assoc.* 83:596–610, 1988.

[99] W. Cleveland and C. Loader. Smoothing by local regression: Principles and methods. Technical Report, AT&T Bell Laboratories, 1995.
[100] T. J. Hastie and C. Loader. *Statist. Sci.* 8:120–143, 1993.
[101] J. N. Morgan and J. A. Sonquist. *J. Amer. Statist. Assoc.* 58:415–434, 1963.
[102] R. A. Becker, J. M. Chambers, and A. R. Wilks. *The NEW S Language.* Chapman & Hall, New York, 1988.
[103] J. M. Chambers and T. J. Hastie (Eds.). *Statistical Models in S*. Chapman & Hall, New York, 1992.
[104] W. N. Venables and B. D. Ripley. *Modern Applied Statistics with S-Plus.* Springer-Verlag, New York, 1994.
[105] L. A. Clark and D. Pregibon. In *Statistical Models in S* (J. M. Chambers and T. J. Hastie, Eds.), Chap. 9. Chapman & Hall, New York, 1992.
[106] J. H. Friedman. *Ann. Statist.* 19:1–141, 1991.
[107] P. Craven and G. Wahba. *Numer. Math.* 31:317–403, 1979.
[108] T. Hastie, R. Tibshirani, and A. Buja. *J. Amer. Statist. Assoc.* 89:1255–1270, 1994.
[109] E. Fix and J. L. Hodges. Discriminatory analysis—nonparametric discrimination: Consistency properties. Technical Report Number 4, Project Number 21-49-004, USAF School of Aviation Medicine, Randolph Field, TX, 1951.
[110] T. M. Cover and P. E. Hart. *IEEE Trans. Inform. Theory* 13:21–27, 1967.
[111] G. T. Toussaint, E. Backer, P. Devijver, K. Fukunaga, and J. Kittler. In *Pattern Recognition Theory and Applications; Proceedings of the NATO Advanced Study Institute* (J. Kittler, K. S. Fu, and L. F. Pau, Eds.), pp. 569–572. D. Reidel, Dordrecht, 1982.
[112] D. L. Wilson. *IEEE Trans. Systems, Man, Cybernet.* 2:408–420, 1972.
[113] P. E. Hart. *IEEE Trans. Inform. Theory* 14:515–516, 1968.
[114] J. Laaksonen and E. Oja. In *Proceedings of the International Conference on Neural Networks*, Vol. 3, pp. 1480–1483. Washington, DC, 1996.
[115] S. Watanabe, P. F. Lambert, C. A. Kulikowski, J. L. Buxton, and R. Walker. In *Computer and Information Sciences II* (J. Tou, Ed.). Academic Press, New York, 1967.
[116] J. Laaksonen and E. Oja. In *Proceedings of the International Conference on Artificial Neural Networks*, pp. 227–232. Bochum, Germany, 1996.
[117] W. S. McCulloch and W. Pitts. *Bull. Math. Biophys.* 5:115–133, 1943.
[118] F. Rosenblatt. *Psychol. Rev.* 65:386–408, 1958.
[119] F. Rosenblatt. *Principles of Neurodynamics: Perceptrons and the Theory of Brain Mechanisms.* Spartan Books, Washington, DC, 1961.
[120] T. J. Sejnowski and C. R. Rosenberg. *J. Complex Syst.* 1:145–168, 1987.
[121] Y. LeCun, B. Boser, J. S. Denker, D. Henderson, R. E. Howard, W. Hubbard, and L. D. Jackel. *Neural Comput.* 1:541–551, 1989.
[122] T. Kohonen. *Computer* 21:11–22, 1988.
[123] C. J. Stone. *J. Roy. Statist. Soc. Ser. B* 36:111–147, 1974.
[124] L. Holmström, S. Sain, and H. Miettinen. *Computer Phys. Commun.* 88:195–210, 1995.
[125] J. Geist, R. A. Wilkinson, S. Janet, P. J. Grother, B. Hammond, N. W. Larsen, R. M. Klear, M. J. Matsko, C. J. C. Burges, R. Creecy, J. J. Hull, T. P. Vogl, and C. L. Wilson. The second census optical character recognition systems conference. Technical Report NISTIR 5452, National Institute of Standards and Technology, 1992.
[126] M. P. Perrone. In *Neural Information Processing Systems 6* (J. D.Cowan, G. Tesauro, and J. Alspector, Eds.), pp. 1188–1189. Morgan Kaufmann, San Francisco, CA, 1994.
[127] L. Xu, A. Krzyżak, and C. Y. Suen. *IEEE Trans. Systems, Man, Cybernet.* 22:418–435, 1992.
[128] T. K. Ho, J. J. Hull, and S. N. Srihari. In *Proceedings of SPIE Conference on Machine Vision Applications in Character Recognition and Industrial Inspection* (D. P. D'Amato, W.-E. Blanz, B. E. Dom, and S. N. Srihari, Eds.), no. 1661 in SPIE, pp. 137–145. SPIE, 1992.

[129] T. K. Ho, J. J. Hull, and S. N. Srihari. *IEEE Trans. Pattern Anal. Machine Intell.* 16:66–75, 1994.

[130] Y. S. Huang, K. Liu, and C. Y. Suen. *Intern. J. Pattern Recog. Artif. Intell.* 9:579–597, 1995.

[131] L. Xu, A. Krzyżak, and C. Y. Suen, In *Proceedings of 1991 International Joint Conference on Neural Networks*, Vol. 1, pp. 43–48. Seattle, WA, 1991.

[132] C. Nadal, R. Legault, and C. Y. Suen. In *Proceedings of the 10th International Conference on Pattern Recognition*, pp. 443–449. Atlantic City, NJ, 1990.

[133] F. Kimura and M. Shridhar. *Pattern Recog.* 24:969–983, 1991.

[134] C. Y. Suen, C. Nadal, R. Legault, T. A. Mai, and L. Lam. *Proceedings of the IEEE* 80:1162–1180, 1992.

[135] L. Lam and C. Y. Suen. In *Proceedings of 12th International Conference on Pattern Recognition*, Vol. II, pp. 418–420. IEEE Computer Society Press, Jerusalem, 1994.

[136] L. Lam and C. Y. Suen. *Pattern Recog. Lett.* 16:945–954, 1995.

[137] L. Xu and M. I. Jordan. In *Proceedings of the World Congress on Neural Networks*, Vol. IV, pp. 227–230, 1993.

[138] L. Xu, M. I. Jordan, and G. E. Hinton. In *Neural Information Processing Systems 7* (G. Tesauro, D. S. Touretzky, and T. K. Leen, Eds.), pp. 633–640. MIT Press, Cambridge, MA, 1995.

[139] M. I. Jordan and R. A. Jacobs. *Neural Comput.* 6:181–214, 1994.

[140] J. Franke and E. Mandler. In *Proceedings of the 11th International Conference on Pattern Recognition*, Vol. II, pp. 611–614. The Hague, 1992.

[141] R. A. Jacobs. *Neural Comput.* 7:867–888, 1995.

[142] V. Tresp and M. Taniguchi. In *Neural Information Processing Systems 7* (G. Tesauro, D. S. Touretzky, and T. K. Leen, Eds.), pp. 419–426. MIT Press, Cambridge, MA, 1995.

[143] M. P. Perrone and L. N. Cooper. In *Artificial Neural Networks for Speech and Vision* (R. J. Mammone, Ed.), pp. 126–142. Chapman & Hall, London/New York, 1993.

[144] L. K. Hansen and P. Salamon. *IEEE Trans. Pattern Anal. Machine Intell.* 12:993–1001, 1990.

[145] A. Krogh and J. Vedelsby. In *Neural Information Processing Systems 7* (G. Tesauro, D. S. Touretzky, and T. K. Leen, Eds.), pp. 231–238. MIT Press, Cambridge, MA, 1995.

[146] D. H. Wolpert. *Neural Networks* 5:241–259, 1992.

[147] F. Śmieja. The pandemonium system of reflective agents. Technical Report 1994/2, German National Research Center for Computer Science (GMD), 1994.

[148] Y. Idan and J.-M. Auger. In *Proceedings of SPIE Conference on Neural and Stochastic Methods in Image and Signal Processing* (S.-S. Chen, Ed.), no. 1766 in SPIE, pp. 437–443, SPIE, 1992.

[149] H. Drucker, R. Schapire, and P. Simard. *Internat. J. Pattern Recog. Artif. Intell.* 7:705–719, 1993.

[150] H. Drucker, C. Cortes, L. D. Jackel, Y. LeCun, and V. Vapnik. *Neural Comput.* 6:1289–1301, 1994.

[151] G. E. Hinton, M. Revow, and P. Dayan. In *Neural Information Processing Systems 7* (G. Tesauro, D. S. Touretzky, and T. K. Leen, Eds.), pp. 1015–1022. MIT Press, Cambridge, MA, 1995.

[152] H. Schwenk and M. Milgram. In *Neural Information Processing Systems 7* (G. Tesauro, D. S. Touretzky, and T. K. Leen, Eds.), pp. 991–998. MIT Press, Cambridge, MA, 1995.

[153] P. Sollich and A. Krogh. In *Neural Information Processing Systems 8* (D. S. Touretzky, M. C. Mozer, and M. E. Hasselmo, Eds.). MIT Press, Cambridge, MA, 1995.

[154] L. Prechelt. A study of experimental evaluations of neural network learning algorithms: Current research practice. Technical Report 19/94, Fakultät für Informatik, Universität Karlsruhe, D-76128 Karlsruhe, Germany, 1994.

[155] W. F. Schmidt, D. F. Levelt, and R. P. W. Duin. In *Pattern Recognition in Practice IV* (E. S. Gelsema and L. S. Kanal, Eds.), Vol. 16 of *Machine Intelligence and Pattern Recognition*. Elsevier Science, New York, 1994.

[156] J. L. Blue, G. T. Candela, P. J. Grother, R. Chellappa, and C. L. Wilson. *Pattern Recog.* 27:485–501, 1994.

[157] L. Holmström, P. Koistinen, J. Laaksonen, and E. Oja. *IEEE Trans. Neural Networks* 8, 1997.

[158] R. P. W. Duin. *Pattern Recog. Lett.* 17:529–536, 1996.

[159] L. Devroye. *IEEE Trans. Pattern Anal. Machine Intell.* 10:530–543, 1988.

[160] SIENA—Stimulation Initiative for European Neural Applications, ESPRIT Project 9811. Available at <http://www.mbfys.kun.nl/snn/siena>.

[161] M. Zhang and J. Fulcher. *IEEE Trans. Neural Networks* 7:555–567, 1996.

[162] W. Konen, T. Maurer, and C. von der Malsburg. *Neural Networks* 7:1019–1030, 1994.

[163] W. Konen, S. Fuhrmann, M. Hormel, and A. Flügel. In *Proceedings of the Industrial Conference "Applications in Industrial & Service Sectors" in the International Conference on Artificial Neural Networks, ICANN'95*, 1995.

[164] F. Takeda and S. Omatu. *IEEE Trans. Neural Networks* 6:73–77, 1995.

[165] Y. Qi and B. R. Hunt. *IEEE Trans. Image Process.* 4:870–874, 1995.

[166] C. S. Cruz *et al.* Hybrid neural methods in classification problems. Technical Report IIC 9501, Instituto de Ingenieria del Conocimiento, Universidad Autonoma, 28049 Madrid, 1995.

[167] A. Hogervorst *et al.* In *Neural Networks: Artificial Intelligence and Industrial Applications. Proceedings of the 3rd Annual SNN Symposium on Neural Networks*. Springer-Verlag, London, 1995.

[168] R. Nekovei and Y. Sung. *IEEE Trans. Neural Networks* 6:64–72, 1995.

[169] G. Chiou and J.-N. Hwang. *IEEE Trans. Image Process.* 4:1407–1416, 1995.

[170] S.Chakrabarti, N. Bindal, and K. Theaghadran. *IEEE Trans. Neural Networks* 6:760–766, 1995.

[171] A. Waxman, M. Seibert, A. Gove, D. Fay, A. Bernardon, C. Lazott, W. Steele, and R. Cunningham. *Neural Networks* 8:1029–1051, 1995.

[172] M. W. Roth. *IEEE Trans. Neural Networks* 1:1990, 1990.

[173] R. Bailey and M. Srinath. *IEEE Trans. Pattern Anal. Machine Intell.* 18:389–399, 1996.

[174] W. Weideman, M. Manry, H.-C. Yau, and W. Gong. *IEEE Trans. Neural Networks* 6:1524–1530, 1995.

[175] S. Lee and J. C.-J. Pan. *IEEE Trans. Neural Networks* 7:455–474, 1996.

[176] A. Waibel *et al. IEEE Trans. Acoustics, Speech, Signal Process.* 37:328–339, 1989.

[177] Y. L. Cun, J. Boser, J. Denker, D. Henderson, R. Howard, W. Hubbard, and L. Jackel. In *Neural Networks, Current Applications* (P. Lisboa, Ed.), pp. 185–195. Chapman & Hall, London, 1992.

[178] Y. Le Cun, J. Denker, S. Solla, R. Howard, and L. Jackel. In *Neural Information Processing Systems* (D. Touretzky, Ed.), Vol. 2. Morgan Kaufman, San Mateo, CA, 1989.

[179] Y. Le Cun, L. Jackel, L. Bottou, C. Cortes, J. Denker, H. Drucker, I. Guyon, U. Müller, E. Säckinger, P. Simard, and V. Vapnik. In *Proceedings of the International Conference on Artificial Neural Networks ICANN'95*, Vol. 1, pp. 53–60, Paris, France, 1995.

[180] K. Fukushima, S. Miyake, and T. Ito. *IEEE Trans. Systems, Man, Cybernet.* 13:826–834, 1983.

[181] D. Hubel and T. Wiesel. *J. Physiol.* 160:106–154, 1962.

[182] K. Fukushima. In *Artificial Neural Networks* (T. Kohonen, K. Mäkisara, J. Kangas, and O. Simula, Eds.), Vol. 1, pp. 105–110. North-Holland, Amsterdam, 1991.

[183] J. Lampinen. In *Applications of Artificial Neural Networks II, Proc. SPIE 1469* (S. K. Rogers, Ed.), pp. 832–842, 1991.

[184] L. Lampinen and S. Smolander. *Internat. J. Pattern Recog. Artif. Intell.* 10:97–113, 1996.

[185] J. Daugman. *J. Opt. Soc. Amer. (A)* 2:1160–1169, 1985.

[186] J. Lampinen and E. Oja. *J. Math. Imag. Vision* 2:261–272, 1992.

[187] J. Lampinen, S. Smolander, and M. Korhonen. In *Proceedings of the Industrial Conference "Technical Diagnosis & Nondestructive Testing" in the International Conference on Artificial Neural Networks, ICANN'95*, 1995.

Comparison of Statistical and Neural Classifiers and Their Applications to Optical Character Recognition and Speech Classification

Ethem Alpaydın
Department of Computer Engineering
Boğaziçi University
TR-80815 İstanbul, Turkey

Fikret Gürgen
Department of Computer Engineering
Boğaziçi University
TR-80815 İstanbul, Turkey

I. INTRODUCTION

Improving person–machine communication leads to a wider use of advanced information technologies. Toward this aim, character recognition and speech recognition are two applications whose automatization allows easier interaction with a computer. As they are the basic means of person-to-person communication, they are known by everyone and require no special training. Speech in particular is the most natural form of human communication and writing is the tool by which humanity has stored and transferred its knowledge for many millennia.

In a typical pattern recognition system (Fig. 1), the first step is the acquisition of data. This raw data is preprocessed to suppress noise and normalize input. Features are those parts of the signal that carry information salient to its identity and their extraction is an abstraction operation where the important is extracted and the irrelevant is discarded. Classification is the assignment of the input as an element of one of a set of predefined classes.

Copyright © 1998 by Academic Press. All rights of reproduction in any form reserved.

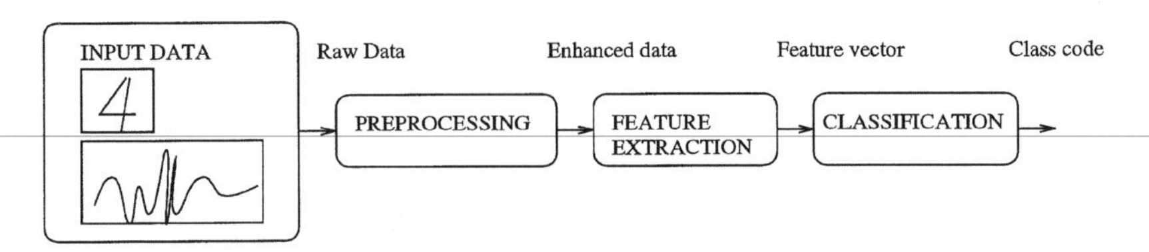

Figure 1 A pattern recognition system where input is an image, as in optical character recognition, or a time series, as in speech classification.

The rules for classification are generally not known exactly and thus are estimated. A classifier is written as a parametric model whose parameters are computed using a given training sample to optimize a given error criterion. Different approaches for classification differ in their assumptions about the model, in the way parameters are computed, or in the error criterion they optimize.

Statistical classifiers model the class-conditional densities and base their decisions on the posteriors which are computed using the class-conditional likelihoods and the priors. Likelihoods are assumed to either come from a given probability density family, e.g., normal, come from a mixture of such densities, or be written in a completely nonparametric way. Bayes decision theory then allows us to choose the class that minimizes the decision risk. The parameters of the densities are estimated to maximize the likelihood of the given sample for that class.

This contrasts with approaches where the discriminants are directly estimated. Neural networks are such approaches and their outputs can be converted directly to posteriors, eliminating the need of assuming a statistical model. From a statistical perspective, a multilayer network is a linear sum of nonlinear basis functions. In the neural network terminology, the nonlinear basis functions are called hidden units and the parameters are called connection weights. In a training process, given a training sample, the weights that minimize the difference between network outputs and required outputs are computed.

This chapter has the aim of comparing these two approaches and extends a previous study [1]. In Section II, we define the two applications that we are concerned with in this study, namely, optical character recognition and speech recognition. We show that these two applications have many common subproblems and quite similar approaches have been used in the past to implement them, both statistical and neural. Section III details how, in the two applications, data are acquired and preprocessed before they can be fed to the classifier. In Section IV, we define formally the problem from a statistical point of view and explain the three approaches of parametric, nonparametric, and semiparametric estimation. In Section V, we discuss the neural approaches such as simple and multilayer perceptrons and radial basis function networks. A literature survey for the two applications is given in Section VI. In Section VII, we give simulation results on two data sets. We conclude in Section VIII.

II. APPLICATIONS

Character recognition is of two forms. In *printed* character recognition, any character image is one of a predefined number of styles which are called *fonts*. Printed character recognition systems generally work by storing templates of character images for all fonts and matching the given image against these stored images to choose the best match. This contrasts with *handwritten* character recognition where there are practically infinite ways of writing a character. It is this latter that we are interested in. In handwritten character recognition, the medium may be two sorts. In *optical* character recognition, the writer writes generally on paper by using a marker of different brightness. The contrast is acquired optically through a scanner or camera and a two-dimensional image is formed. Because recognition is done generally long after the writing is done, it is named *off-line*. In *pen-based* character recognition, the writing is done on a touch-sensitive pad using a special pen. While it is moved over the pad, the coordinates of the pen-tip are returned at each sampling, leading to a sequence of pen-tip positions for each written character. Recognition is done while writing takes place and is called *on-line*. A special journal issue on different approaches to character recognition, edited by Pavlidis and Mori [2], recently appeared.

In speech recognition, there is no analogy to printed character recognition as the sound to be recognized is natural, never synthetic. Speech is captured using a microphone and is digitized to get a set of samples in time from the sound waveform. The sound spectrogram is a three-dimensional representation of the speech intensity, in different frequency bands, over time. Another way of representing speech is by modeling the human vocal tract as a tube whose resonances produce speech; they are called *formants* and they represent the frequencies that pass the most acoustic energy; typically there are three. For a more detailed analysis, refer to the book by Rabiner and Juang [3].

The two tasks of character recognition and speech recognition have a number of common problems. One is the segmentation of characters or phonemes from a stream of text image or speech. This is especially a problem when cursive handwriting or continuous phrases is the case; the system, before recognizing individual characters or phonemes, should determine the boundaries. To facilitate recognition, some systems require that inputs be isolated before recognition, by providing separate boxes for different characters or slow, careful articulation of words.

Another problem is that of being independent from writer or speaker; a recognizer that can recognize inputs from a small set of people is rarely useful. But the recognizer should also have the ability to adapt to a particular user if required so as to be able to recognize that user's handwriting/speech with higher accuracy.

Recognition is also dependent on the domain. If only postal codes are to be optically recognized, then in most countries only digits are valid classes; a voice-controlled system may have a small set of commands to be recognized. To minimize errors and rejects, more complicated tasks require also the integration of a vocabulary so as to be able to use lexical information to aid in recognition. This creates the problems of storing a lexicon and accessing it efficiently.

The best existing systems perform well only on artificially constrained tasks. Performance is better if samples are provided for all speakers/writers, when words are spoken/letters are written in isolation, when the vocabulary size is small, and when restricted language models are used to constrain allowable sequences [2, 4].

III. DATA ACQUISITION AND PREPROCESSING

A. Optical Character Recognition

In optical character recognition, preprocessing should be done before individual character images can be fed to the classifier.

Depending on how small the characters are, a suitable resolution should be chosen first for scanning. Typically a resolution of 300 dpi (dots per inch) is used. With smaller resolutions and smaller character images, dots and accents may be lost or images may get connected.

In most applications, characters are written in special places on preprinted forms [2, 5]. So the first step after scanning is registration, which is the determination of how the original form is translated and rotated to get the final image. This is done by matching a number of points in the input form to the original blank form. Then given the coordinates of the input fields, their coordinates in the input form can be computed and the input field images extracted. The characters in a field are then segmented and individual character images are obtained. These are then size-normalized to scale all characters to the same height and width and slant-normalized to reduce the slant variations in order to be left only with shape differences.

The so-formed bitmap can be fed as input to the classifier, or features may be extracted from the image to get invariances and/or reduce dimensionality. One common approach is low-pass filtering the image, which gives invariance to small translations. Easy-to-extract features in character images are the ratio of width to height, number of *on* pixels, number of line crossings along certain horizontal and vertical directions, etc. A more expensive preprocessing is to have small local kernels with which all parts of the image are convolved, to detect line segments, corners, etc.

B. Speech Recognition

Digitized speech samples are obtained by an antialiasing filter and an analog-to-digital converter (A/D) [3, 6]. A low-pass antialiasing filter should be set below the Nyquist frequency (half of the sample rate) so that the Fourier transform of the signal will be bandlimited. An A/D converter commonly consists of a sampling circuit and a hold circuit. A 10–12-kHz sampling frequency usually includes the first five formants for most talkers, but may not capture all the unvoiced energy such as the /s/ phoneme. An 8-kHz sampling frequency can be selected to be used in a 4-kHz telephone channel.

Once the speech signal has been digitized, the discrete-time representation is usually analyzed within short-time intervals. Depending on the application, an analysis window of 5–25 ms is chosen in which it is assumed that the speech signal is time-invariant or quasi-stationary. Time-invariant analysis is essential since parameter estimation in a time-varying (nonstationary) system is a much more difficult problem.

The utterances of speakers are recorded in a soundproof room or in certain environmental conditions such as no-noise with a microphone at a certain bandwidth. The required segments of each utterance are manually, or by use of an accurate segmentation algorithm, endpointed and processed into frames.

A linear predictive coding (LPC)-based analysis procedure [3, 6, 7] can be used to obtain desired features such as cepstrum coefficients or fast Fourier transform (FFT) coefficients. Recent feature extraction methods concentrate also on auditory modeling and time-frequency representation of speech signals.

IV. STATISTICAL CLASSIFIERS

In pattern recognition, we are asked to assign a multidimensional input $\boldsymbol{x} \in \Re^d$ to one of a set of classes C_j, $j = 1, \ldots, m$ [8, 9]. Sometimes the additional action of *reject* is added to choose when no class or more than one class is probable. A *classifier* is then a mapping from $\Re^d$ to $\{C_1, \ldots, C_m, C_{rej}\}$.

In statistical decision theory, actions have costs and the decision having the minimum cost is made [10]. Assuming that correct decisions have no cost, all incorrect decisions have equal cost, and no rejects, for minimum expected cost (or risk), the so-called *Bayes' decision rule* states that a given input should be assigned to the class with the highest *posterior* probability [8–10]:

$$c = \arg\max_j P(C_j|\boldsymbol{x}). \tag{1}$$

The posteriors are almost never exactly known and thus are estimated. We are given a sample $\mathcal{X} = \{\boldsymbol{x}_i, y_i\}$, $i = 1, \ldots, n$ and $y \in \{C_1, \ldots, C_m\}$. In statistical

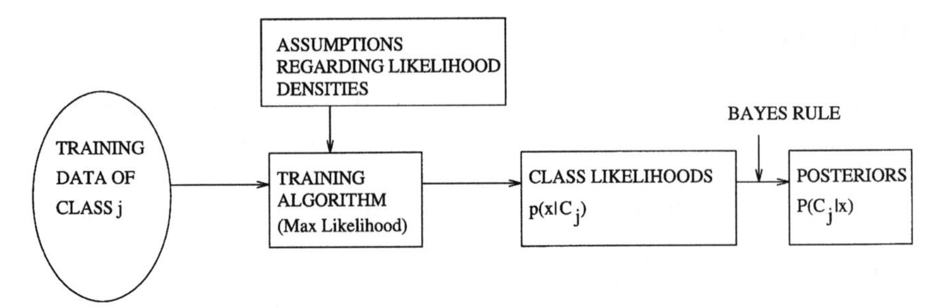

Figure 2 Building a statistical classifier.

pattern recognition theory (Fig. 2), Bayes rule factors the posterior into a *prior* class probability $P(C_j)$ and a class-conditional density or *likelihood* $p(\boldsymbol{x}|C_j)$:

$$P(C_j|\boldsymbol{x}) = \frac{p(\boldsymbol{x}|C_j)P(C_j)}{p(\boldsymbol{x})} = \frac{p(\boldsymbol{x}|C_j)P(C_j)}{\sum_k p(\boldsymbol{x}|C_k)P(C_k)}. \tag{2}$$

The estimate of the prior is just the fraction of samples belonging to that class. If n_j is the number of samples belonging to class j, $\sum_j n_j = n$, then

$$\widehat{P}(C_j) = n_j/n. \tag{3}$$

The real problem is that of estimating the class-conditional likelihood densities $p(\boldsymbol{x}|C_j)$. There are three approaches:

1. *Parametric Methods.* These assume that class-conditional densities have a certain parametric form, e.g., normal, whose sufficient statistics are estimated from the data [8, 9]. These methods generally reduce to *distance-based* methods where, depending on assumptions made on the data, the good distance metric is chosen.
2. *Nonparametric Methods.* When no such assumptions can be made, the densities need to be estimated directly from the data. These are also known as *kernel-based estimators* [8, 11, 12].
3. *Semiparametric Methods.* The densities are written as a mixture model whose parameters are estimated [8, 13–16]. In the case of normal mixtures, this approach is equivalent to cluster-based classification strategies such as LVQ of Kohonen [17] and is similar to Gaussian radial basis function networks [18].

A decision rule as given in Eq. (1) has the effect of dividing the input space into mutually exclusive regions called the *decision regions* where each region is assigned to one of the m classes. Bounding these decision regions are the decision boundaries or *discriminants* that separate the examples of one class from others.

Pattern classification may equally well be thought of as defining appropriate discriminant functions, $g_j(\boldsymbol{x})$, and we assign feature vector $\boldsymbol{x}$ to class c if

$$g_c(\boldsymbol{x}) = \max_j g_j(\boldsymbol{x}). \tag{4}$$

An immediate discriminant function is the posterior probability, or its variants. The following are all equivalent:

$$\begin{aligned} g_j(\boldsymbol{x}) &= P(C_j|\boldsymbol{x}), \\ g'_j(\boldsymbol{x}) &= p(\boldsymbol{x}|C_j)P(C_j), \\ g''_j(\boldsymbol{x}) &= \log p(\boldsymbol{x}|C_j) + \log P(C_j). \end{aligned} \tag{5}$$

A. Parametric Bayes Classifiers

The shape of decision regions defined by a Bayes classifier depends on the assumed form for $p(\boldsymbol{x}|C_j)$ [8, 9]. Most frequently, it is taken to be multivariate normal which assumes that examples from a class are noisy versions of an ideal class member. The ideal class prototype is given by the class mean $\boldsymbol{\mu}_j$, and the characteristics of the noise appear as the covariance matrix Σ_j. When $p(\boldsymbol{x}|C_j) \sim \mathcal{N}(\boldsymbol{\mu}_j, \Sigma_j)$, it is written as

$$\hat{p}(\boldsymbol{x}|C_j) = \frac{1}{(2\pi)^{d/2}|\Sigma_j|^{1/2}} \exp\big[-(1/2)(\boldsymbol{x} - \boldsymbol{\mu}_j)^T \Sigma_j^{-1}(\boldsymbol{x} - \boldsymbol{\mu}_j)\big]. \tag{6}$$

This leads to the following discriminant function [ignoring the common term of $-(d/2)\log 2\pi$]:

$$g_j(\boldsymbol{x}) = -(1/2)\log|\Sigma_j| - (1/2)(\boldsymbol{x} - \boldsymbol{\mu}_j)^T \Sigma_j^{-1}(\boldsymbol{x} - \boldsymbol{\mu}_j) + \log P(C_j). \tag{7}$$

When $\boldsymbol{x}$ is d-dimensional, the free parameters are d for the mean vector and $d(d+1)/2$ for the (symmetric) covariance matrix. This latter is $O(d^2)$ which is disturbing if d is large. A large number of free parameters both makes the system more complex and requires larger training samples for their reliable estimation. Thus assumptions are made to keep the number of parameters small, which has the effect of *regularization*.

1. Independent Features of Equal Variances

When the dimensions of the feature vector $\boldsymbol{x}$ are independent, i.e., $\mathrm{Cov}(x_k, x_l) = 0,\ k \neq l,\ \forall k,\ l = 1, \ldots, d$, and have equal variances $\mathrm{Var}(x_k) = \sigma^2,\ \forall k = 1, \ldots, d$, then $\Sigma_j = \Sigma = \sigma^2 I$. Because the covariance matrices are equal, the first term of Eq. (7) can be ignored. $\Sigma^{-1} = (1/\sigma^2)I$ and we obtain

$$g_j(\boldsymbol{x}) = -\big(1/2\sigma^2\big)\|\boldsymbol{x} - \boldsymbol{\mu}_j\|^2 + \log P(C_j). \tag{8}$$

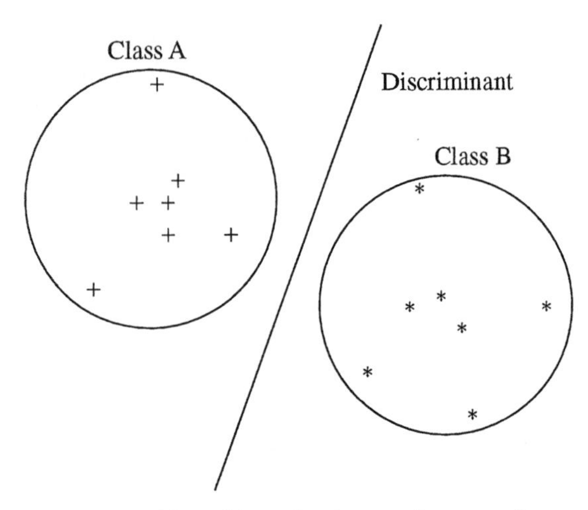

Figure 3 Example two-class problem with equal variances and zero covariances and the linear discriminant.

Assuming equal priors, this reduces to assigning input to the class with the nearest mean. If the priors are not equal, the discriminant is moved toward the less likely mean. Geometrically speaking, class densities are hyperspherical with $\boldsymbol{\mu}_j$ as the center and σ^2 defining the radius (Fig. 3). It can easily be shown that the discriminants $\{\boldsymbol{x} | g_i(\boldsymbol{x}) = g_j(\boldsymbol{x}),\ i \neq j\}$ are linear. The number of parameters for m classes is $m \cdot d$ for the means and 1 for σ^2.

2. Independent Features of Unequal Variances

If features are independent and the variances along different dimensions vary: $\mathrm{Var}(x_k) = \sigma_k^2,\ \forall k = 1, \ldots, d$, but are equal for all classes, we obtain

$$g_j(\boldsymbol{x}) = -\sum_{k=1}^{d} \frac{(x_k - \mu_{jk})^2}{2\sigma_k^2} + \log P(C_j). \tag{9}$$

This also assigns the input to the class of the nearest mean but now Euclidean distance is generalized to *Mahalanobis distance* taking also differences in variances into account. Class densities now are axis-aligned hyperellipsoids and the discriminants they lead to are linear (Fig. 4). The number of parameters is $m \cdot d + d$.

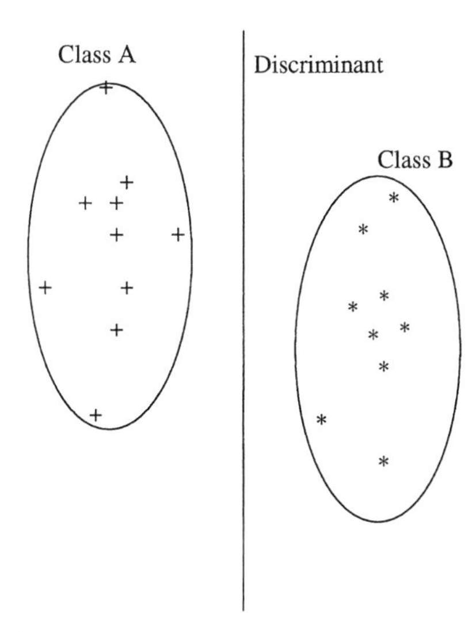

Figure 4 Example two-class problem with different variances and zero covariances and the linear discriminant.

3. Arbitrary Covariances

We are not interested in estimating full covariance matrices as $O(d^2)$ parameters require quite large training samples for accurate estimation. It is known that when classes have arbitrary but equal covariances, the discriminants are linear, and we will be considering linear discriminants in more detail in Section V.A.

When classes have different and full covariances, discriminants can be quadratic. The total number of parameters to be estimated is $m(d + d(d + 1)/2)$ and this can only be feasible with quite small d and/or a very large number of samples as otherwise we may have ill-conditioned covariance matrices. It may thus be preferable to use a linear discriminant, e.g., Fisher's linear discriminant, when we do not have enough data even if we know that the two covariance matrices are different and the discriminant is quadratic. When a common covariance matrix is assumed, this introduces an effect of regularization. Friedman's *regularized discriminant analysis* writes the covariance matrix of a class as a weighted sum of the estimated covariance matrix of that class and the covariance matrix common to all classes, the relative weight of two being estimated using cross-validation. The common covariance matrix can even be forced to be diagonal if there is even less data available, providing further regularization [9].

There are also techniques to decrease the dimensionality. One is *subset selection* which means choosing the most important p dimensions from d, ignoring the $d - p$ dimensions. *Principal component analysis* (PCA) chooses the most important p linear combinations of the d dimensions.

B. Nonparametric Kernel-Based Density Estimators

In parametric estimation, we assume the knowledge of a certain form of density family for the likelihood whose parameters are estimated from the data. In the nonparametric case, we directly estimate the entire density function. Then we need a large sample for our estimate not to be biased by the particular sample we use. The *kernel estimator* is given as

$$\hat{p}(\boldsymbol{x}|C_j) = \frac{1}{n_j h^d} \sum_{i=1}^{n_j} K\left(\frac{\boldsymbol{x} - \boldsymbol{x}_i}{h}\right). \tag{10}$$

h is the *window width* or the *smoothing parameter*. Depending on the shape of K, one can have various estimators [11].

One disadvantage of kernel estimators is the requirement of storing the whole sample. One possibility is to selectively discard patterns that do not convey much information [19]. Another is to cluster data and keep reference vectors that represent clusters of patterns instead of the patterns themselves. The semiparametric mixture models discussed in Section IV.C correspond to this idea.

1. K-Nearest Neighbor (k-NN)

Let us denote the kth nearest sample to $\boldsymbol{x}$ as $\boldsymbol{x}^{[k]}$ and let $V^k(\boldsymbol{x})$ be the volume of the d-dimensional sphere of radius $r^k \equiv \|\boldsymbol{x} - \boldsymbol{x}^{[k]}\|$; thus $V^k(\boldsymbol{x}) = r^k c_d$, where c_d is the volume of the unit sphere in d dimensions, e.g., $c_1 = 2$, $c_2 = \pi$, $c_3 = 4\pi/3$, etc. If out of the k neighbors, k^j of them are labelled ω_j, then the k-nearest neighbor estimate is (Fig. 5)

$$\widehat{P}(\omega_j|\boldsymbol{x}) = k^j/k. \tag{11}$$

2. Parzen Windows

For $\hat{p}$ to be a legitimate density function, K should be nonnegative and integrate to 1. For a smooth approximation, K is generally taken to be the normal density:

$$K(\mathbf{u}) = \left(\frac{1}{\sqrt{2\pi}}\right)^d \exp\left[-\frac{\|\mathbf{u}\|^2}{2}\right]. \tag{12}$$

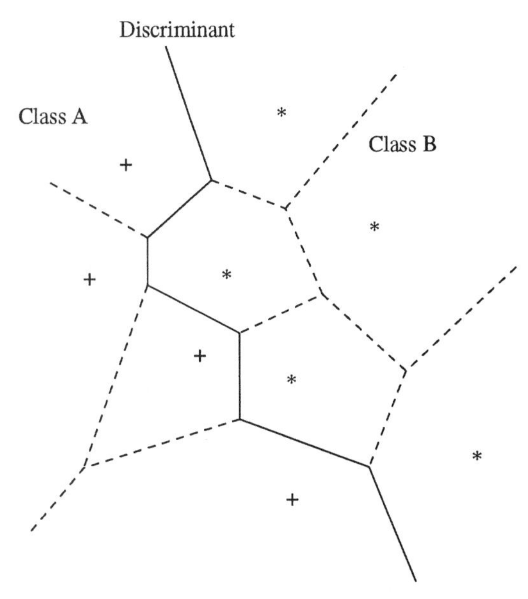

Figure 5 Example two-class problem with sample points and the arbitrary discriminant found by one nearest neighbor. The dotted lines show the Voronoi tesselation.

Here the kernel estimator is a sum of "bumps" placed at the observations. K determines the shape of the bumps while h determines their widths. When spheric bumps with equal h in all dimensions are used, this corresponds to using Euclidean distance. If this assumption of equal variances (and independent dimensions) is not valid, then different variances (and covariances) can be estimated and Mahalanobis distance can be used instead. This also applies to k-NN.

3. Choosing h or k

In kernel-based estimation, the correct choice of the spread parameter (k or h) is critical. If it is large, then even distant neighbors affect the density at $\boldsymbol{x}$ leading to a very smooth estimate. If it is small, $\hat{p}$ is the superposition of n sharp pulses centered at the samples and is a "noisy" estimate.

With Parzen windows, when h is small with a small sample, it is possible that no samples fall in the kernel, leading to a zero estimate; k-NN guarantees that k samples fall in the kernel. Small or large h leads to a decrease in success. When h is small, there are few samples, and when it is large, there are too many. The good h value is to be found using cross-validation on a separate set.

For the k-nearest neighbor, it has been shown [8] that the performance of the 1-nearest neighbor in classification is never worse than twice the Bayesian risk where complete knowledge of the distributions is assumed. It can thus be said that at least half of this knowledge is provided by the nearest neighbor. The performance can be improved by increasing the number of neighbors considered, in which case the error asymptotically converges to the Bayes risk.

When the samples are noisy, we expect k-NN with small k not to perform well. Large k takes into account the effect of very distant samples and thus is not good either. When h (or k) is decreased to decrease bias, variance increases and vice versa. This can intuitively be explained as follows [20]. When h (or k) is large, $\hat{p}$ is the weighted average of many samples and thus does not change much from one sample set to another. The variance contribution is small; however, response is biased toward the population response. (In the extreme case, when $h \to \infty$, $\hat{p}$ is the sample average and is independent of the input $\boldsymbol{x}$.) On the other extreme, when h is small, there is small bias but the response is dependent on the particular sample used; therefore, the variance contribution is high. The same argument holds with k-NN, with k in place of h. Choosing h or k implies a trade-off between bias (systematic error) and variance (random error).

C. Semiparametric Mixture Models

The parametric approach assumes that examples of a class are corrupted versions of an ideal class prototype. This ideal prototype is estimated by the mean of the examples and examples are classified depending on how similar they are to these prototypes. In certain cases, for a class, it is not possible to choose one single prototype; instead there may be several. For example, in character recognition, while writing “7” one prototype may be a seven with a horizontal middle bar (European version) and one without (American version). A *mixture density* defines the class-conditional density as a sum of a small number of densities [8, 9, 15]:

$$\hat{p}(\boldsymbol{x}|C_j) = \sum_{h=1}^{h_j} p(\boldsymbol{x}|\omega_{jh}, \Phi_j)P(\omega_{jh}), \tag{13}$$

where the conditional densities $p(\boldsymbol{x}|\omega_{jh}, \Phi_j)$ are called the *component densities* and the prior probabilities $P(\omega_{jh})$ are called the *mixing parameters*. Note that here we have one mixture model for each class leading to an overall *mixture of mixtures* [16].

We want to estimate the parameters Φ_j, that include the sufficient statistics of the component densities, and the mixing proportions, that maximize the likelihood of a given iid sample $\mathcal{X}_j$ of class j:

$$\mathcal{L}(\Phi_j|\mathcal{X}_j) = \sum_{i=1}^{n_j} \log p\big(\boldsymbol{x}_i|\Phi_j\big) = \sum_i \log \sum_h p\big(\boldsymbol{x}_i|\omega_{jh}, \Phi_j\big) P(\omega_{jh}). \tag{14}$$

This does not have an analytical solution but an iterative procedure exists based on the expectation-maximization (EM) algorithm [13, 21]. In the expectation (E) step of the algorithm, using the current set of parameters, we compute the probability that the sample $\boldsymbol{x}_i$ is generated by component h of class j:

$$P(\omega_{jh}|\boldsymbol{x}_i, \Phi_j) = \frac{p(\boldsymbol{x}_i|\omega_{jh}, \Phi_j)P(\omega_{jh})}{\sum_l p(\boldsymbol{x}_i|\omega_{jl}, \Phi_j)P(\omega_{jl})} \equiv \tau_{jhi}. \tag{15}$$

Assuming Gaussian components, i.e., $p(\boldsymbol{x}|\omega_{jh}, \Phi_j) \sim \mathcal{N}(\mu_{jh}, \Sigma_{jh})$, we have

$$\tau_{jhi} = \frac{P(\omega_{jh})|\boldsymbol{\Sigma}_{jh}|^{-1/2} \exp[-(1/2)(\boldsymbol{x}_i - \boldsymbol{\mu}_{jh})^T \boldsymbol{\Sigma}_{jh}^{-1}(\boldsymbol{x}_i - \boldsymbol{\mu}_{jh})]}{\sum_l P(\omega_{jl})|\boldsymbol{\Sigma}_{jl}|^{-1/2} \exp[-(1/2)(\boldsymbol{x}_i - \boldsymbol{\mu}_{jl})^T \boldsymbol{\Sigma}_{jl}^{-1}(\boldsymbol{x}_i - \boldsymbol{\mu}_{jl})]}. \tag{16}$$

In the M step, we update the component parameters Φ_j based on the probabilities computed in the E step:

$$\begin{aligned}
\widehat{P}(\omega_{jh}) &= (1/n_j) \sum_i \tau_{jhi}, \\
\boldsymbol{\mu}_{jh} &= \frac{\sum_i \tau_{jhi} \boldsymbol{x}_i}{\sum_i \tau_{jhi}}, \\
\Sigma_{jh} &= \frac{\sum_i \tau_{jhi}(\boldsymbol{x}_i - \boldsymbol{\mu}_{jh})(\boldsymbol{x}_i - \boldsymbol{\mu}_{jh})^T}{\sum_i \tau_{jhi}}.
\end{aligned} \tag{17}$$

As in the parametric approaches, simplifying assumptions can be made on the covariance matrices for regularization. Assuming equal hyperspheric densities, one uses (squared) Euclidean distance in computing the posteriors. If we merely compute the Euclidean distance $\|\boldsymbol{x}_i - \boldsymbol{\mu}_{jh}\|^2$, find the nearest mean $\boldsymbol{\mu}_{jc}$ nearest to $\boldsymbol{x}$, and set its τ to 1 and all others to zero, and only update $\boldsymbol{\mu}_{jc}$ for that example, we get the k-means procedure. The *on-line* version of the same algorithm updates the mean after each pattern. For each pattern $\boldsymbol{x} \in C_j$, we find $\boldsymbol{\mu}_{jc}$ such that

$$\|\boldsymbol{x} - \boldsymbol{\mu}_{jc}\| = \min_h \|\boldsymbol{x} - \boldsymbol{\mu}_{jh}\|, \tag{18}$$

and then do the update immediately:

$$\Delta\boldsymbol{\mu}_{jc} = \eta(\boldsymbol{x} - \boldsymbol{\mu}_{jc}), \tag{19}$$

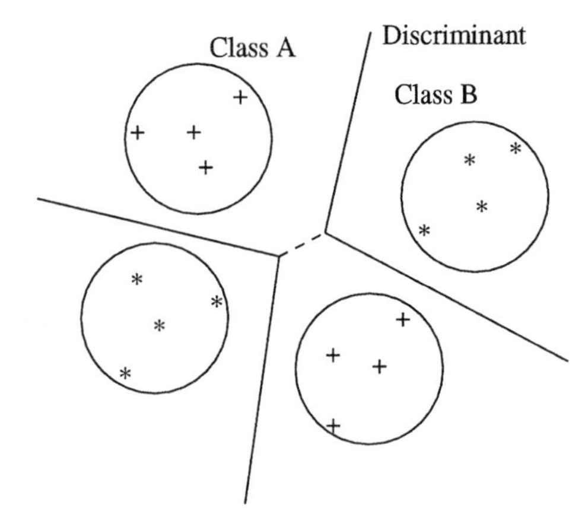

Figure 6 Example two-class problem with two reference vectors per class and the arbitrary discriminant found by LVQ. The dotted lines show the Voronoi tesselation.

where η is a learning factor that is gradually decreased toward zero for convergence. The rationale of this method is that by moving the mean closer to the sample we increase the likelihood of seeing that sample.

1. Learning Vector Quantization

Kohonen [17] proposed learning vector quantization which also moves means of wrong classes away. For a given input $\boldsymbol{x}$, we find the closest mean $\boldsymbol{\mu}_{jc}$ among all classes (Fig. 6):

$$\|\boldsymbol{x} - \boldsymbol{\mu}_{jc}\| = \min_{k,h} \|\boldsymbol{x} - \boldsymbol{\mu}_{kh}\|,$$

and then we move the mean toward the input if the classes of the mean and the input agree and we move the mean away from the input if they disagree:

$$\Delta\boldsymbol{\mu}_{jc} = \begin{cases} +\eta(\boldsymbol{x} - \boldsymbol{\mu}_{jc}), & \text{if } \boldsymbol{x} \in C_j, \\ -\eta(\boldsymbol{x} - \boldsymbol{\mu}_{jc}), & \text{otherwise.} \end{cases} \tag{20}$$

V. NEURAL CLASSIFIERS

An artificial neural network is a network of simple processing units that are interconnected through weighted connections [17, 22, 23]. The interconnection topology between the units and the weights of the connections define the operation of the network. We are generally interested in feedforward networks where a set of units are designated as the input units through which input features are fed to the network. There is then a layer of *hidden* units that extract features from the input. This is followed by the layer of output units where in classification each output corresponds to one class.

There are a number of advantages to using neural network-type classifiers for pattern recognition [24]:

1. They can learn, i.e., given a sufficiently large labelled training set, the parameters can be computed to optimize a given error criterion.
2. They can generate any kind of nonlinear function of the input.
3. Because they are capable of incorporating multiple constraints and finding optimal combinations of constraints for classification, features do not need to be treated as independent. More generally, there is no need for strong assumptions about the statistical distributions of the input features (as is usually required in Bayes classifiers).
4. Artificial neural networks are highly parallel and regular structures which makes them especially amenable to high-performance parallel architectures and hardware implementations.

Statistical pattern recognition assumes a certain model for the densities and, using Bayes decision rule, we see what type of discriminant functions they lead to. The neural approach is to assume a certain model for the discriminants (posteriors) directly, as defined by the network operation (Fig. 7). For simplicity $g_j(\boldsymbol{x})$ can be assumed to be linear in $\boldsymbol{x}$:

$$g_j(\boldsymbol{x}) = \boldsymbol{W}^T \boldsymbol{x}' = \sum_{k=1}^{d} W_{jk} x_k + W_{j0}, \tag{21}$$

where $\boldsymbol{x}'$ is $(\boldsymbol{x}, 1)^T$, augmented to include also an intercept (or bias) term. This is a neural network called a *perceptron* where units in the input layer take the

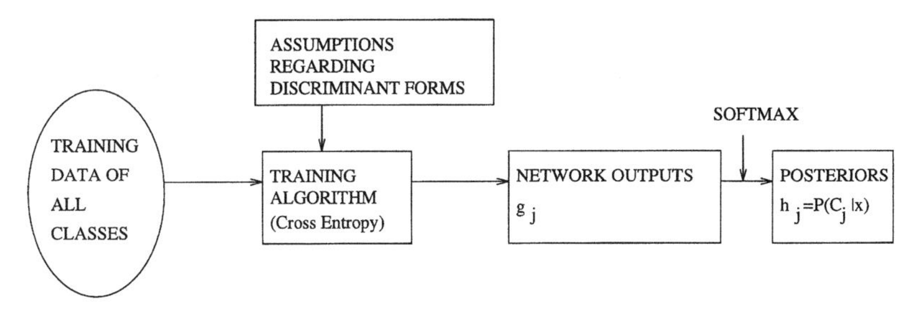

Figure 7 Building a neural classifier.

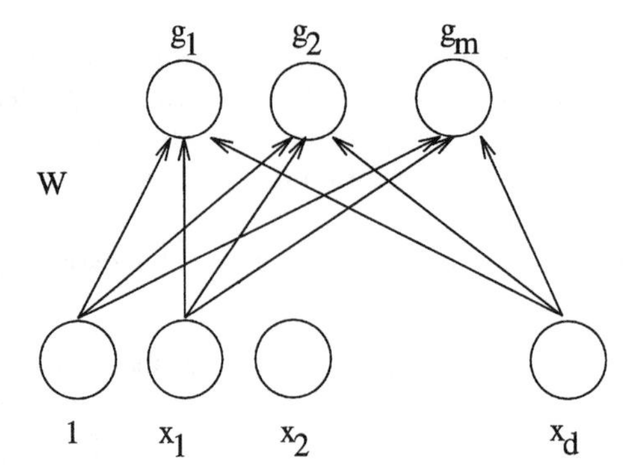

Figure 8 A linear classifier realized as a perceptron neural network.

value $\boldsymbol{x}$ and units in the output layer take $g_j(\boldsymbol{x})$. The weights of the connections between are $\boldsymbol{W}$ (Fig. 8).

Discriminants in real life are rarely linear so one way to approximate nonlinear functions is by estimating them as a linear sum of a number of nonlinear basis functions (Fig. 9):

$$g_j(\boldsymbol{x}) = \sum_{h=1}^{H} W_{jh}\phi_h(\boldsymbol{x}) + W_{j0}. \tag{22}$$

In neural network terminology, the basis functions, $\phi_h(\cdot)$, are called the hidden units, and if the basis function is Gaussian this approach is called a *radial basis function network*; it is called a *multilayer perceptron* if it is a sigmoid. The well-known statistical technique of *projection pursuit regression* has the difference that basis functions need not be fixed identical but are estimated in a nonparametric manner.

In classification, we know that outputs are probabilities and that they sum up to 1. This can be enforced using the softmax model [23]:

$$h_j = \frac{\exp g_j}{\sum_k \exp g_k}, \tag{23}$$

and the error measure to be minimized is the cross-entropy between the two distributions [9, 23]:

$$E = -\sum_i \sum_j r_{ij} \log h_j(\boldsymbol{x}_i), \tag{24}$$

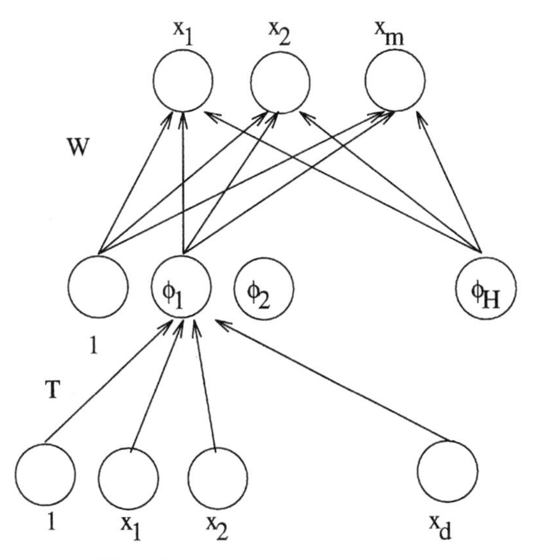

Figure 9 A multilayer neural network.

where $r_{ij} = 1$ if $\boldsymbol{x}_i \in C_j$ and 0 otherwise. $\boldsymbol{W}_j$, $j = 1, \ldots, m$, can be optimized using gradient descent:

$$\Delta W_{jh} = -\eta \frac{\partial E}{\partial W_{jh}}. \tag{25}$$

Internal parameters of the basis functions, i.e., weights from the input layer to the hidden layer, can also be trained similarly if $\phi(\cdot)$ is differentiable. This technique is called *back-propagation* of errors [22].

Note that because of the dependence introduced through softmax, a given pattern is used to train the discriminants of all classes. This contrasts with the statistical approach where a pattern affects the likelihood of one class only.

A. Simple Perceptrons

A perceptron as defined in Eq. (21) defines a linear discriminant and works if samples from a class can be separated linearly from samples of all other classes where $\boldsymbol{W}_j$ defines the position and orientation of the separating hyperplane. This model is attractive due to a number of reasons. It is optimal when classes are normal and share a common covariance matrix. It has a small number of parameters and thus does not require large amounts of memory, and it is simple to implement.

B. Multilayer Perceptrons

In a multilayer perceptron as defined in Eq. (22), there is also a hidden layer whose units correspond to the basis functions, $\phi_h(\cdot)$. They extract nonlinear input combinations to be able to define nonlinear discriminants [22]. Usually, the hidden units implement perceptrons passed through the sigmoid function:

$$\phi_h(\boldsymbol{x}) = \phi\left(T_h^T \boldsymbol{x}'\right) = \frac{1}{1 + \exp[-T_h^T \boldsymbol{x}']}. \tag{26}$$

Connection weights of both layers, $\boldsymbol{T}$ and $\boldsymbol{W}$, are trained in a supervised manner by gradient descent over a cost function like the cross-entropy.

It has been shown [25, 26] that this type of a neural network is a universal approximator, i.e., can approximate any continuous function with desired accuracy. It has also been shown [27] that in the large sample case, multilayer perceptrons estimate posterior probabilities, thus building a link between multilayer networks and statistical classifiers. These theorems do not tell how many hidden units are necessary, so one should test several alternatives on a cross-validation set and choose the best.

C. Radial Basis Functions

A radial basis function (RBF) network [18, 28] is another type of feedforward, multilayer network where the basis function is a Gaussian:

$$\phi_h(\boldsymbol{x}) = \phi\left(\|T_h - \boldsymbol{x}\|\right) = \exp\left[-\frac{\|T_h - \boldsymbol{x}\|^2}{2\sigma_h^2}\right]. \tag{27}$$

Sometimes $\phi(\cdot)$ are normalized to sum up to 1. RBF is also a universal approximator. Unlike Parzen windows where we have one Gaussian for each sample, in RBF networks we have less. Means of Gaussians may be seen as reference vectors in vector quantization or components in mixture models, the difference being that in the latter cases, a reference vector or a component belongs to one class only whereas here, class discriminants are defined as a linear combination of them.

Training can be done in one of two ways. In the uncoupled version, originally proposed by Moody and Darken [18], the Gaussian centers are trained in an unsupervised manner, e.g., using k-means. σ, the spread of Gaussians, is computed as a factor of the average of intercenter distances. The second layer of $\boldsymbol{W}$ is a single-layer perceptron and is trained using gradient-descent rule in a supervised manner. In the coupled version, all parameters are trained in a supervised manner together, using back-propagation.

Because units have local responses, only a small number of Gaussians are active for each input, thus one generally needs many more Gaussians than sigmoids, but learning is faster when only a few units need to be updated for an input. The generalization ability of RBF can be extended by having the weight of each hidden unit, W_{jh}, not a scalar but a linear function of the input [29]. This corresponds to a piecewise linear approximation of the discriminant instead of a piecewise constant approximation.

VI. LITERATURE SURVEY

A. Optical Character Recognition

Optical character recognition is one of the most popular pattern recognition applications and many systems have been proposed in the past toward this aim; see the special journal issue edited by Pavlidis and Mori for a review [2]. This is because it is a significant application of evident economic utility and also because it is a test bed before more complicated visual pattern recognition applications are attempted.

One of the earliest neural network-based systems for handwritten character recognition is the Neocognitron of Fukushima [30]. A significant amount of work on optical recognition of postal ZIP codes was done by a group at AT&T Bell Labs by Le Cun and others [31, 32]. The system uses a multilayered network with local connections and weight sharing trained with back-propagation for classification. This implements a hierarchical cone where simpler local features are extracted in parallel which combine to form higher-level, less local features and which finally define the digits. An extensive study of back-propagation for optical recognition of both handwritten letters and digits is given by Martin and Pittman [33].

Keeler, Martin, and others at MCC worked on combining segmentation and recognition in one integrated system [34, 35]. This is necessary if characters are touching in such a way that they cannot be segmented by a straightforward segmentation procedure.

Several comparative studies have also been done, either by fixing the data set and varying the methods or by also using a number of data sets. Guyon *et al.* [36] is an early reference where simple and multilayer perceptrons are compared with statistical distance-based classifiers such as k-NN in recognizing handwritten digits for automatic reading of postal codes. A comparison of k-NN, multilayer perceptron, and radial basis functions in recognizing handwritten digits is given by Lee [37]. A review of the task and several neural and conventional approaches is given by Senior [38]. Comparison of distance-based classifiers, single and multilayer perceptrons, and radial basis function networks is given in [39].

For the task of optical handwritten character recognition, a significant step was the production of a CDROM (Special Database 3) by the National Institute of Standards and Technology (NIST) [5] which includes a large set of digitized character images and computer subroutines that process them. This allowed many researchers a common test bed of significant size and quality on which to compare their approaches. Many of the above-mentioned works use this data set or its predecessor. It is available by writing to NIST. A comparison of four statistical and three neural network classifiers is given by Blue *et al.* [40] for optical character recognition and a similar task, fingerprint recognition (for which a similar CDROM was also made available by NIST). Researchers from NIST made several studies using this data set and technical reports can be accessed over the Internet.

Recently with the reduction of cost of computing power and memory, it has been possible to have multiple systems for the same task which are then combined to improve accuracy [41, 42]. One approach is to have parallel models and then take a vote. Another approach is to have models cascaded where simpler models are used to classify simpler images and complex methods are used to classify images of poorer quality.

B. Speech Recognition

In speech recognition, the input is dynamic, i.e., changes in time. Classifiers we have considered up to now are static, i.e., assume that the whole input feature vector is available for classification. To use a static classifier for a dynamic task, a *time delay approach* is used [43]. This uses an input layer with tapped delay lines and can be used if the input buffer is large enough to accommodate the longest possible sequence or if a resampling is done to normalize length. This basically is mapping time into space by having multiple copies of the input units.

If the classifier is to accept input vectors sequentially, the classifier should have some kind of internal state that is a function of the current input and the previous internal state. In the neural network terminology, these are named *recurrent* networks which contrast with feedforward networks by having also connections between units in the same layer or connections to units of a preceding layer [22]. For short sequences, a recurrent network can be converted into an equivalent feedforward network by *unfolding* it over time. This is another way of mapping time into space, the difference being that now copies of the whole network are done. In some recurrent architectures, a separate set of units are designated as *feedback units* containing the hidden or output values generated by the preceding input. In theory, the current state of the whole network will nonlinearly depend on a combination of the previous network state and the current input [24]. A comparison of different recurrent architectures and learning rules is given in [44].

Furui [45] discusses various methods for speech recognition. Lippmann [4] and Waibel and Hampshire [46] give two reviews on using neural networks for speech recognition. Early work used recurrent neural networks for representation of temporal context but after the introduction of time delay neural networks by Waibel *et al.* [43], feedforward networks were also used for phoneme recognition. Lee and Lippmann [47] and Ng and Lippmann [48] for the same two artificial and two speech tasks compare a large number of conventional and neural pattern classifiers. Comparison of distance-based classifiers, single and multilayer perceptrons, and radial basis function networks for phoneme recognition is given in [49]. The recent book by Bourlard and Morgan [24] discusses in more detail neural approaches to speech classification. Currently the most efficient approach for speech recognition is accepted to be *hidden Markov models* (HMMs) [24]. An HMM models speech as a sequence of discrete stationary states with instantaneous transition between states. At any state, there is a stochastic output process that describes the probability of occurrence of some feature vectors and a stochastic state-transition matrix conditioned on the input. It is called "hidden" because the sequence of states is not directly observable but is apparent only from the observed sequence of events. Generally there is one HMM for every word and states correspond to phonemes, syllables, or demi-syllables. HMMs are also used to recognize individual phonemes where states correspond to substructures. Bourlard and Morgan [24] give a detailed discussion of HMM models and their use in speech recognition. They also show [7] how HMMs and multilayer networks can be combined for continuous speech recognition where the network estimates the emission probabilities for HMMs. A recent reference on current speech recognition methodologies is [50].

VII. SIMULATION RESULTS

For optical character recognition (OCR) experiments, we used the set of programs recently made available by NIST [5] to generate a data base on which to test the algorithms we discuss. Forty-four people have filled in forms which are scanned and processed to get 32×32 matrices of handwritten digits. These matrices are then low-pass filtered and undersampled to 8×8 to decrease dimensionality. Each element is in the range 0–16. These 44 forms are divided into two clusters randomly as 30 forms in one side and 14 forms in the other. From the first 30, we generated three sets: training set, cross-validation set, and the writer-dependent test set. The training set has 1934 digits. The cross-validation set contains 946 digits and is used to choose the best set of parameters during training, e.g., number of basis functions, point to stop training, neighborhood size, etc. The writer-dependent test set is used for testing after training and has 943 digits. The remaining 14 forms have no relationship with those used in training and they constitute the writer-independent test set containing 1797 digits. We make sure that all sets contain approximately equal numbers of examples from each class.

Table I

Properties of the Data Sets Used

	Number of features	Input data type	Number of classes	Number of training examples	Number of cross-validation examples	Number of test examples	Number of indep. test examples
OCR	64	int: 0..16	10	1934	946	943	1797
SR	112	real: 0..1	6	600	300	300	683

For /b,d,g,m,n,N/ speech phoneme recognition (SR) experiments, the data base contains 5240 Japanese isolated words and phrases. Two hundred samples for each class are taken from the even-numbered and odd-numbered words. Six hundred samples are used for training, 300 for cross-validation, and 300 for testing. A further 683 phrases are used only for testing after having trained on isolated words. As is known, the speaking style and speed of phrases differ from the isolated words. Phonemes are extracted from hand-labelled discrete word utterances and phrase utterances which have a sampling frequency of 12 KHz. Seven speech frames (each 10 ms) are used as input. For each 10-ms frame, 16 Mel-scaled FFT coefficients are computed as feature values. The final input fed to the classifier has 112 dimensions. Properties of the data sets are summarized in Table I.

Results with various algorithms on the two sets are given in Table II. For each data set, the first column contains results on the test set that is generated in an

Table II

Results on the Two Applications

	OCR[a]		SR[a,b]	
Method	Test	Independent test	Test	Phrases
Bayes	90.77	89.43	63.33 (82.00)	36.75 (58.86)
k-NN	97.56	97.61	87.67 (96.33)	62.52 (72.91)
Parzen	97.99	97.44	90.00 (95.67)	67.50 (72.91)
LVQ	96.48, 0.34	96.42, 0.32	83.43, 1.83 (92.73, 2.10)	61.93, 2.71 (67.57, 2.88)
SP	96.06, 0.44	93.85, 0.32	92.10, 0.75	56.41, 2.59
MLP	97.51, 0.41	94.78, 0.41	93.83, 0.82	64.86, 2.44
RBF	98.11, 0.17	95.41, 0.31	91.90, 1.51	57.13, 5.54 (60.38, 5.59)

[a] Values reported are average and standard deviations of ten independent runs (when applicable).

[b] Values in parentheses for SR are improved results by allowing different variances for different features.

identical manner with the training set, i.e., the same writers or the same articulation. The second column contains data that are taken from different writers or continuous-speech phrases and thus are more natural than the first, and it is actually the success in these columns that matters.

A visual analysis of the results is possible through Figs. 10 and 11. The two axes are accuracy on the independent test set and phrases for OCR and SR respectively and memory requirement. We assumed each real-valued parameter to require 32 bits of storage. Input features require 4 bits for OCR (a value in the range 0 . . . 16) and 16 bits for SR (32,768 discrete levels). The number of training epochs is also marked for each technique.

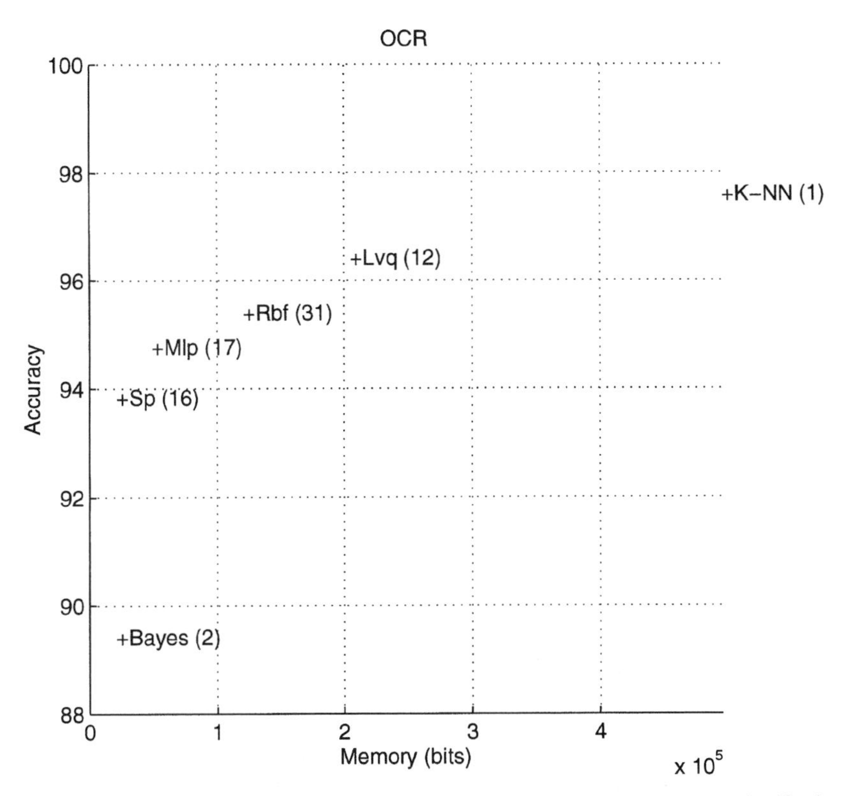

Figure 10 Results on the optical digit recognition data set. Accuracy is percent correct classification on writer-independent test set and memory is the number of bits required to store the free parameters of the system. Each pattern value (in the range 0 . . . 16) requires four bits and connection weights are assumed to require 32 bits. Values in parentheses are the number of epochs required for calculating the parameters.

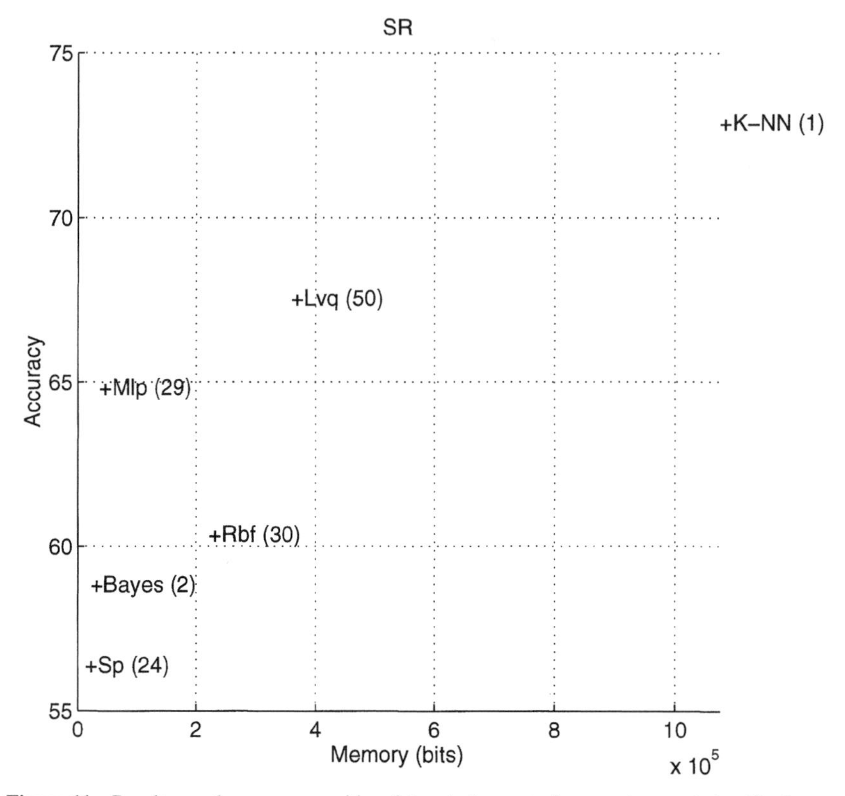

Figure 11 Results on phoneme recognition data set. Accuracy is percent correct classification on phrase set and memory is the number of bits required to store the free parameters of the system. Each pattern value is assumed to require 16 bits and connection weights are assumed to require 32 bits. Values in parentheses are the number of epochs required for calculating the parameters.

In SR, using different variances for different features leads to a big improvement, whereas in OCR it does not. This information is also used in k-NN, Parzen windows, and LVQ, where it improves accuracy; note the large difference in accuracy between the two percentages in the fourth column of Table II. Knowing that variances differ with RBF, we allowed Gaussians to have different spreads along different dimensions. A similar method that can be used with any classifier, and not only distance-based ones, is *z-normalization* where each feature value is normalized to have zero mean and unit variance. Note that this assumes that all samples for a feature are drawn from one unimodal normal distribution and thus may fail if this assumption does not hold. For example, with the multilayer perceptron on SR, though success on the test set increases after z-normalization, success on phrases decreases; this shows that feature values for phrases obey a different distribution.

In both applications, k-NN (or Parzen windows) has the highest accuracy. It is also the most expensive technique in terms of computation and memory. This however is no longer a serious drawback as the cost of storage and computation is getting cheaper. LVQ uses less storage by clustering data but has lower accuracy. RBF requires less storage than LVQ by sharing clusters between classes. A multilayer perceptron (MLP) generalizes better than a single-layer perceptron (SP), indicating that discriminants are not linear. Parametric Bayes classifiers that assume independent features do not perform well, indicating that the input features are highly correlated.

VIII. CONCLUSIONS

The similarity between statistical and neural techniques is greater than generally agreed. Many of the neural techniques are either identical or bear much resemblance to previously proposed statistical techniques. For a good discussion of neural networks from statisticians' point of view and vice versa, see the collection of articles in [51]. The recent interest in neural networks did much to revive interest in the old field of statistical pattern recognition [23].

Omohundro [52] discusses how nonparametric kernel estimators can be implemented as neural networks (by representing each sample with a Gaussian centered at the sample) and also discusses efficient data structures for the purpose. One example is the *probabilistic neural network* of Specht [12] which is a neural network implementation of Parzen windows. This approach is also known as *memory-based* as it can be seen as interpolating from a table of stored samples, and is called *lazy* in machine-learning literature as there is no learning process but the computation is deferred up until recognition is done.

Neural networks based on mixture models have also been proposed. Nowlan [15] considers them as "soft" variants of competitive approaches when used for vector quantization. Traven [14] proposes to use a mixture of Gaussians and calls this a "neural network approach" and uses EM to optimize parameters without saying so.

Statistics can also be used to improve the performance of neural techniques. Analysis of variances and use of a preprocessing such as z-normalization or principal component analysis improve accuracy considerably in practice. The quality of the training sample is perhaps the most important factor, as with an unrepresentative sample any statistic would be wrong.

Known statistical techniques such as k-NN also provide a benchmark against which more complex approaches such as multilayer perceptrons can be compared. Simple methods such as k-NN generally perform quite well and much of the functionality of neural networks such as parallel distributed processing can be obtained from such distance-based methods without requiring complicated computation, precise weights, and lengthy error-minimization techniques.

We have also reached the conclusion that generally there is not one method that is significantly superior to all others in all respects of generalization accuracy, learning time, memory requirements, and implementation complexity. The relative importances of these four factors differ from one application to another and thus in choosing one method, all of these should be taken into account, and not only generalization accuracy as has frequently been done in the past.

ACKNOWLEDGMENTS

This work is supported by Grant EEEAG–143 from TÜBİTAK, the Turkish Scientific and Technical Research Council and Grant 95HA0108 from Boğaziçi University Research Funds. The OCR data set has been collected and processed by C. Kaynak using programs made available by NIST. Thanks to F. Masulli and S. Furui for comments on this chapter.

REFERENCES

[1] E. Alpaydın and F. Gürgen. Comparison of kernel estimators, perceptrons and radial-basis functions for OCR and speech classification. *Neural Comput. Appl.* 3:38–49, 1995.

[2] T. Pavlidis and S. Mori. Special issue on optical character recognition. *Proc. IEEE* 80(7), 1992.

[3] L. Rabiner and B.-H. Juang. *Fundamentals of Speech Recognition*. Prentice-Hall, Englewood Cliffs, NJ, 1993.

[4] R. P. Lippmann. Review of neural networks for speech recognition. *Neural Comput.* 1:1–38, 1989.

[5] M. D. Garris, J. L. Blue, G. T. Candela, D. L. Dimmick, J. Geist, P. J. Grother, S. A. Janet, and C. L. Wilson. NIST form-based handprint recognition system. NISTIR 5469, National Institute of Standards and Technology, Computer Systems Laboratory, Advanced Systems Division, Gaithersburg, MD, 1994.

[6] D. P. Morgan and L. S. Christopher. *Neural Networks and Speech Processing*. Kluwer Academic, Dordrecht/Norwell, MA, 1991.

[7] N. Morgan and H. Bourlard. Continuous speech recognition: An introduction to the hybrid HMM/connectionist approach. *IEEE Signal Process. Mag.* 12:25–42, 1995.

[8] R. O. Duda and P. E. Hart. *Pattern Classification and Scene Analysis*. Wiley, New York, 1973.

[9] G. J. McLachlan. *Discriminant Analysis and Statistical Pattern Recognition*. Wiley, New York, 1992.

[10] J. O. Berger. *Statistical Decision Theory and Bayesian Analysis*, 2nd ed. Springer-Verlag, Berlin/New York, 1980.

[11] B. W. Silverman. *Density Estimation for Statistics and Data Analysis*. Chapman & Hall, London, 1986.

[12] D. F. Specht. Probabilistic neural networks. *Neural Networks* 3:109–118, 1990.

[13] R. A. Redner and H. F. Walker. Mixture densities, maximum likelihood and the EM algorithm. *SIAM Rev.* 26:195–239, 1984.

[14] H. G. C. Traven. A neural network approach to statistical pattern classification by 'semiparametric' estimation of probability density functions. *IEEE Trans. Neural Networks* 2:366–377, 1991.
[15] S. J. Nowlan. Soft competitive adaptation: Neural network learning algorithms based on fitting statistical mixtures. Ph.D. Thesis, School of Computer Science, Carnegie Mellon University, 1991.
[16] R. L. Streit and T. E. Luginbuhl. Maximum likelihood training of probabilistic neural networks. *IEEE Trans. Neural Networks* 5:764–783, 1994.
[17] T. Kohonen. *Self-Organization and Associative Memory*. Springer-Verlag, Berlin/New York, 1988.
[18] J. Moody and C. J. Darken. Fast learning in networks of locally-tuned processing units. *Neural Comput.* 1:281–294, 1989.
[19] E. Alpaydın. GAL: Networks that grow when they learn and shrink when they forget. *Internat. J. Pattern Recog. Artif. Intell.* 8:391–414, 1994.
[20] S. Geman, E. Bienenstock, and R. Doursat. Neural networks and the bias/variance dilemma. *Neural Comput.* 4:1–58, 1992.
[21] A. P. Dempster, N. M., Laird, and D. B. Rubin. Maximum likelihood from incomplete data via the EM algorithm. *J. Roy. Statist. Soc. B* 39:1–38, 1977.
[22] J. Hertz, A. Krogh, and R. Palmer. *An Introduction to the Theory of Neural Computation*. Addison-Wesley, Reading, MA, 1991.
[23] B. D. Ripley. Neural networks and related methods for classification. *J. Roy. Statist. Soc. B* 56:409–456, 1994.
[24] H. A. Bourlard and N. Morgan. *Connectionist Speech Recognition: A Hybrid Approach*. Kluwer Academic, Dordrecht/Norwell, MA, 1994.
[25] K. Funahashi. On the approximate realization of continuous mapping by neural networks. *Neural Networks* 2:183–192, 1989.
[26] K. Hornik, M. Stinchcombe, and H. White. Multilayer feedforward networks are universal approximators. *Neural Networks* 2:359–366, 1989.
[27] D. W. Ruck, S. K. Rogers, M. Kabrisky, M. E. Oxley, and B. W. Suter. The multi-layer perceptron as an approximation to a bayes optimal discriminant function. *IEEE Trans. Neural Networks* 1:296–298, 1990.
[28] T. Poggio and F. Girosi. Networks for approximation and learning. *Proc. IEEE* 78:1481–1497, 1990.
[29] E. Alpaydın and M. I. Jordan. Local linear perceptrons for classification. *IEEE Trans. Neural Networks* 7:788–792, 1996.
[30] K. Fukushima. Neocognitron: A hierarchical neural network capable of visual pattern recognition. *Neural Networks* 1:119–130, 1988.
[31] Y. Le Cun, B. Boser, J. S. Denker, D. Henderson, R. E. Howard, W. Hubbard, and L. D. Jackel. Handwritten digit recognition with a back-propagation network. In *Advances in Neural Information Processing Systems 2* (D. Touretzky, Ed.), pp. 396–404. Morgan Kaufmann, San Mateo, CA, 1990.
[32] O. Matan, H. S. Baird, J. Bromley, C. J. C. Burges, J. S. Denker, L. D. Jackel, Y. Le Cun, E. P. D. Pednault, W. D. Satterfield, C. E. Stenard, and T. J. Thompson. Reading handwritten digits: A zip code recognition system. *IEEE Computer* 25:59–62, 1992.
[33] G. L. Martin and J. A. Pittman. Recognizing hand-printed letters and digits using backpropagation learning. In *Advances in Neural Information Processing Systems 2* (D. Touretzky, Ed.), pp. 405–414. Morgan Kaufmann, San Mateo, CA, 1990.
[34] J. Keeler and D. E. Rumelhart. A self-organizing integrated segmentation and recognition neural net. In *Advances in Neural Information Processing Systems 4* (J. E. Moody, S. J. Hanson, R. P. Lippmann, Eds.), pp. 496–503. Morgan Kaufmann, San Mateo, CA, 1992.

[35] G. L. Martin and M. Rashid. Recognizing overlapping hand-printed characters by centered-object integrated segmentation and recognition. In *Advances in Neural Information Processing Systems 4* (J. E. Moody, S. J. Hanson, R. P. Lippmann, Eds.), pp. 504–511. Morgan Kaufmann, San Mateo, CA, 1992.

[36] I. Guyon, I. Poujoud, L. Personnaz, G. Dreyfus, J. Denker, and Y. Le Cun. Comparing different neural architectures for classifying handwritten digits. In *International Joint Conference on Neural Networks 1989*, Vol. 2, pp. 127–132, Washington, DC, 1989.

[37] Y. Lee. Handwritten digit recognition using K-nearest-neighbor, radial-basis function, and backpropagation neural networks. *Neural Comput.* 3:440–449, 1991.

[38] A. W. Senior. Off-line handwriting recognition: A review and experiments. CUED/F-INFENG/TR 105. Cambridge University Engineering Department, 1992.

[39] E. Alpaydın, S. Aratma, and M. Yağcı. Recognition of handwritten digits using neural networks. *ELEKTRİK, Turk. J. Elect. Engin. Computer Sci.* 2:20–31, 1994.

[40] J. L. Blue, G. T. Candela, P. J. Grother, R. Chellappa, and C. L. Wilson. Evaluation of pattern classifiers for fingerprint and OCR applications. *Pattern Recogn.* 27:485–501, 1994.

[41] S. N. Srihari. High-performance reading machines. *Proc. IEEE* 80:1120–1132, 1992.

[42] C. Y. Suen, C. Nadal, R. Legault, T. A. Mai, and L. Lam. Computer recognition of unconstrained handwritten numerals. *Proc. IEEE* 80:1162–1180, 1992.

[43] A. Waibel, T. Hanazawa, G. Hinton, K. Shikano, and K. J. Lang. Phoneme recognition using time-delay neural networks. *IEEE Trans. Acoustics, Speech, Signal Process.* 37:328–339, 1989.

[44] F. Gürgen, M. Şıhmanoğlu, and E. Alpaydın. Learning speech dynamics by neural networks with delay elements. In *ICT'96, International Conference on Telecommunications* (B. Sankur, Ed.), pp. 156–161, Boğaziçi University Press, Istanbul, 1996.

[45] S. Furui. *Digital Speech Processing, Synthesis and Recognition.* Marcel Dekker, New York, 1989.

[46] A. Waibel and J. B. Hampshire II. Neural network applications to speech. In *Neural Networks: Concepts, Applications, and Implementations 1* (P. Antognetti & V. Milutinović, Eds.), pp. 54–76. Prentice-Hall, Englewood Cliffs, NJ, 1991.

[47] Y. Lee and R. Lippmann. Practical characteristics of neural network and conventional pattern classifiers on artificial and speech problems. In *Advances in Neural Information Processing Systems 2* (D. Touretzky, Ed.), pp. 168–177. Morgan Kaufmann, San Mateo, CA, 1990.

[48] K. Ng and R. Lippmann. Practical characteristics of neural network and conventional pattern classifiers. In *Advances in Neural Information Processing Systems 3* (R. Lippmann, J. Moody, D. Touretzky, Eds.), pp. 970–976. Morgan Kaufmann, San Mateo, CA, 1991.

[49] F. Gürgen, R. Alpaydın, U. Ünlüakın, and E. Alpaydın. Distributed and local neural classifiers for phoneme recognition. *Pattern Recogn. Lett.* 15:1111–1118, 1994.

[50] C.-H. Lee, F. K. Soong, and K. K. Paliwal. *Automatic Speech and Speaker Recognition: Advanced Topics.* Kluwer Academic, Dordrecht/Norwell, MA, 1996.

[51] V. Cherkassky, J. H. Friedman, and H. Wechsler. *From Statistics to Neural Networks: Theory and Pattern Recognition Applications*, NATO ASI Series F, Vol. 136. Springer-Verlag, Berlin/New York, 1994.

[52] S. M. Omohundro. Efficient algorithms with neural network behavior. *Complex Syst.* 1:273–347, 1987.

Medical Imaging

Ying Sun
Department of Electrical and
Computer Engineering
University of Rhode Island
Kingston, Rhode Island 02881

Reza Nekovei
Remote Sensing Laboratory
University of Rhode Island
Bay Campus
Narragansett, Rhode Island 02882

I. INTRODUCTION

The purpose of this chapter is twofold. First, we report the findings of a literature search for applications of artificial neural networks (ANNs) in medical imaging. Based on the literature search we review the current status of ANN techniques in the medical imaging area. Second, using an example of detecting blood vessels in angiograms we show the formulation and performance of a feedforward back-propagation (BP) network as well as a self-adaptive (SA) network for image segmentation. The example illustrates the use of both supervised and unsupervised ANN classifiers for feature extraction at the lower (pixel) level of processing medical images. We also compare the ANNs with the more conventional classifiers in terms of their classification performance.

The chapter is organized into six sections. In Section I, we introduce the various modalities of medical imaging used in modern hospitals nowadays. This section is intended to review the basic physics of the medical imaging. In Section II, we review the recent research efforts of ANN applications in medical imaging. The intention here is to give a general, collective view of the past and ongoing researches on the relevant topics. In Section III, we state our own research problem, i.e., the identification of blood vessels in X-ray angiograms. In Section IV, we present the result of applying a feedforward back-propagation network to the blood-vessel segmentation problem. With this problem, we demonstrate the use of ANN for supervised feature extraction and discuss the important issues related to the network configuration and training parameters. In Section V, using the same segmentation problem we show the formulation and performance of a self-adaptive network, which represents an unsupervised ANN approach to this

Image Processing and Pattern Recognition
Copyright © 1998 by Academic Press. All rights of reproduction in any form reserved.

problem. In Section VI, we draw conclusions based on our experimental results and discuss the implications of our study for the general applications of neural networks in medical imaging.

A. Medical Imaging

The history of medical imaging began a century ago. The landmark discovery of X-rays by Wilhelm Conrad Röntgen in 1895 ushered in the development of noninvasive methods for visualization of internal organs. The birth of the digital computer in 1946 brought medical imaging into a new era of computer-assisted imagery. During the second half of the twentieth century the medical imaging technologies have diversified and advanced at an accelerating rate.

Today, clinical diagnostics rely heavily on the various medical imaging systems. In addition to the conventional X-ray radiography, computer-assisted tomography (CAT) and magnetic resonance imaging (MRI) produce two-dimensional (2D) cross sections and three-dimensional (3D) imagery of the internal organs, drastically improving our capability to diagnose various diseases. X-ray angiography used in the cardiac catheterization laboratory allows us to detect stenoses in the coronary arteries and guide the treatment procedures such as balloon angioplasty and cardiac ablation. Ultrasonography has become a routine procedure for fetus examination. Two-dimensional echocardiography combined with color Doppler flow imaging has emerged as a powerful and convenient tool for diagnosing heart valve abnormalities and for assessing cardiac functions. In the area of nuclear medicine, the scintillation gamma camera provides 2D images of pharmaceuticals labeled by radioactive isotopes. Single photon emission computed tomography (SPECT) and positron emission tomography (PET) further allow for 3D imaging of radioactive tracers.

Whereas a detailed study of medical imaging is beyond the scope of this chapter, introducing some background knowledge of the routinely used medical imaging systems should help us better understand the nature of the problems under investigation. We approach this by first studying the different media used in medical imaging. Then, we summarize the physics involved in the various imaging modalities. For a detailed treatment of this subject the readers are referred to the book by Webb [1].

B. Media Used for Medical Imaging

1. X-Rays

X-rays are electromagnetic waves generated by the X-ray tube. The wavelength of the X-rays is between 0.1 and 100 angstroms (Å), where $1\text{Å} = 10^{-10}$ m. The wavelength of X-rays is much shorter than that of visible light, which is between 3800 Å (violet) and 7600 Å (red). The energy of the X-ray photons is on the order of 0.1 to 100 KeV. Energy, frequency, and wavelength are related by Einstein's photon formula: $E = h\nu = (hc)/\lambda$, where E is photon energy, ν is frequency, λ is wavelength, h is the Planck constant (6.626×10^{-34} J · s or 4.1375×10^{-15} eV · s), and c is the speed of light (3×10^{8} m/s). By substituting the constants into the equation, E (eV) and λ (m) are related by $E = 1.24 \times 10^{-6}/\lambda$. A shorter wavelength corresponds to a higher photon energy and a higher degree of penetration through the human tissue.

Figure 1a shows the arrangement of X-ray source, X-ray detector, and subject under examination in an X-ray imaging system. The X-ray image is a 2D projection of the spatial distribution of the X-ray absorption coefficient within the subject. The parameters in an X-ray imaging system are adjusted such that a suitable trade-off between image contrast and X-ray dose is made. To minimize the X-ray dose given to the patient, the X-ray exposure should be set at a minimal level but enough to produce a sufficient image contrast for the intended diagnostic purpose. The energy range for diagnostic X-rays is between 10 and 150 KeV. Within this range the human tissue appears to be semitransparent to the X-rays. X-rays with energy below this range are mostly absorbed by the tissue and X-rays with energy beyond this range mostly penetrate through the tissue; neither would produce an adequate contrast in the X-ray image.

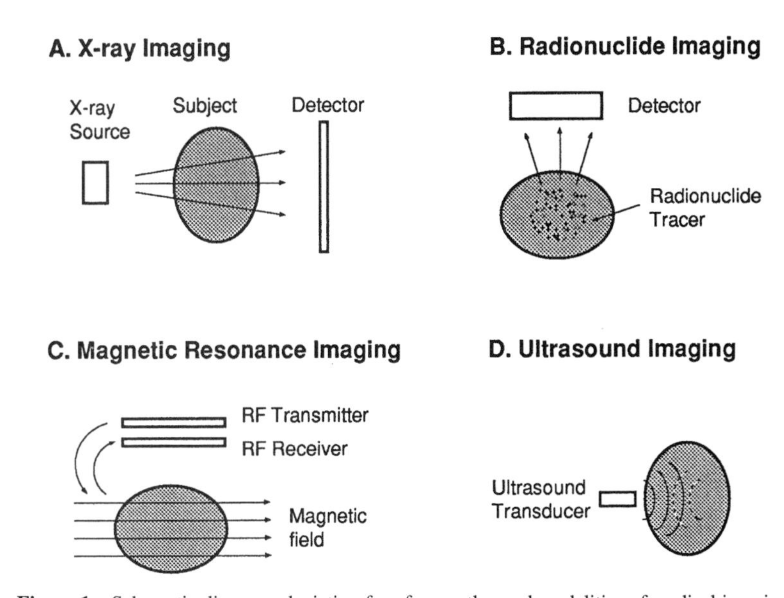

Figure 1 Schematic diagrams depicting four frequently used modalities of medical imaging.

In conventional X-ray radiography, the 3D function of absorption coefficient is projected onto the 2D image plane. In computer-assisted tomography (CAT), X-ray projections are acquired around the subject. Then, the 2D cross-sectional slices of the subject are reconstructed from the projection data. The mathematical relationship between the projections and the reconstructed slice was first studied by Johann Radon in 1907 [2]. The modern methods for tomographic reconstruction include the filtered-backprojection, which is based on the inverse Radon transform, and the algebraic reconstruction technique, which is an iterative numerical approach.

2. γ-Rays

Radioactive isotopes emitting γ-rays can be combined with appropriate pharmaceuticals and used as tracers for radionuclide imaging. Because the radioactive tracer is injected into the human body, as shown by the sketch in Fig. 1b, it is preferable to have a somewhat higher photon energy allowing for better penetration of the radiations from inside the body. However, photons with too high an energy can penetrate the components of the imaging system as well, resulting in a low detection efficiency. The most frequently used radionuclide for medical imaging is the Technitium in the metastable state ($^{99}Tc^{m}$) which has a decay half-life of 6.02 hours and emits γ-rays with a photon energy of 140 KeV. At this energy photons penetrate well through the tissue and can still be effectively detected. Detection of γ-rays is typically accomplished by the sodium iodide (NaI) crystals in the scintillation gamma camera, which produces 2D images of the radioactive tracer. For single photon emission computed tomography (SPECT), the gamma camera is mounted on a rotational gantry and used as an area detector. The acquired data are tomographically reconstructed to produce 2D slices. The 3D imagery can also be rendered from the 2D slices.

The radionuclides used in positron emission tomography (PET) emit positrons instead of γ-rays. A positron has the same rest mass as an electron (9.11×10^{-31} Kg), but carries a positive charge. A positron does not travel far in the tissue before it encounters and annihilates with an electron. The annihilation creates two photons traveling in opposite directions. The energy of each photon can be computed from the ubiquitous Einstein's equation: $E = mc^2$. By substituting the mass of the positron and the speed of light into the above equation and applying the conversion factor of 1 eV = 1.6×10^{-19} J, the photon energy is determined to be 511 KeV. Detection of the 511-KeV photons cannot be effectively achieved by the standard gamma camera because of the high photon energy. It is typically accomplished with a ring-shaped array of bismuth germanate (BGO) crystals [3] or two xenon-filled multiwire chambers positioned at the opposite sides of the subject.

Positron-emitting radionuclides are proton-rich isotopes prepared by bombarding specimens with accelerated protons in a cyclotron. Gallium 68 is a frequently used tracer for PET scan, and has a half-life of 68 minutes. The short half-life of ^{68}Ga requires that the PET system be installed in the vicinity of a cyclotron such that the radionuclides can be prepared and applied to patients within a sufficiently short period of time.

3. Magnetic Resonance

The principle of nuclear magnetic resonance was discovered in 1946 and has been successfully applied to identifying chemical compounds and molecular structures since then. The development of the commercial systems for magnetic resonance imaging (MRI) began in the late 1970s. For the past two decades MRI has rapidly emerged as an important diagnostic tool in many areas of medicine such as neurology, oncology, and sports medicine.

Although in principle MRI is capable of imaging the distribution of different types of molecules in the tissue, clinical MRI systems nowadays are designed to image the distribution of the H_2O molecules which constitute over 80% of the total body weight. The H_2O molecules are randomly oriented in the tissue. Under the influence of a strong magnetic field all the H_2O molecules orient themselves in the direction of the magnetic field and spin at a specific angular frequency. This angular frequency, called Larmor angular frequency, is directly proportional to the strength of the magnetic field. The magnetic field strength required for diagnostic MRI is around 1 Tesla, which is on the order of 10,000 times stronger than the earth's magnetic field. The corresponding resonance frequency is in the megahertz range.

The basic concept of MRI can be represented by Fig. 1c, although the actual MRI system is far more sophisticated. Once the water molecules are aligned by the magnetic field, additional energy can be introduced by using a radio-frequency (RF) transmitter. The electromagnetic wave generated by the RF transmitter is polarized and tuned to the resonance frequencies of the molecules. A short burst of the RF electromagnetic wave (pulse) is sent and pushes the water molecules off their original axis. As the water molecules return to their original orientation, energy is released also in the form of an RF signal. The magnitude of the received RF signal is proportional to the amount of the water molecules. The time constant for the molecules to recover their orientation is the T_1 time constant, which can be extracted by use of special pulse sequences. The received RF signal shows an exponential decay. The decay time constant, called the T_2 time constant, is related to the *dephasing*. As the water molecules gradually recover their orientation, they release the RF energy and spin out of phase, resulting in signal cancellation at the receiver. The T_2 decay is related to the inhomogeneity of the water molecules in their surrounding environment.

The spatial information in MRI is encoded by applying a small magnetic field gradient across the imaging plane. Each point on the imaging plane has a unique magnetic field strength corresponding to a unique resonance frequency. Thus, the MR images can be reconstructed from the Fourier transform of the received RF signals.

4. Ultrasound

Sound wave is transmitted by propagating the vibration from molecules to molecules. The velocity of sound in tissue is on average 1540 m/s, varying over a range of $\pm 6\%$ for the different types of tissue. At the interface of two different types of tissue a portion of the wave energy is reflected back. Medical ultrasonography nowadays predominantly exploits the reflected echoes. The appropriate frequency for diagnostic sound wave is in the megahertz range, beyond the human audible range (20 Hz to 20 KHz).

As shown in Fig. 1d, an ultrasound probe is used for both transmitting and receiving the ultrasound waves. The probe usually consists of an array of piezoelectric transducers. The beam-forming technique can be used to steel the ultrasound wave and scan the beam over a fan-shaped sector. This is accomplished by transmitting and receiving phase-shifted signals across the array. For each angle a burst of ultrasound is transmitted. Then, its echoes are recorded over a time period and converted to image intensity along the scan line.

Two-dimensional echocardiography provides dynamic imaging of the cardiovascular system. The velocity of blood flow contributes to a frequency shift in the returned echoes, i.e., the well-known Doppler effect. The frequency shift can be used to estimate the velocity of blood flow in the direction of the incident wave. The Doppler flow information is often coded with pseudocolors and overlapped with the 2D echocardiogram shown in grayscale.

5. Other Media

Visible light and infrared light have been used for medical imaging. However, their applications are limited because of the low degree of penetration through the tissue. For electrical impedance tomography low-level electrical currents can be injected into the body via multiple electrodes to measure the distribution of impedance. The formidable problem in electrical impedance tomography is related to the fact that the electrical current follows the least-resistance path, not necessarily a straight line.

II. REVIEW OF ARTIFICIAL NEURAL NETWORK APPLICATIONS IN MEDICAL IMAGING

A. Model for Medical Image Processing

Detecting/classifying patterns is one area where ANNs have made significant contributions. For medical diagnostics, detecting abnormalities and associating them with the possible causes are the two fundamental tasks. From this point of view, the diagnostic problems in medicine lend themselves to neural network computing. Medical diagnostics rely mainly on:

1. input data—patient history, symptoms, and test results;
2. knowledge—cumulative experiences in medical diagnostics; and
3. analysis—medical expert's interpretation of data based on his/her knowledge.

To apply an ANN to a medical diagnostic problem, the relevant diagnostic knowledge can be used in training. The trained ANN takes the patient's data as input and generates diagnostic output, which can be compared to the medical expert's diagnostics for the purpose of verification.

The interpretation of diagnostic medical images, however, is usually quite sophisticated and involves multiple levels of processing. To provide a common platform for studying the various problems of medical-imaging-based diagnostics, we employ a three-level model as shown in Fig. 2. At the lowest level, images are formed. Some imaging modalities, such as the conventional X-rays, do not require

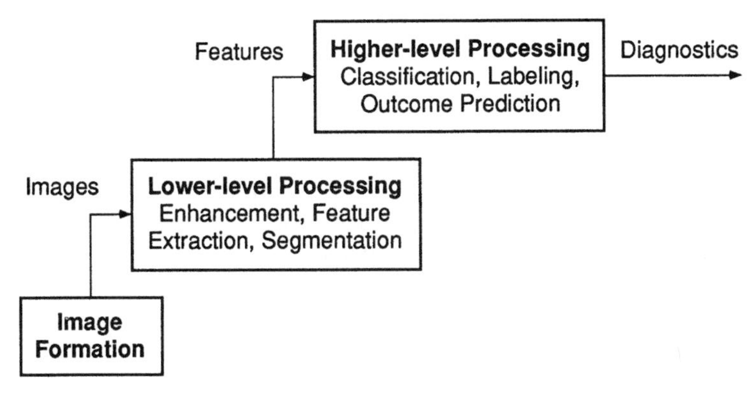

Figure 2 Model for diagnostic system using medical images.

any computation, whereas others, such as CAT scan, require extensive computation for reconstructing images from projections. Image processing is separated into two levels: the lower-level processing and the higher-level processing. The lower-level processing takes image pixels as input and performs tasks such as image enhancement, feature extraction, and image segmentation. The higher-level processing takes the output from the lower-level processing as input and generates output related to medical diagnostics. Tasks accomplished in the higher-level processing include classification of features, detection of specific lesions, and diagnosis for various abnormalities.

B. Review of Recent Literature

Based on the three-level model discussed previously we now review the recent research works involving neural network applications in medical imaging. Obviously, not every research in this area can be properly pigeonholed into one of the three levels in our model, i.e., image formation, lower-level processing, and higher-level processing. Nevertheless, we will attempt to categorize these works so that the review can be conducted in a more coherent manner and with a focus on the neural network system techniques. The scope of the literature search is limited to published journal articles in the past five years, between 1992 and 1996. The search is by no means exhaustive but should reflect the current state of ANN applications in the medical imaging area. The review is focused on the problems intended and the techniques applied. We do not include the results from the individual studies because a simplified presentation of the data without detailed discussion may sometimes be misleading. The interested readers are referred to the original articles. For related research works prior to 1992 the readers are referred to the paper by Miller *et al.* [4] who conducted a comprehensive review of ANN applications in medical imaging as well as medical signal processing.

1. Image Formation

In the SPECT system, the tomograms are reconstructed from the planar data which are acquired by use of a gamma camera rotating around the subject. Kerr and Bartlett [5] showed that this tomographic reconstruction problem can be solved by using a standard back-propagation (BP) ANN trained on either a set of simulated images or a series of rudimentary geometric SPECT scans; the performance can be further improved by employing a statistically tailored BP ANN. Munley *et al.* [6] used a supervised ANN to perform the SPECT reconstruction and to simultaneously compensate for collimator, attenuation, and scatter effects.

For MRI, the design of the various pulse sequences and the processing of MR signals remain important research areas. Cagnoni *et al.* [7] trained an ANN to synthesize a spin echo multiecho sequence for each slice of a multislice sequence for improved signal-to-noise ratio. Yan and Mao [8] used a BP ANN to reduce the artifact caused by the truncation of high-frequency MR signals; their method was improved upon by Hui and Smith [9].

For electrical impedance tomography, Adler and Guardo [10] showed that the reconstruction can be conducted on a finite element model using an ANN trained by the Widrow-Hoff learning rule.

2. Lower-Level Processing

The MRI is particularly capable of differentiating soft tissues such as gray matter, white matter, and cerebrospinal fluid in the brain. Computer algorithms for automated segmentation and labeling of MRI brain scans are useful for quantifying tissue volumes. Raff *et al.* [11] employed an ANN to determine the appropriate threshold between the gray matter and the white matter; the BP ANN was trained by the bimodal histogram of the remaining image with the cerebrospinal-fluid regions removed. Li *et al.* [12] developed an automated system for segmentation and labeling of the MRI brain scan based on two Boolean neural networks which have binary inputs, binary outputs, and integer weights. Özkan *et al.* [13] approached this segmentation problem by applying a supervised ANN to multimodal images including T_1-weighted, T_2-weighted, and proton-density-weighted MRI brain scans and CT scans. Unsupervised ANNs were also employed: Cheng *et al.* [14] and Lin *et al.* [15] approached the problem of medical image segmentation by using a Hopfield neural network with a winner-takes-all learning mechanism.

Three-dimensional imagery of internal anatomical structures can be generated from 2D MRI or CT scans by a 3D rendering algorithm. Coppini *et al.* [16] developed an ANN-based system for automated segmentation and recognition of 3D structures from a set of 2D slices; they employed two supervised ANNs trained by back-propagation.

Radionuclide imaging with a gamma camera has limited accuracy in quantitative applications due to the scatter effect and the loss of photons that penetrate through the detector. Qian *et al.* [17] studied the restoration of gamma-camera images by combining an order statistic filter and an unsupervised Hopfield neural network.

In the area of chest radiography Lo *et al.* [18] applied a 2D convolution BP ANN to lung nodule detection.

In the area of X-ray angiography Nekovei and Sun [19] applied a BP ANN to segmentation of vascular structures in coronary arteriograms and systematically studied the effects of various ANN parameters on training and generalization.

3. Higher-Level Processing

Cerebral perfusion has been routinely studied with brain SPECT scans by using appropriate radiopharmaceuticals, such as $^{99}\mathrm{Tc}^{m}$-HMPAO, that can pass through the blood–brain barrier. Chan *et al.* [20] trained a BP ANN to discriminate normal from abnormal perfusion patterns with inputs from 120 standard cortical regions. DeFigueiredo *et al.* [21] used a supervised ANN to discriminate elderly normal, Alzheimer disease, and vascular dementia subjects based on intensities averaged over various regions defined by suitable masks. Page *et al.* [22] used the theoretical profiling technique to extract cortical perfusion patterns which were then input to an ANN for diagnosing Alzheimer disease.

Myocardial perfusion has been routinely studied with cardiac SPECT scans by using appropriate radiopharmaceuticals such as ^{201}Tl-chloride. Fujita *et al.* [23] employed a BP ANN to diagnose coronary artery disease with 256 inputs representing the perfusion patterns in SPECT. Hamilton *et al.* [24] trained an ANN to predict lesion presence without the need to compare the SPECT data with a normal data base.

Ventilation-perfusion lung scans are simultaneous radionuclide images of lung ventilation distribution and pulmonary blood perfusion. Scott *et al.* [25] trained an ANN to diagnose pulmonary embolism with 28 inputs representing the ventilation-perfusion findings; training was based on 100 consecutive ventilation-perfusion scans with angiographic correlation. Tourassi *et al.* [26] employed a supervised ANN to predict pulmonary embolism by using both ventilation-perfusion scans and chest radiographs as inputs. Fisher *et al.* [27] found that brief training (50–100 iterations) was suitable for an ANN that predicted pulmonary embolism from ventilation-perfusion features; further training diminished network performance.

MRI scans have also been studied by employing ANNs for higher-level processing. Kischell *et al.* [28] extracted a comprehensive feature vector from MRI brain scans which was used as input to an ANN for classifying brain compartments and head injury lesions; they studied two ANNs involving supervised training (back-propagation and counter-propagation) as well as two unsupervised ANNs (Kohonen learning vector quantifier and analog adaptive resonance theory). Azhari *et al.* [29] studied myocardial motions by using tagged MRI in dogs; a supervised ANN was used to map acute ischemic regions with features obtained from 24 cuboids from the 3D MRI images of the left ventricle.

In the area of coronary angiography Suzuki *et al.* [30] employed a BP ANN to estimate the percent diameter stenosis with inputs from a vessel tracking algorithm.

In the area of X-ray mammography Zheng *et al.* [31] detected microcalcifications by employing an ANN trained by back-propagation with Kalman filtering; inputs to their neural network were spatial and spectral features extracted with a preprocessing stage. Sahiner *et al.* [32] used a convolution ANN classifier to classify regions of interest on mammograms as either mass or normal tissue. Baker *et al.* [33] used standardized mammographic descriptors and patient histories as inputs to an ANN for predicting the outcome of breast biopsy.

Ultrasonography has also been used to diagnose breast tumors. Goldberg *et al.* [34] used an ANN to improve the specificity of detecting malignant breast lesions based on selected texture features from the ultrasonograms.

III. SEGMENTATION OF ARTERIOGRAMS

A. BACKGROUND

An angiogram is a time sequence of X-ray images of blood vessels or cardiac chambers infused with an X-ray contrast agent. Angiography is used during cardiac catheterization for various diagnostic purposes [35] and for guiding treatment procedures such as coronary angioplasty [36]. An angiogram of the arteries is termed an arteriogram. The arteriogram can be used to study the artery's lumen geometry, dimensions, and blood flow; however, extracting such information is not a trivial task because of the following problems. First, the signal-to-noise ratio of the arteriogram is generally low due to the need for minimizing the X-ray dose and the dosage of the X-ray contrast agent given to the patient. Second, the complex imaging chain of the angiographic system contributes to the presence of various types of noise in the images [37]. Third, segmentation of the vascular structures is complicated by the overlapping of vessel branches and the interference from irrelevant anatomical structures. Fourth, analysis of the arteriogram is further complicated by the dynamics from motions of the heart, blood flows, and infusion of the X-ray contrast agent.

Segmentation of arteriograms can be accomplished by use of a vessel tracking algorithm such as the recursive tracking algorithm that we developed previously [38, 39]. This algorithm begins with a user-defined root node, tracks one vessel segment at a time, and identifies the entire vascular tree structures in a recursive fashion. Figure 3 shows two examples of coronary arteriograms and the results of the vessel tracking algorithm. The tracking approach produces a segment-by-segment description of the vascular network which is useful for applications such as 3D reconstruction of coronary arteries from biplane angiograms [40]. However, the tracking approach may not be suitable for some other applications because of the following drawbacks. User intervention such as specifying the root node, although minimal, is nonetheless required. The tracking approach is based on sequential search that does not take advantage of distributive parallel processing.

The tracking algorithm is also susceptible to noise and background variations. For example, in Fig. 3 the segmentation of the bottom image is better than that of the top image. The top image is a digitized cineangiogram (DCA) originally

a. Digitized cineangiogram of left coronary artery and tracking result

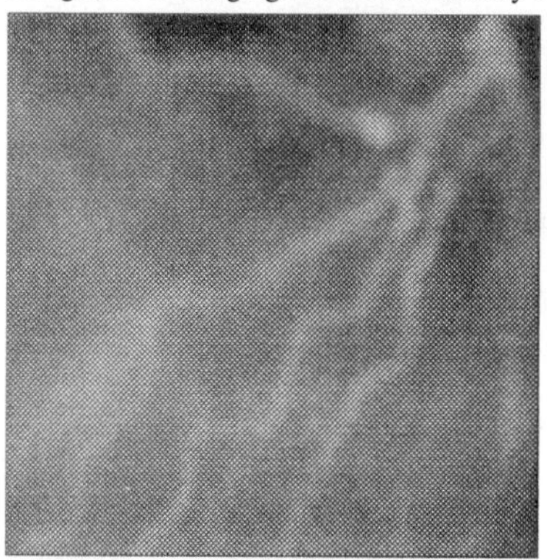

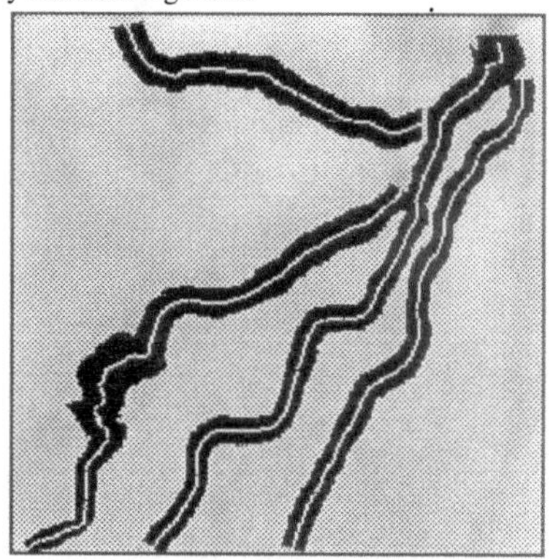

b. Digital substration angiogram of right coronary artery and tracking result

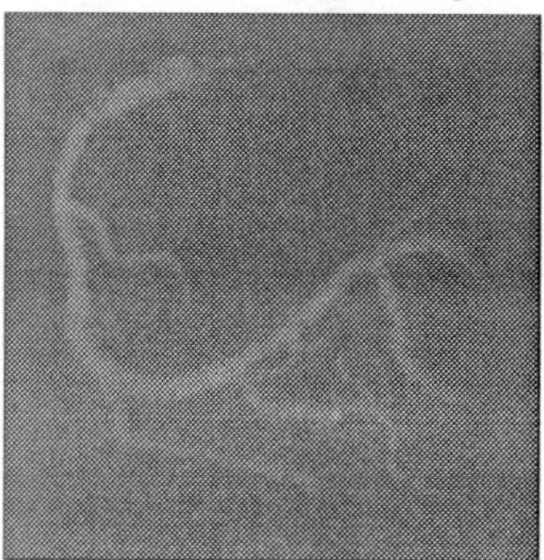

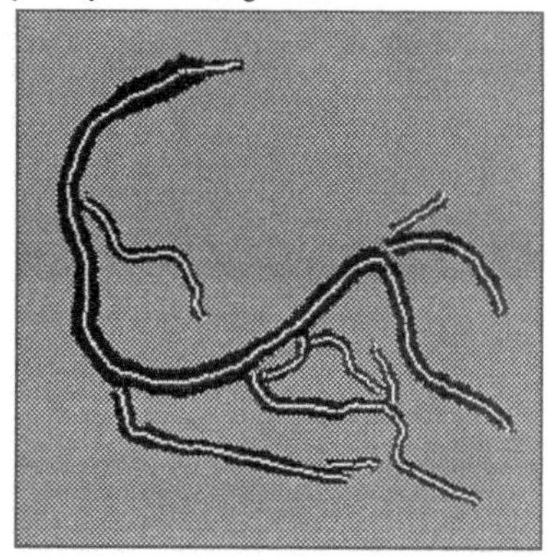

Figure 3 Two arteriograms and segmentations by vessel tracking.

recorded on 35-mm film, whereas the bottom image is a digital subtraction angiogram (DSA) with improved signal-to-noise ratio. DSA [41] is obtained by digitally subtracting two frames: one during and one before the injection of the X-ray contrast agent. The two frames correspond to the same point of the cardiac cycle by synchronization with respect to the R wave of the electrocardiogram. In addition to DCA and DSA, a third type of angiograms used in this study is the direct video angiogram (DVA). The DVA is digitized on-line via a video camera focused on the X-ray image intensifier. In our study, the DCA contains the highest level of noise due to the involvement of the complex imaging chain. The DSA has the highest signal-to-noise ratio (SNR) but may contain subtraction artifacts caused by miss-registration between the two frames during subtraction. The DVA has an intermediate image quality, between those of DCA and DSA.

In another previous research [42], we studied the problem of arteriogram segmentation by an approach based on the pixel grayscale. We developed an iterative ternary classification (ITC) algorithm which used two grayscale thresholds to classify each pixel to one of three classes, i.e., artery, background, and undecided. By iterating on the undecided class the two thresholds are brought closer together and the output converges to a two-class segmentation. The result from the ITC algorithm will be compared with the ANN result as demonstrated later.

B. Problem Statement

In this study, the problem we attempt to solve is the segmentation of arteriograms. The arteriogram is to be segmented into two classes, i.e., vessel and background, with the ANN approach. The ANN-based segmentation is conducted at the lower-level processing. Image pixel values are used as direct input to the ANN. Because we are particularly interested in ANN's capability of extracting features from the *raw* image data, we do not consider the possibility of using a separate non-ANN stage for preprocessing. However, a postprocessing stage may be employed if necessary.

The purpose of this study is twofold: (1) to develop practical ANN-based classifiers for the segmentation of arteriograms, and (2) using the arteriogram segmentation problem as an example, to study the neural network system techniques in terms of network topology, training parameters, generalization capability, supervised versus unsupervised trade-off, and mechanisms for self-organization. In the following two sections, we discuss two ANN classifiers. In Section IV, we review a BP ANN classifier developed in a previous study [19]. In Section V, we derive and evaluate an unsupervised ANN classifier that employs a self-adaptive mechanism for grayscale thresholding on a pixel-by-pixel basis.

IV. Back-Propagation Artificial Neural Network for Arteriogram Segmentation: A Supervised Approach

A. Overview of the Feedforward Back-Propagation Neural Network

Multilayer perceptron with back-propagation learning [43] is perhaps the most common paradigm for supervised neural network computing to date. This has been observed in the medical imaging area (see Section II) as well as many other pattern recognition areas. In a multilayer feedforward network the neurons are

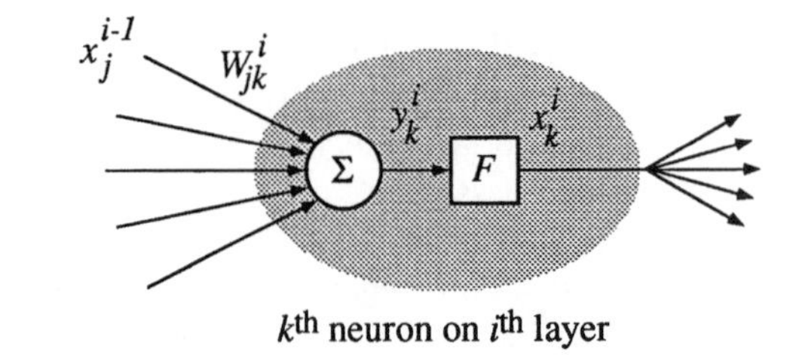

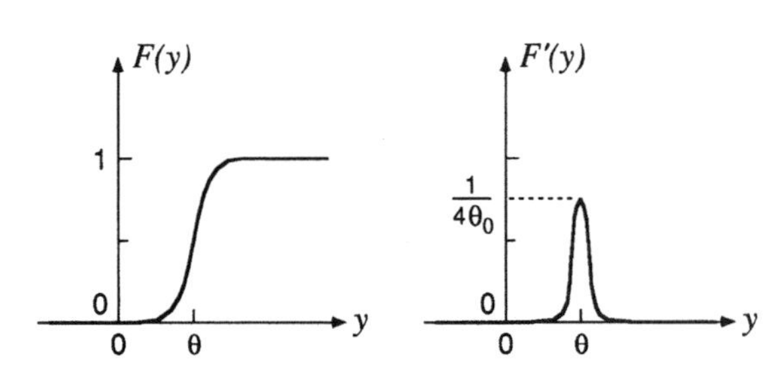

Figure 4 Neuron model with sigmoid activation function.

fully connected in the sense that a neuron on a layer other than the input layer receives signals from all neurons on the previous layer, but from no other. Figure 4 shows the standard neuron model, representing the kth neuron on the ith layer of a feedforward network. The summation operator produces the linear combination of the weighted outputs from all neurons on the previous $(i-1)$th layer:

$$y_k^i = \sum_{\text{all } j} W_{jk}^i x_j^{i-1}, \tag{1}$$

where W_{jk}^i is the weight associated with the link that connects the jth neuron on the $(i-1)$th layer to the kth neuron on the ith layer. The nonlinearity associated with each neuron is an important element, without which the multilayer structure would collapse down to a single-layer linear perceptron [44]. In order to propagate the learning information backward and through the nonlinearity, the nonlinear function needs to be differentiable. The sigmoidal function has frequently been used for this purpose. The output from the nonlinearity is given by

$$x \equiv \mathcal{F}(y) = \frac{1}{1 + e^{(-y+\theta)/\theta_0}}. \tag{2}$$

$\mathcal{F}(y)$ is between 0 and 1; θ is the activation point where $\mathcal{F}(\theta) = (1/2)$. The nonlinearity parameter θ_0 controls the slope of the transition. A lower θ_0 results in a steeper transition. The sigmoidal function approaches to the hard-limiter as

θ_0 approaches to zero. This function has an advantage that its derivative can be easily computed:

$$\mathcal{F}'(y) = \frac{1}{\theta_0}\mathcal{F}(y)\big[1 - \mathcal{F}(y)\big]. \tag{3}$$

The sigmoid function and its derivative are shown in Fig. 4.

The back-propagation learning employs a gradient descent method to train the network weights such that the mean squared error between the actual network output vectors and the desired output vectors is minimized. The back-propagation learning algorithm, often referred to as the generalized delta rule, was elegantly derived by Rumelhart *et al.* [43]. The amount of weight adjustment at each iteration is proportional to the input and the associated δ which can be computed in a *back-propagation* fashion. Let p be the iteration number. At the pth iteration the weight adjustment is according to

$$\Delta W^i_{jk}(p+1) \equiv W^i_{jk}(p+1) - W^i_{jk}(p) = \beta \cdot \delta^i_k(p) \cdot x^{i-1}_j + \alpha \cdot \Delta W^i_{jk}(p), \tag{4}$$

where β is an empirical parameter controlling the rate of learning. The second term on the right-hand side is the momentum term which improves stability and accuracy by slowing the learning process near convergence. The δ function is updated for each neuron at each iteration according to

$$\delta^i_k = \begin{cases} \mathcal{F}'(y^i_k)(d_k - x^i_k), & \text{for output layer,} \\ \mathcal{F}'(y^i_k)\sum_n \delta^{i+1}_n W^{i+1}_{kn}, & \text{otherwise,} \end{cases} \tag{5}$$

where d_k is the labeled output for the kth neuron on the output layer. Because the δs on the ith layer can be determined only when the δs on the $(i+1)$th layer are known, the learning must be carried out in the *backward* direction, i.e., from the output layer toward the input layer.

The weights are typically initialized to small random values before the back-propagation learning commences. For a training set consisting of N pairs of input and labeled output and for an output layer containing M neurons, we define the system error (E) as

$$E = \frac{1}{MN}\sum_{k=1}^{M}\sum_{n=1}^{N}\big[d_k(n) - x_k(n)\big]^2. \tag{6}$$

The training process iterates on computing the δs and updating the weights. The process terminates upon the satisfaction of a stopping criterion, e.g., when the system error is below an acceptable threshold or when the number of iterations exceeds a predetermined threshold.

B. Back-Propagation Artificial Neural Network Classifier for Arteriogram Segmentation

We developed a classifier based on the standard feedforward back-propagation ANN to segment the arteriograms. The structure of this supervised classifier is shown in Fig. 5. The neural network takes image grayscale values as direct inputs. The grayscale values are taken from a window centered about the pixel to be classified. The output layer contains two neurons—one represents vessel and the other represents background—and whichever outputs the larger value prevails. The feedforward network classifies one pixel at a time. Segmentation of the vascular structures is accomplished by scanning the window over the entire image.

In contrast to the elegant derivation of the back-propagation learning, the theories for configuring the neural networks and selecting training parameters are relatively weak. We therefore conducted a systematic study on the various configurations and training parameters for this problem. We attempted to answer questions such as:

- Given a fixed complexity in terms of the total number of weights in the network, what is the most suitable network topology for our segmentation problem? What is the optimal number of hidden layers? How should the neurons be distributed among the input and hidden layers? Do the deep, shallow, and bottleneck network topologies [45] perform differently?
- How should the training set be defined? How many test samples should be included? Should the test samples be hand-picked or randomly selected?
- Does the initial random weight pattern affect the result of learning?

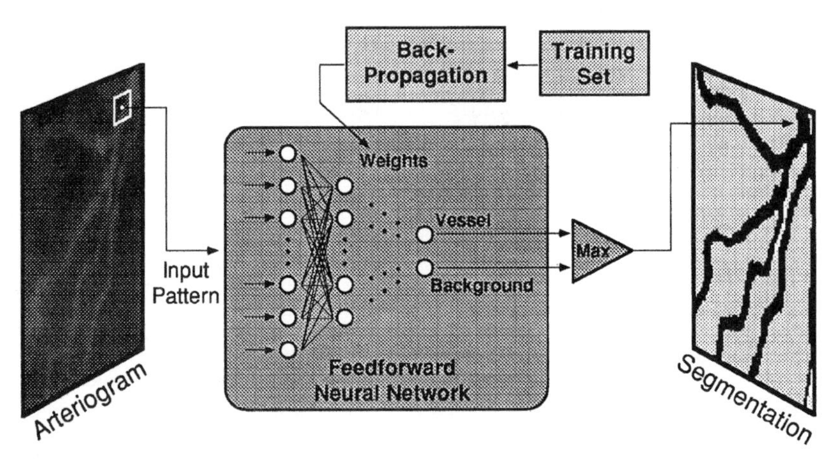

Figure 5 Back-propagation ANN classifier for arteriogram segmentation.

- What values should be used for the learning rate (β) and the momentum rate (α)?
- How many iterations of the learning process should be allowed to run? Does *overlearning* have a negative effect on generalization?

The study addressing these questions has recently been published [19] and is not repeated here. The interested readers are referred to the original publication for the details. In the following we summarize the important findings from that study which are relevant to the present discussion.

We implemented the BP ANN classifiers in the C language for the VAX 11/780 or compatible machines (Digital Equipment Corporation, Maynard, Massachusetts). The training for each network took between 2 and 10 CPU hours, depending on the number of iterations required to reach the specified system error. A systematic study was conducted on the various combinations of network configurations and parameters. The combined computational time for the entire study was on the order of 5000 CPU hours using several networked VAX systems. A topology that yielded the optimal performance was identified, as shown in Fig. 6. This feedforward network consisted of 121 neurons on the input layer to receive grayscale values from an 11×11 window, 17 neurons in the hidden layer, and 2 neurons on the output layer. The total number of weights for the neural network was 2091 ($121 \times 17 + 17 \times 2$). This classifier is referred to as "121-17-2" in the following discussion.

The selection of the training samples had a significant effect on the performance of the classifier. Random selection of samples over the entire image resulted in a training set containing many more background pixels than vessel pixels. A training set consisting of carefully chosen pixels at various parts of the background, edges, and centers of the vessels gave the best performance. The coronary arteriogram shown in Fig. 6 was used to provide the training data. This image contained 256×256 pixels with 8-bit grayscale. The arteriogram was segmented by a human operator to produce a target image. The 75 samples marked by crosses in the arteriogram defined the training set for this study. The BP ANN classifier was repetitively training over these 75 samples until either the system error was less than 0.15 or the total number of iterations reached 3500.

The 121-17-2 classifier was considered converged after 764 iterations during training; it correctly classified 65 samples, i.e., 87% of the 75 training samples corresponding to a system error of 0.13. It generalized quite well. For the remaining 60,441 pixels of the test angiogram—excluding the 75 training samples and the 5-pixel-wide borders that cannot be reached by the center of the 11×11 window—the classification accuracy was 92%. The generalization performance was even better than the training performance because the training set was chosen to represent the *problematic* cases. The 121-17-2 classifier also generalized well for other arteriograms including the DCA, DVA, and DSA types. These results

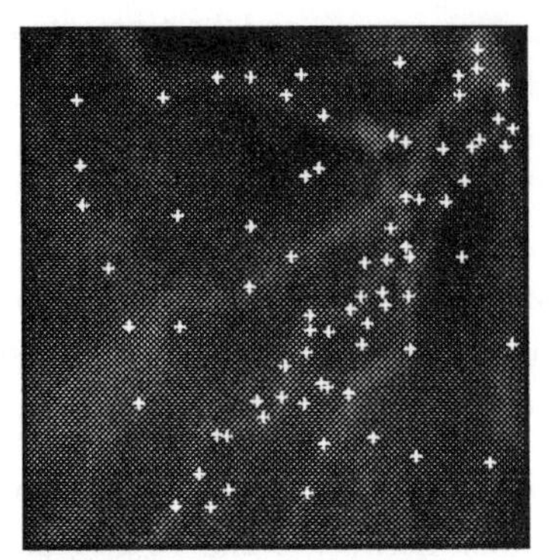

Digitized Cineangiogram and 75 Selected Training Samples

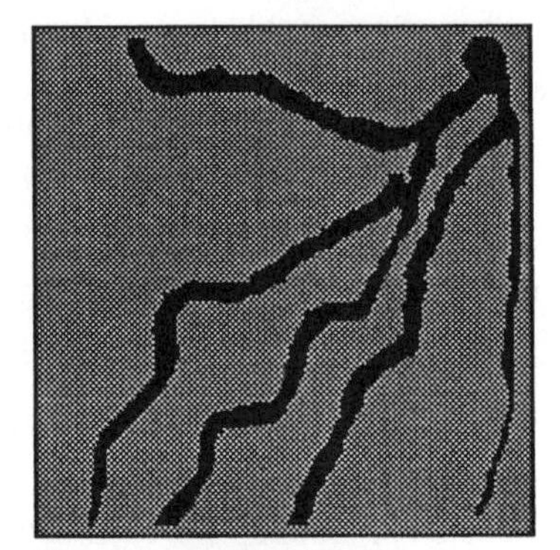

Target Image Obtained by Manual Segmentation

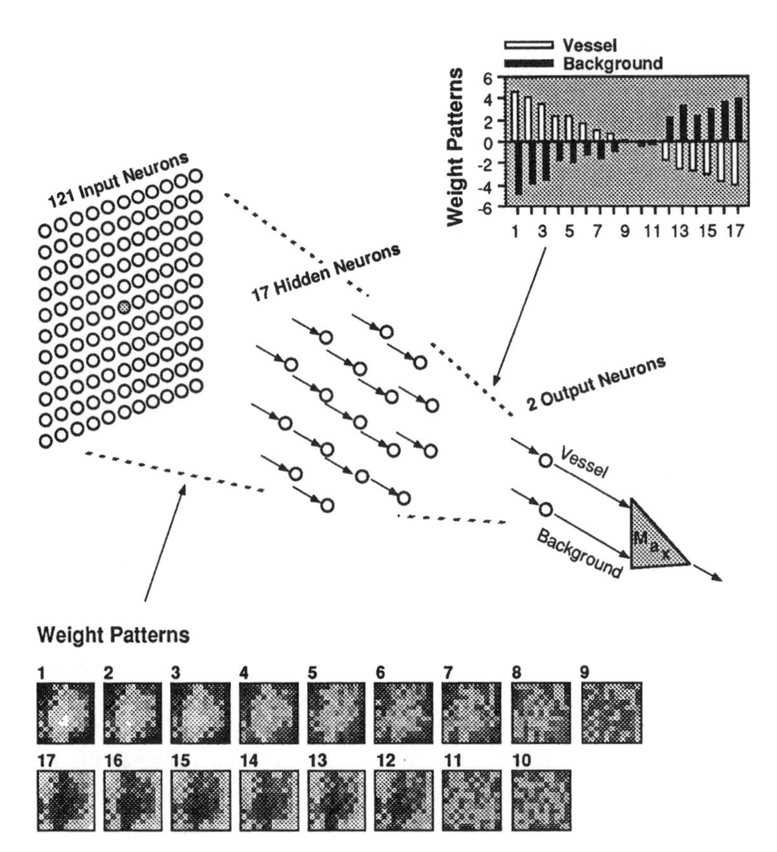

Figure 6 Training data and weight patterns for 121-17-2 back-propagation classifier.

will be presented and compared with the results from an unsupervised classifier in the next section.

To gain an insight into the classification mechanisms of the 121-17-2 classifier, in Fig. 6 the weight patterns between input and hidden layer are displayed as image templates, and the weight patterns between hidden and output layer are plotted in a bar graph. The ANN classifier first acts as a matched filter—the 17 weight templates are convolved over the image to search for specific vessel patterns. Templates 1–6 and templates 12–17 show well-structured patterns and there seems to be a complement relationship between the two sets of templates. Notice that, because the vessels may appear in any orientation, these patterns are more or less radially symmetric. The 17 hidden neurons are activated when the corresponding patterns are sufficiently matched. The weights connecting to the *vessel* output neuron vary systematically from positive to negative, indicating some form of spatial differentiation. As expected, the weights connecting to the *background* output neuron show exactly the complement of those connecting to the vessel neuron. Thus, we conclude that the trained BP ANN classifier behaves as a matched filter followed by a nonlinear decision tree.

V. SELF-ADAPTIVE ARTIFICIAL NEURAL NETWORK FOR ARTERIOGRAM SEGMENTATION: AN UNSUPERVISED APPROACH

A. Adaptive Systems and Gradient Search Method

An adaptive system is a system capable of altering its internal structure to improve its performance by means of an iterative learning algorithm. It is typically a nonlinear system which produces the desirable output by manipulating the input signals through a set of adjustable variables (weights). The weight adjustment is accomplished through an optimization procedure based on a certain performance criterion.

The adaptive system approach is attractive for classification tasks due to its self-organizing, generalizable, and fault-tolerant characteristics. However, the adaptive system is generally difficult to analyze and to control because of its complex implicit mechanisms for decision making. The nonlinear elements in the system also make it difficult to back-track the cause when an erroneous decision is made by the system.

The adaptive systems can be classified in terms of open-loop adaptation and closed-loop adaptation [46]. The open-loop adaptive system adjusts its weights solely based on its input, whereas the closed-loop adaptation is based on both input and feedback from the output. The closed-loop adaptive system has proven to be by far the more powerful model, especially for nonlinear, time-varying, and/or nonstationary processes.

During adaptation the weights are adjusted in such a way that the output is brought closer to the desired response. In contrast to the supervised neural network, the adaptive system does not rely on user-defined training data. The desired response is guided by an internal mechanism designed to solve the specific classification problem. The adaptation process is accomplished by means of minimizing an error signal (ξ), which is usually based on a distance measure between the desired response and the actual response. For a given set of input, the error signal forms a performance surface in the multivariate space defined by the weights. An adaptation algorithm (or learning algorithm) adjusts the weights to move the operating point down the performance surface until the minimum is reached. For most practical applications it is impossible to derive an analytical expression of the performance surface, nor is it possible to conduct an exhaustive search for the global minimum over the multivariate space due to the large number of weights in the system. The adaptation algorithm must be designed to find an optimal or near-optimal solution via a step-by-step search based solely on the local behavior of the performance surface.

The estimate of the local gradient can be used to guide the search toward the minimum on the performance surface. A widely used gradient search method is the steepest descent method since it has fewer restrictions on data and system characteristics than other adaptation algorithms. Steepest descent search is an iterative method in which all the system weights are modified in the direction of the negative gradient. The search begins with an initial weight vector, usually arbitrarily selected. At the kth iteration the new weight vector is determined from the present weight vector $\mathbf{W}_k$ and the gradient ∇_k according to

$$\mathbf{W}_{k+1} = \mathbf{W}_k + \beta(-\nabla_k), \tag{7}$$

where β is the learning rate that controls the stability and the rate of convergence. The gradient defined by

$$\nabla_k = \left. \frac{d\xi}{d\mathbf{W}} \right|_{\mathbf{W}=\mathbf{W}_k} \tag{8}$$

needs to be estimated at each step of the iteration. The search terminates when the gradient is a null vector, or $\mathbf{W}_{k+1} = \mathbf{W}_k$, indicating a minimum on the performance surface is reached.

For a linear system the performance surface based on the mean squared error is shaped like a bowl (a hyperparaboloid for more than two weights) and has only a single minimum. Therefore, it is guaranteed to converge to the optimal solution. Although the linear adaptive classifier has proven to be a statistically optimal classifier [47–49], it is only applicable to the linearly separable problems such as detecting a signal in the presence of white Gaussian noise. For more complicated problems, such as arteriogram segmentation with the presence of background variation and other types of noise, it is necessary to consider a nonlinear adaptive classifier.

For a nonlinear system, however, the performance surface may embody a combination of steep and flat regions [50]. Hence, it is possible that the search is guided to a local minimum and terminates prematurely before the global minimum is reached. The search may also become unstable at a steep part of the surface especially when the learning rate (β) is not sufficiently small. These problems arise as the consequence of forcing the search always in the *downhill* direction on an ill-conditioned performance surface. Fortunately, it has been shown that in a variety of practical applications the system is quasi-linear [43]. The performance surface for a quasi-linear system is differentiable and nondecreasing in all directions away from the global minimum. Thus, if the system is quasi-linear, the performance surface does not contain local minima to trap a gradient-based search.

B. Derivation of the Self-Adaptive Classifier

1. Architecture

An intuitive approach to the arteriogram segmentation problem is to apply a grayscale threshold on the arteriogram—assume that the pixel values of the vessel are generally higher than the pixel values of the background over the entire image. If the histogram of the arteriogram is bimodal showing a peak for vessel and a peak for background, the appropriate value for the threshold can be either manually selected or statistically determined [51]. The single-threshold approach may work for a digital subtraction angiogram with the background properly removed. Unfortunately, the histogram of an unprocessed arteriogram is almost never bimodal. Due to the large background variation, segmentation based on a single threshold usually performs poorly. To improve the segmentation a variable threshold can be used. The idea of variable thresholding is demonstrated with the intensity profile of a scan line across an arteriogram as shown in Fig. 7. Notice that the background intensity is significantly increased on the right side. The threshold is adapted for each pixel based on statistics extracted from the neighborhood of the pixel.

In the following, we derive a self-adaptive (SA) classifier for arteriogram segmentation. The SA classifier employs a variable threshold in conjunction with an adaptation algorithm to segment an arteriogram into the vessel class and the background class. The classification is achieved through an iterative process in which the expected input is estimated from the system output and compared to the actual

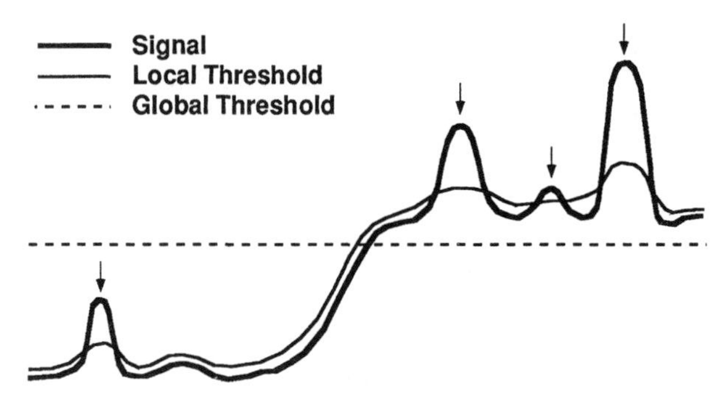

Figure 7 Graphic illustration of fixed versus variable thresholding for intensity profile of scan line across arteriogram. Arrows indicate locations of vessels.

input. The comparison produces an error signal which controls the thresholding parameters.

The variable threshold (local threshold) for each pixel (i, j) in the image is determined according to

$$T_{ij} = \mu_{ij} + W_{ij}\sigma_{ij}. \tag{9}$$

W_{ij} is the weight that controls the threshold for pixel (i, j). μ_{ij} is the mean in a neighborhood of $\omega_0 \times \omega_0$ pixels centered about (i, j):

$$\mu_{ij} = \frac{1}{\omega_0^2} \sum_{m=(1-\omega_0)/2}^{(\omega_0-1)/2} \sum_{n=(1-\omega_0)/2}^{(\omega_0-1)/2} x_{i+m,j+n}, \tag{10}$$

where x is the pixel grayscale value. σ_{ij} is a measure of scatter about the mean (standard deviation) in the neighborhood:

$$\sigma_{ij} = \left[\frac{1}{\omega_0^2 - 1} \sum_{m=(1-\omega_0)/2}^{(\omega_0-1)/2} \sum_{n=(1-\omega_0)/2}^{(\omega_0-1)/2} (x_{i+m,j+n} - \mu_{ij})^2 \right]^{1/2}. \tag{11}$$

Once the local thresholds are computed, the entire image is segmented to create a binary image:

$$y_{ij} = \begin{cases} 0, & \text{if } x_{ij} \le T_{ij},\ x_{ij} \in \text{background}, \\ 1, & \text{if } x_{ij} > T_{ij},\ x_{ij} \in \text{vessel}. \end{cases} \tag{12}$$

In selecting the window size ω_0, there exists a trade-off between rejecting noise and retaining threshold locality. As the window size increases, local thresholding acts more like global thresholding. As the window size decreases, statistics estimated from the neighborhood become less reliable. Although local thresholding favors a small window size in general, the low signal-to-noise ratio in the arteriogram requires a sufficiently large window size to provide reliable statistics for estimating the local threshold. To circumvent the difficulty of the window-size trade-off, we use a one-layer self-adaptive network to control the weight (W_{ij}) applied to the standard deviation for each pixel. The SA classifier performs variable thresholding for each pixel with a threshold computed from the estimates of the neighborhood's mean plus the weighted standard deviation.

In most adaptive systems the error signal is the difference between the desired output and the actual output. In our case, however, we do not have the desired output because the information about the vessel location is not available *a priori* in the unsupervised situation. Thus, instead of comparing the outputs, we compare the inputs. The adaptive system presented here obtains its error signal from the distance between the actual input x_{ij} and the estimated input $\hat{x}_{ij}$. Figure 8 demonstrates the overall architecture for the SA classifier.

Another important feature of the present system is that, instead of using Eq. (12) to perform thresholding, the hard-limiter can be replaced by a soft-limiter. When the soft-limiter is used, the system output (y_{ij}^k) comes from the sigmoid function:

$$y_{ij}^k = \mathcal{F}(x_{ij} - \mu_{ij} - W_{ij}^k \sigma_{ij}), \tag{13}$$

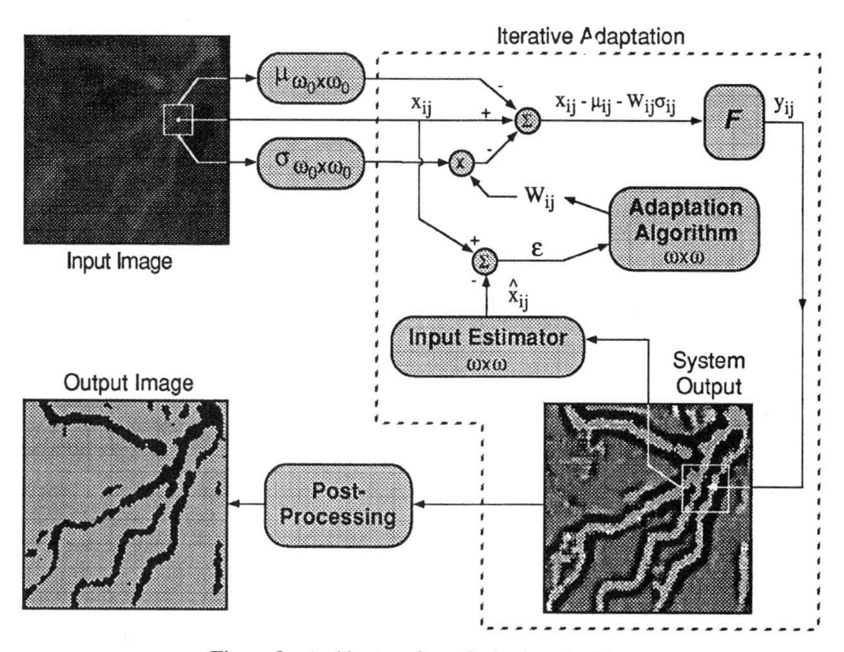

Figure 8 Architecture for self-adaptive classifier.

where the sigmoid function $\mathcal{F}$ is defined by Eq. (2). The sigmoid function is continuous and varies monotonically from 0 to 1. The output y_{ij}^k is not exactly binary; it can be considered as the probability that pixel (i, j) belongs to the vessel class. The derivative of the sigmoid function exists, as defined by Eq. (3), allowing us to carry the adaptation process through the nonlinearity as discussed in the following.

2. Estimation of the Error Signal

At the kth iteration the error signal (ξ^k) is defined as the distance between the actual input (x_{ij}) and an estimated input ($\hat{x}^k$) from the present output within an $\omega \times \omega$ window:

$$\xi^k = \|x_{ij} - \hat{x}_{ij}^k\|_2 = \left[a \sum_{\omega} \left(x_{ij} - \hat{x}_{ij}^k\right)^2\right]^{1/2}, \tag{14}$$

where a is a normalizing constant; the notation $\sum_{\omega}$ denotes the summation over the $\omega \times \omega$ pixels centered about pixel (i, j). Notice that this window size ω is not necessarily the same as the window size ω_0 which was used in the previous section for obtaining the means and standard deviations from the input image. To simplify the derivation of the algorithm, we choose the error measure (ε^k) as the sum of squared errors:

$$\varepsilon^k = \left(\xi^k\right)^2 = a \sum_{\omega} \left(x_{ij} - \hat{x}_{ij}^k\right)^2. \tag{15}$$

The estimate of the input signal is based on the mean value of each class (vessel or background) in the moving window. At each iteration, the estimated input for a vessel (background) pixel is set to the mean value of all detected vessel (background) pixels within the $\omega \times \omega$ window.

First, let us assume that the system's nonlinearity function is the hard-limiter. In this case, the system output is a binary image consisting of ones and zeros. The input estimate is given by

$$\hat{x}_{ij}^k = \dot{\mu}^k y_{ij}^k + \ddot{\mu}^k (1 - y_{ij}^k), \tag{16}$$

where $\dot{\mu}^k$ and $\ddot{\mu}^k$ are the means for the vessel class and the background class, respectively. That is,

$$\hat{x}_{ij}^k = \begin{cases} \dot{\mu}^k, & \text{if } y_{ij}^k = 0, \\ \ddot{\mu}^k, & \text{if } y_{ij}^k = 1. \end{cases} \tag{17}$$

The mean of each class can be estimated by

$$\dot{\mu}^k = \frac{\sum_\omega x_{mn} y_{mn}^k}{\sum_\omega y_{mn}^k}, \tag{18}$$

$$\ddot{\mu}^k = \frac{\sum_\omega x_{mn}(1 - y_{mn}^k)}{\sum_\omega (1 - y_{mn}^k)}. \tag{19}$$

Our adaptation algorithm described in the next section requires differentiability along the signal path. The adaptation process would be blocked by the hard-limiter because its derivative does not exit. Substituting the hard-limiter by the soft-limiter for the above input estimator will introduce some error. This error, however, should be negligibly small, especially for a soft-limiter that has a low θ_0 corresponding to an abrupt transition between 0 and 1. Notice that as θ_0 approaches to zero, the soft-limiter approaches to the hard-limiter. The error also diminishes as the output pixel values converge to either 0 or 1.

By substituting Eqs. (18) and (19) into Eq. (16), the input estimate is given by

$$\hat{x}_{ij}^k = \left[\frac{\sum_\omega x_{mn} y_{mn}^k}{\sum_\omega y_{mn}^k} y_{ij}^k\right] + \left[\frac{\sum_\omega x_{mn}(1 - y_{mn}^k)}{\sum_\omega (1 - y_{mn}^k)}(1 - y_{ij}^k)\right]. \tag{20}$$

3. Adaptation Algorithm

The adaptation algorithm developed in this section is analogous to the steepest descent method in the sense that the operating point descends on the performance surface toward the minimum. The weights are initialized to small random values. At the kth iteration the weights are adjusted in the direction opposed to the gradient of the error signal ε:

$$W_{ij}^{k+1} = W_{ij}^k - \beta \left.\frac{\partial \varepsilon^k}{\partial W_{ij}}\right|_{W_{ij}^k}, \tag{21}$$

where β is the adaptation coefficient or learning rate that regulates the speed and stability of the system. The partial derivative $\partial\varepsilon^k/\partial W_{ij}$ can be evaluated using the chain-rule:

$$\frac{\partial \varepsilon^k}{\partial W_{ij}} = \left(\frac{\partial \varepsilon^k}{\partial \hat{x}_{ij}^k}\right)\left(\frac{\partial \hat{x}_{ij}^k}{\partial W_{ij}}\right). \tag{22}$$

Substituting Eq. (15) into the first term on the right-hand side, we have

$$\frac{\partial \varepsilon^k}{\partial \hat{x}_{ij}^k} = a \sum_{(m,n)\in\omega} \frac{\partial (x_{mn} - \hat{x}_{mn}^k)^2}{\partial \hat{x}_{ij}^k}. \tag{23}$$

The only nonzero term in the summation is at $(m, n) = (i, j)$, thus

$$\frac{\partial \varepsilon^k}{\partial \hat{x}_{ij}^k} = -2a\left(x_{ij} - \hat{x}_{ij}^k\right). \tag{24}$$

To solve for the second term on the right-hand side of Eq. (22), we apply the chain-rule again:

$$\frac{\partial \hat{x}_{ij}^k}{\partial W_{ij}} = \left(\frac{\partial \hat{x}_{ij}^k}{\partial y_{ij}^k}\right)\left(\frac{\partial y_{ij}^k}{\partial W_{ij}}\right). \tag{25}$$

The second term on the right-hand side can be determined based on the fact that $y_{ij}^k = \mathcal{F}(x_{ij} - \mu_{ij} - W_{ij}^k \sigma_{ij})$, where $\mathcal{F}(\cdot)$ is the sigmoidal function with its derivative defined by Eq. (3). We have

$$\begin{aligned} \frac{\partial y_{ij}^k}{\partial W_{ij}} &= \frac{1}{\theta_0}\mathcal{F}\left(x_{ij} - \mu_{ij} - W_{ij}^k \sigma_{ij}\right)\left[1 - \mathcal{F}(x_{ij} - \mu_{ij} - W_{ij}^k \sigma_{ij})\right] \\ &\quad \times \frac{\partial (x_{ij} - \mu_{ij} - W_{ij}^k \sigma_{ij})}{\partial W_{ij}} \\ &= -\frac{1}{\theta_0} y_{ij}^k \left(1 - y_{ij}^k\right)\sigma_{ij}. \end{aligned} \tag{26}$$

Substituting $\hat{x}_{ij}^k$ given by Eq. (16) into the first term of Eq. (25), we obtain

$$\frac{\partial \hat{x}_{ij}^k}{\partial y_{ij}^k} = \frac{\partial[\dot{\mu}^k y_{ij}^k + \ddot{\mu}^k(1 - y_{ij}^k)]}{\partial y_{ij}^k} \tag{27}$$

$$= \frac{\partial \dot{\mu}^k y_{ij}^k}{\partial y_{ij}^k} + \frac{\partial \ddot{\mu}^k (1 - y_{ij}^k)}{\partial y_{ij}^k} \tag{28}$$

$$= \left[\frac{\partial \dot{\mu}^k}{\partial y_{ij}^k} y_{ij}^k + \dot{\mu}^k\right] + \left[\frac{\partial \ddot{\mu}^k}{\partial y_{ij}^k}(1 - y_{ij}^k) - \ddot{\mu}^k\right]. \tag{29}$$

Now we define

$$\dot{\nu}_{ij}^k = \frac{\partial \dot{\mu}^k}{\partial y_{ij}^k} = \frac{x_{ij}\sum_\omega y_{mn}^k - \sum_\omega x_{mn} y_{mn}^k}{(\sum_\omega y_{mn}^k)^2} \tag{30}$$

and

$$\ddot{\nu}_{ij}^k = -\frac{\partial \ddot{\mu}^k}{\partial y_{ij}^k} = \frac{x_{ij}\sum_\omega (1 - y_{mn}^k) - \sum_\omega x_{mn}(1 - y_{mn}^k)}{[\sum_\omega (1 - y_{mn}^k)]^2}. \tag{31}$$

Therefore,

$$\frac{\partial \hat{x}_{ij}^k}{\partial y_{ij}^k} = \dot{\nu}^k y_{ij}^k + \dot{\mu}^k - \ddot{\nu}^k(1 - y_{ij}^k) - \ddot{\mu}^k. \quad (32)$$

Finally, by combining Eqs. (24), (26), and (32), the weight update Eq. (21) can be solved according to

$$W_{(i,j)\in\omega}^{k+1} = W_{(i,j)\in\omega}^{k} - \beta\big[-2a\big(x_{ij} - \hat{x}_{ij}^k\big)\big]\left[\frac{1}{\theta_0} y_{ij}^k\big(1 - y_{ij}^k\big)\sigma_{ij}\right] \times \big[\dot{\nu}^k y_{ij}^k + \dot{\mu}^k - \ddot{\nu}^k\big(1 - y_{ij}^k\big) - \ddot{\mu}^k\big] \quad (33)$$

$$= W_{(i,j)\in\omega}^{k} - \rho\sigma_{ij}\big(x_{ij} - \hat{x}_{ij}^k\big)y_{ij}^k\big(1 - y_{ij}^k\big) \times \big[\big(\dot{\mu}^k + y_{ij}^k\dot{\nu}_{ij}^k\big) - \big(\ddot{\mu}^k + \big(1 - y_{ij}^k\big)\ddot{\nu}_{ij}^k\big)\big], \quad (34)$$

where $\rho = 2\beta a/\theta_0$.

Due to the complexity of the above expression the mechanisms of the adaptation are not self-evident. In the following, we isolate and study the individual terms in Eq. (34) with the intention to obtain a better insight into the weight update mechanisms. In Fig. 9, we plot the term $y_{ij}(1 - y_{ij})$ versus y_{ij}. This term, affecting the rate of weight adjustment, has the maximum at $y_{ij} = 1/2$, decreases on both sides, and reaches 0 at $y_{ij} = 0$ and $y_{ij} = 1$. This term contributes to the reduction of the learning rate when the output converges to either 0 or 1, and thereby improves accuracy and stability. Thus, it has a similar effect as the momentum term used in supervised BP learning discussed in Section IV.

The amount of weight adjustment is proportional to $(x_{ij} - \hat{x}_{ij}^k)$, which is self-explanatory. The remaining term can be rewritten in the following form:

$$\big(\dot{\mu}^k - \ddot{\mu}^k\big) + \big[\dot{\nu}_{ij}^k y_{ij}^k - \ddot{\nu}_{ij}^k\big(1 - y_{ij}^k\big)\big]. \quad (35)$$

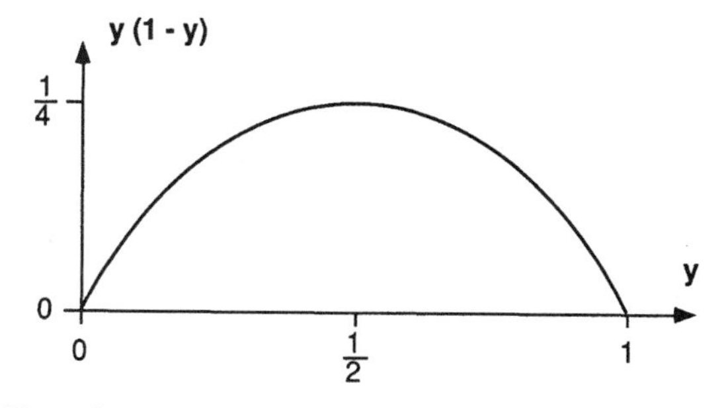

Figure 9 Weight update rate, $y(1 - y)$, plotted versus output, y.

Furthermore, $\dot{v}_{ij}^k$ and $\ddot{v}_{ij}^k$ can be rearranged as

$$\begin{aligned}\dot{v}_{ij}^k &= \frac{1}{\sum_\omega y_{mn}^k}\left[\frac{x_{ij}\sum_\omega y_{mn}^k}{\sum_\omega y_{mn}^k} - \frac{\sum_\omega x_{mn} y_{mn}^k}{\sum_\omega y_{mn}^k}\right] \\ &= \frac{1}{\dot{\kappa}}\left(x_{ij} - \dot{\mu}^k\right),\end{aligned} \tag{36}$$

$$\begin{aligned}\ddot{v}_{ij}^k &= \frac{1}{\sum_\omega (1 - y_{mn}^k)}\left[\frac{x_{ij}\sum_\omega (1 - y_{mn}^k)}{\sum_\omega (1 - y_{mn}^k)} - \frac{\sum_\omega x_{mn}(1 - y_{mn}^k)}{\sum_\omega (1 - y_{mn}^k)}\right] \\ &= \frac{1}{\ddot{\kappa}}\left(x_{ij} - \ddot{\mu}^k\right),\end{aligned} \tag{37}$$

where $\dot{\kappa}$ and $\ddot{\kappa}$ are the pixel counts for the vessel class and the background class, respectively. Now the term can be presented as

$$\left(\dot{\mu}^k - \ddot{\mu}^k\right)\left[\frac{1}{\dot{\kappa}}\left(x_{ij} - \dot{\mu}^k\right)y_{ij}^k - \frac{1}{\ddot{\kappa}}\left(x_{ij} - \ddot{\mu}^k\right)\left(1 - y_{ij}^k\right)\right]. \tag{38}$$

The first part of this term is the difference between the mean vessel intensity and the mean background intensity within the $\omega \times \omega$ window. The second part represents a consistency measure for the present classification of pixel (i, j) within the $\omega \times \omega$ window. For example, if the present classification is vessel ($y_{ij} = 1$), this part is reduced to $(1/\dot{\kappa})(x_{ij} - \dot{\mu}^k)$ which is the difference between the pixel intensity and the mean vessel intensity normalized by the vessel pixel count within the $\omega \times \omega$ window.

4. Postprocessing

Some background variations have features similar to those of vessels. These background variations are incorrectly classified as vessel and result in speckled artifacts scattered over the background area in the segmented arteriogram. Fortunately, these speckled artifacts are easily detectable due to their appearance as isolated small clusters and can be removed by a postprocessing stage. The various filtering techniques based on mathematical morphology [52] seem to be particularly suitable for this purpose.

The following describes one feasible algorithm for postprocessing based on a simple median filter.

Step 1. Make a binary image by assigning all output pixels which have not reached vessel class to the background class. In other words, all the unclassified pixels are absorbed by the background:

$$y_{ij} = \begin{cases} 1, & \text{if } y_{ij} = 1, \\ 0, & \text{if } y_{ij} < 1. \end{cases} \tag{39}$$

Step 2. Remove the speckled artifacts by applying a moving median filter over the binary image. Within a local window the center pixel value y_{ij} is replaced by the median of all pixel values within the window. For a binary image the median filter can be implemented simply by assigning the dominant class within the window to the center pixel.

The median filter is useful here because it reduces single-pixel noise while it preserves edges in the image. The simple median filter can be generalized to an $n \times n$ median filter which can correctly remove a noise pixel as long as the total number of noise pixels with the window is less than $(n^2 + 1)/2$. We have found that a 5×5 median filter provides a satisfactory performance for the post-processing.

C. Performance Evaluation of the Self-Adaptive Classifier

We implemented the SA classifier on a conventional computer to evaluate its properties and classification performance. First, we conducted a systematic study on the effects of various system parameters including input window size ω_0, adaptation window size ω, nonlinearity parameter θ_0, and learning rate β. We should emphasize that it is very important to study the sensitivity of these empirical parameters. Should the performance be very sensitive to certain parameters, the system would not generalize well and the adaptation scheme associated with those parameters should be reevaluated. Next, after the system parameters were properly selected, we applied the SA classifier to arteriograms including the DCA, DVA, and DSA types described in Section III. The segmentation results by the SA classifier were also compared with those by the BP classifier discussed in Section IV.

1. Convergence

As with any adaptive system, a primary concern with the SA classifier is its convergence. In our experiments, the SA classifier converged in practically every case within 10 iterations. The rapid convergence of the system was observed from the weight matrix values, W_{ij}, through the iterations. The weight matrix was initialized with random values. As the iteration commenced, it rapidly organized itself to provide the appropriate local thresholds for the vessel pixels in the arteriogram. As expected, when the weight matrix was shown as an image, it resembled the vascular structure in the input image. In Fig. 10 we demonstrate the convergence of weight matrix by displaying it as an image at the first, third, and tenth iterations (left to right). The weights were mapped into 8-bit grayscale and the resulting image was histogram-equalized to improve visualization.

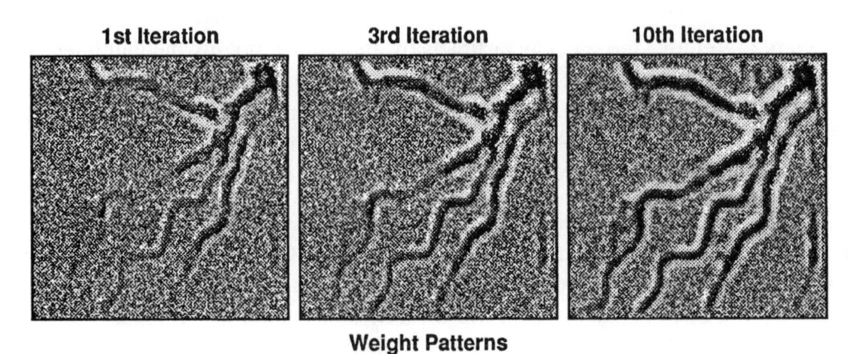

Figure 10 Weights (W_{ij}) in SA classifier shown as images through iterations.

2. Window Effects

The SA classifier employs two moving windows. The input window with size ω_0 is used to estimate the mean and variance around each pixel from the arteriogram. The adaptation window with size ω is used to assess the error signal. The input window is applied only once before the iteration begins, whereas the adaptation window is used at each step of the iteration. The experimental results show that the two moving window sizes have direct but relatively minor effects on the performance of the SA classifier. Figure 11 shows the effects of ω_0 and ω on the segmentation of an arteriogram.

The input window size (ω_0) controls the *smoothness* of the segmented image. A small input window produces less reliable statistics and results in a relatively noisy segmentation. In contrast, a large input window produces a relatively smooth segmentation but has a smearing effect on edges and anatomical details.

The adaptation window size (ω) shows a somewhat greater effect on the performance than the input window size does. The adaptation window size affects the segmentation quality and, to a lesser extent, the convergence rate. A small adaptation window slows the adaptation and can cause premature convergence at a local minimum. An adaptation window significantly larger than the vessel width makes the system behave like a global-thresholding method and reduces the classification accuracy. The best performance is associated with an adaptation window size slightly larger than the average width of the vessels under investigation.

3. Rate Parameters

Referring to Eq. (34), the learning process is affected by three parameters: learning rate β, nonlinearity parameter θ_0, and normalization factor a. The normalization factor is a constant calculated according to the size of the adaptation

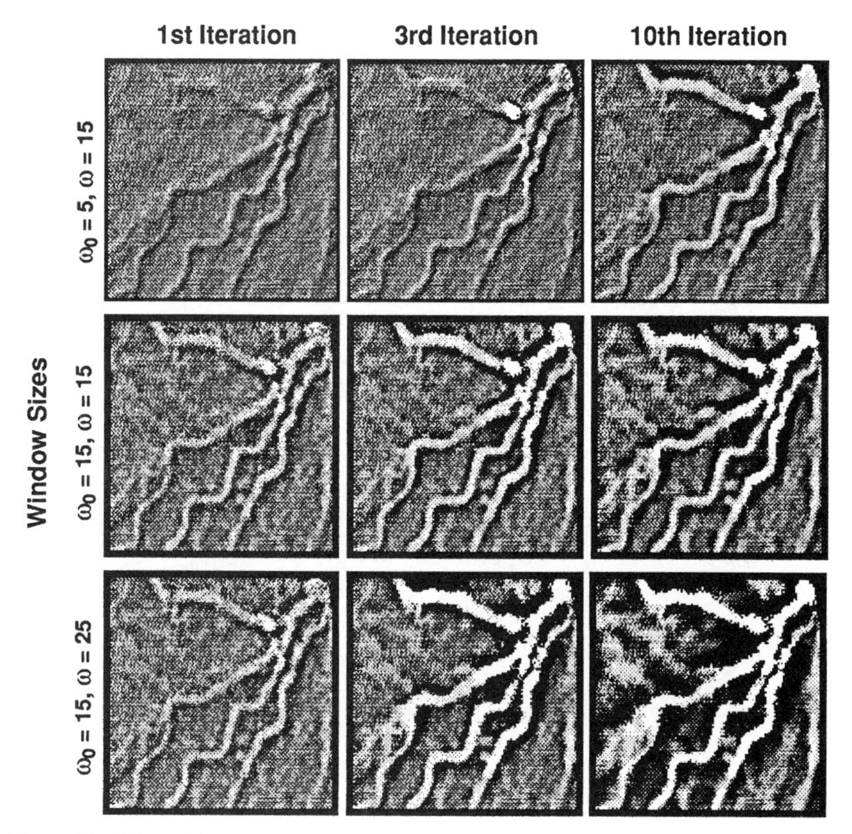

Figure 11 Effects of input window size ω_0 and adaptation window size ω on system output through iterations.

window. This is why the adaptation window size affects the rate of convergence as discussed above. Once the adaptation window size is determined, the learning rate and the nonlinearity parameter are the only two parameters that can control the rate of convergence.

How learning rate should be controlled to achieve the best performance is a common problem to all the steepest-descent-type algorithms. Learning rate controls not only the rate of convergence but also the stability of the system. A high learning rate can result in an unstable system producing noisy and inaccurate outputs. A low learning rate can result in slow convergence or premature convergence to a local minimum. Figure 12 illustrates the effects of learning rate on the test image. The best performance was achieved by choosing β between 0.01 and 0.09.

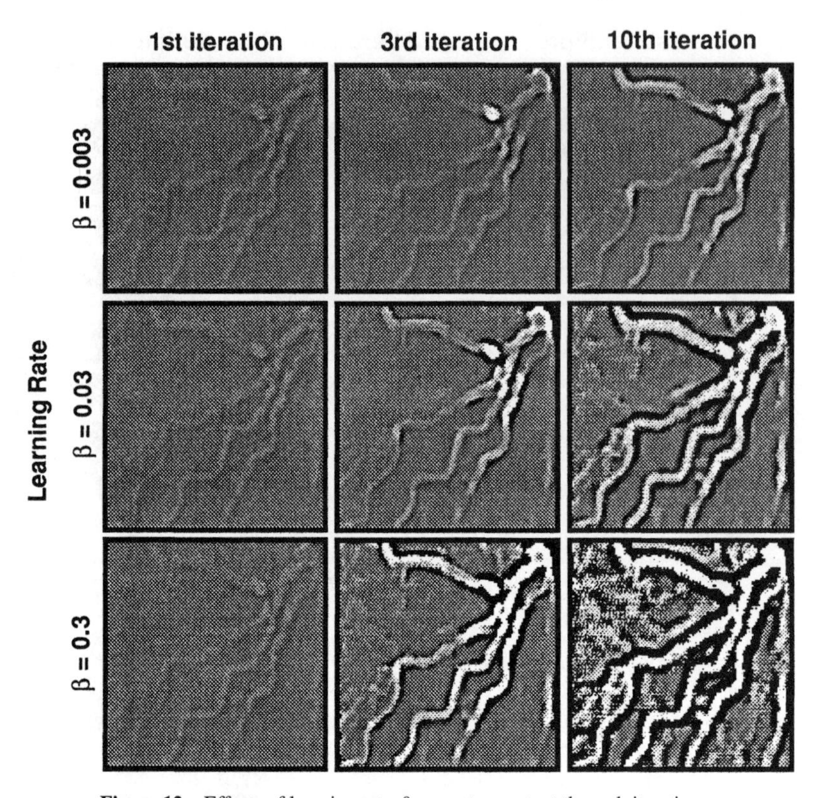

Figure 12 Effects of learning rate β on system output through iterations.

The nonlinear transition of the sigmoid function is affected by θ_0: A small θ_0 results in an abrupt transition and a large θ_0 results in a gradual transition. The adjustment of θ_0 has two effects on the system. Referring to parameter ρ in Eq. (34), θ_0 is combined with β to control the rate of weight update. θ_0 also directly controls the quantization of the system output and affects the amount of information being fed back from output to weight adaptation. A large θ_0 slows the convergence, increases the likelihood of local-minimum convergence, but provides more information (less quantization) for better adaptation. A small θ_0 makes the sigmoid function closer to a hard-limiter. In the extreme case of hard-limiter ($\theta_0 = 0$) the adaptation mechanism stops functioning completely because the derivative of the nonlinearity required by Eq. (26) no longer exists. Figure 13 shows the effects of θ_0 on the segmentation of the test image. According to the experimental results the appropriate range for θ_0 is between 0.1 and 1.0.

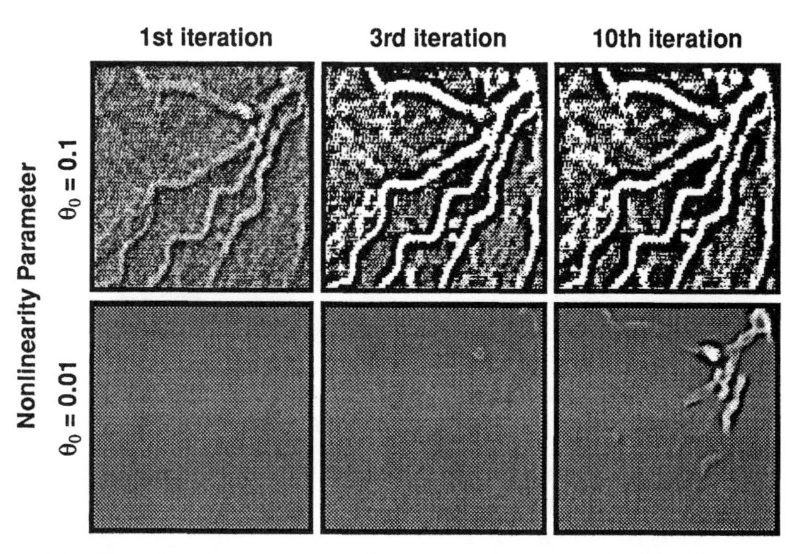

Figure 13 Effects of nonlinearity parameter θ_0 on system output through iterations. Smaller θ_0 corresponds to more abrupt transition in sigmoid function.

4. Segmentation of Arteriograms

We evaluated the performance of the SA classifier with a set of arteriograms representing a broad range of image quality. The supervised BP classifier developed in Section IV was also applied to the same set of arteriograms so that the performance between the unsupervised and supervised classifier can be compared. Figure 14 shows the results for the original arteriogram, which was used by the BP classifier for training and by the SA classifier for parameter optimization. This image is a digitized cineangiogram (DCA) of the left coronary artery. In Fig. 14 the four images are arteriogram (upper-left), segmentation by the BP classifier (upper-right), output of the SA classifier before postprocessing (lower-left), and segmentation after postprocessing (lower-right). In the same format, Fig. 15 shows the results for a different DCA frame that belongs to the same sequence of the original arteriogram shown in Fig. 14. Figure 16 shows the results for a direct video angiogram (DVA) of the right iliac arteries. Finally, Fig. 17 shows the results for a digital subtraction angiogram (DSA) of the right coronary artery.

The results presented above have provided a qualitative comparison between the supervised BP classifier and the unsupervised SA classifier. Generally speaking, the two classifiers are comparable in performing the task of arteriogram segmentation. The SA classifier shows a high sensitivity for detecting smaller vessels, as seen in Figs. 16 and 17. The SA classifier also produces a cleaner background;

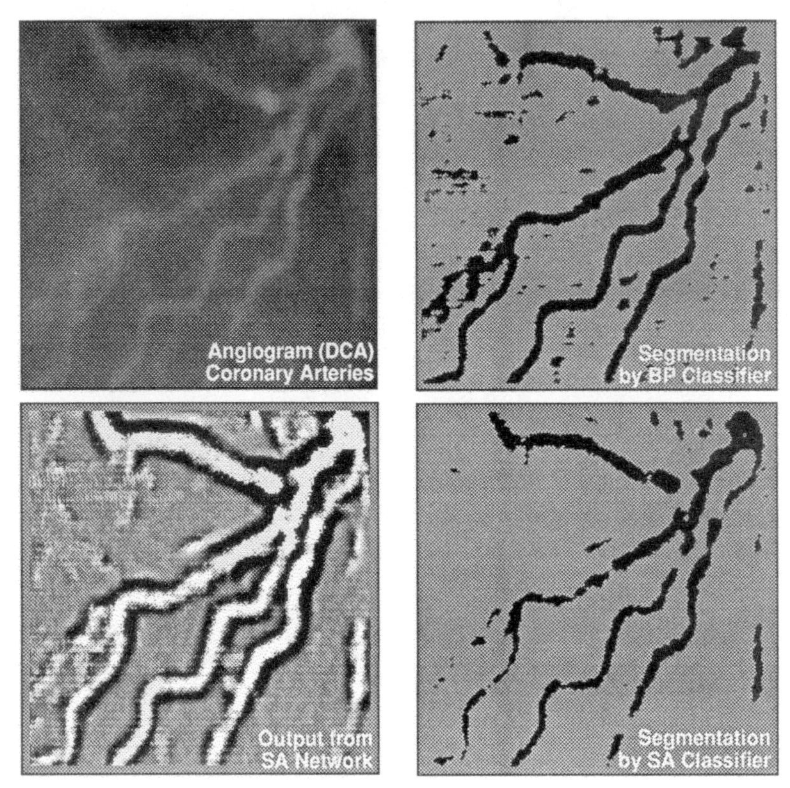

Figure 14 Digitized cineangiogram of left coronary artery (original test image) and segmentation results by BP classifier, SA classifier before postprocessing, and SA classifier after postprocessing.

however, much of that should be attributed to the postprocessing stage. In Fig. 15, a large dark background area can be observed on the left side of the arteriogram. The SA classifier incorrectly extracts the edge of this area as part of the vascular structure. In contrast, the BP classifier correctly ignores this edge. The SA classifier does not contain a mechanism to take advantage of the fact that a vessel segment has two parallel borders. The BP classifier, on the other hand, seems to be well trained to handle this situation.

To further provide a quantitative evaluation of the two classifiers, we use the original DCA image and the target image shown in Fig. 6. The target image defined by a human operator is used as the gold standard. In this comparison, we also include two other classifiers: the iterative ternary classifier (ITC) developed in a previous study [42] and a maximum likelihood estimator (MLE) that computes a global threshold based on the classic Bayesian approach [19]. Figure 18 shows

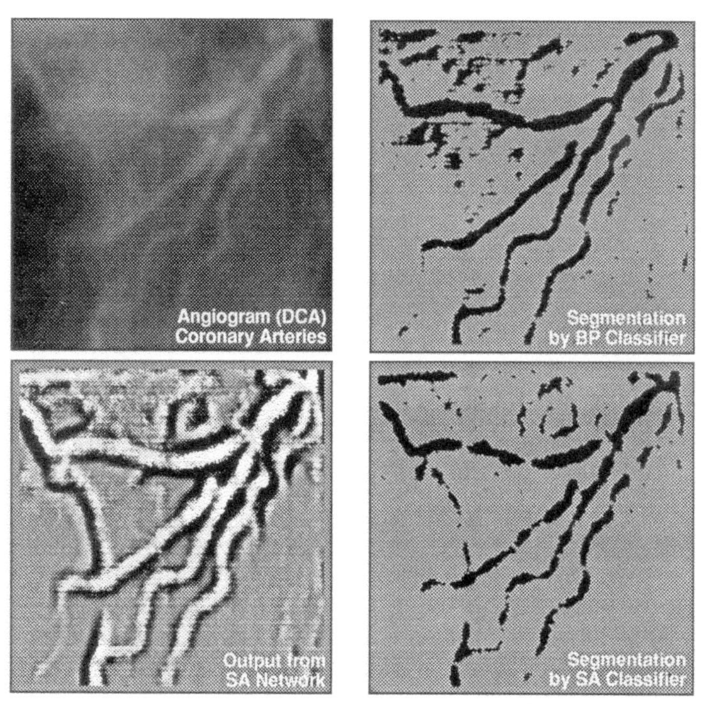

Figure 15 Digitized cineangiogram of left coronary artery and segmentation results by BP classifier, SA classifier before postprocessing, and SA classifier after postprocessing.

the segmentation results from these four classifiers: SA, BP, ITC, and MLE. The performance indexes including classification accuracy, learning time, and classification time are summarized in Table I. The SA classifier showed the best performance with 94% accuracy, closely followed by the BP classifier's 92%. The

Table I
Performance Comparison of Four Classifiers

Algorithm	Accuracy	Learning time (s)	Classification time (s)
Self-adaptive ANN classifier	94%	0	360
Back-propagation ANN classifier	92%	7,150	540
Iterative ternary classifier	83%	0	170
Maximum likelihood estimator	68%	60	350

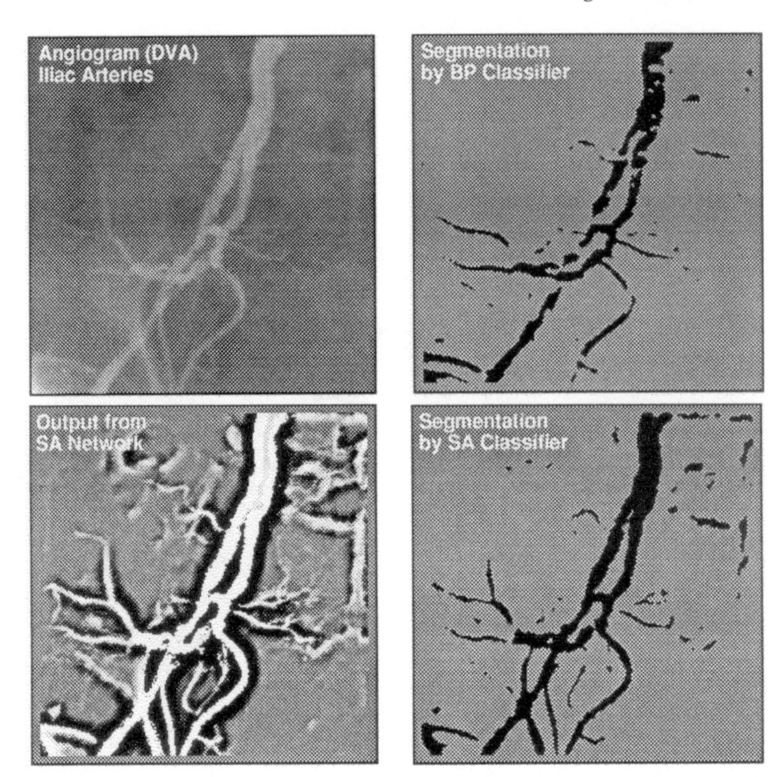

Figure 16 Direct video angiogram of iliac arteries and segmentation results by BP classifier, SA classifier before postprocessing, and SA classifier after postprocessing.

parameters for the SA classifier were: $\omega_0 = 11$, $\omega = 11$, $\theta_0 = 0.1$, and $\beta = 0.03$. The parameters for the BP classifier were: topology = 121-17-2, $\alpha = 0.5$, and $\beta = 0.05$. The two ANN-based classifiers generally performed better than the other two methods. This may be attributed to the ability of ANN to form highly nonlinear decision boundaries and to classify patterns with non-Gaussian distributions.

VI. CONCLUSIONS

A. NEURAL NETWORK APPLICATIONS IN MEDICAL IMAGING

The literature review in Section II—although it was neither exhaustive nor in-depth—should provide a perspective for the trend of ANN applications in the medical imaging area. While technique-oriented researches have been conducted

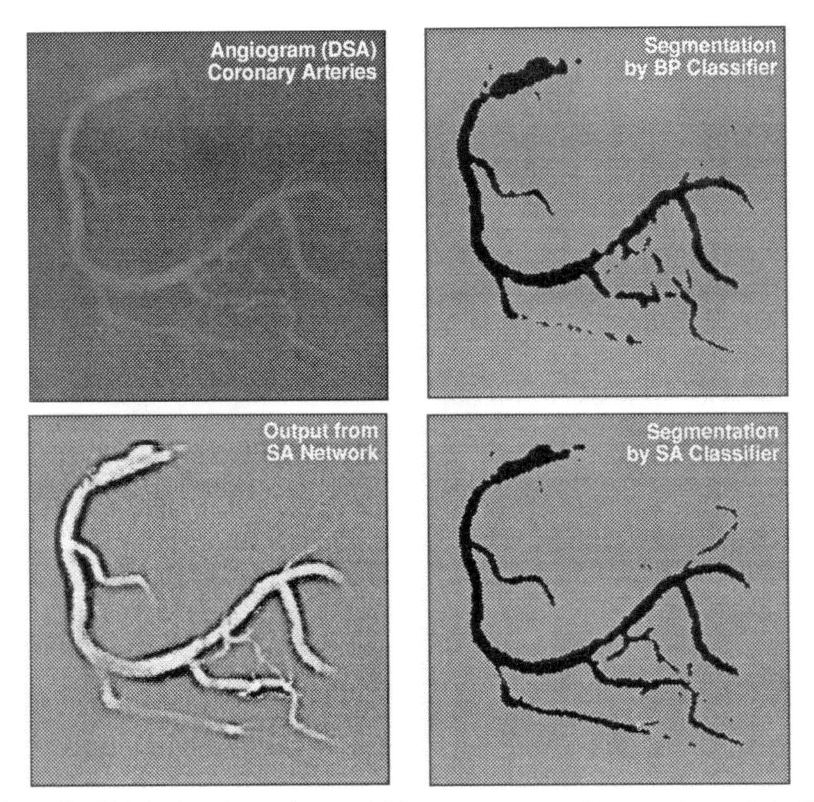

Figure 17 Digital subtraction angiogram of right coronary artery and segmentation results by BP classifier, SA classifier before postprocessing, and SA classifier after postprocessing.

by using ANNs for lower-level processing of medical images, clinical applications have been predominantly for higher-level processing whereby features are first extracted in a preprocessing stage by using more conventional pattern recognition methods. The use of ANNs for higher-level processing is attractive for several reasons. First, the lower-level processing involves a large amount of data from image pixels and usually requires customized software. Second, by incorporating a preprocessing stage for data reduction, it is much easier to adopt a general commercial neural network software for the specific diagnostic problem. Third, medical experts are accustomed to the use of image features extracted by conventional pattern recognition techniques and information from patient history, which are more suitable as inputs to an ANN at the higher processing level. Fourth, the inputs to the higher-level processing are usually more meaningful and bear clinical significance; therefore, it is easier to back-track the problem when the output of the ANN is erroneous.

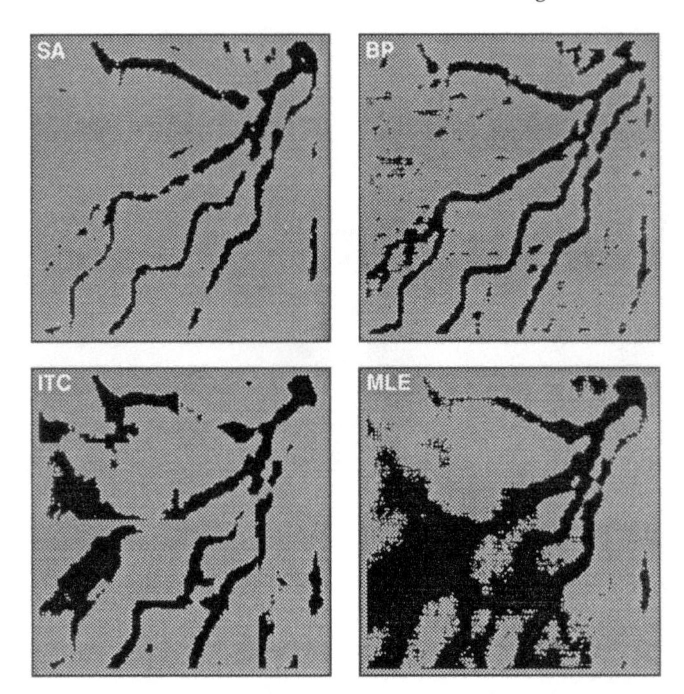

Figure 18 Arteriogram segmentation by self-adaptive classifier, back-propagation classifier, iterative ternary classifier, and maximum likelihood estimator.

If the ANN classifiers can be considered separable from the *conventional* classification methods, it must be due to the distributive parallel processing nature of neural network computing. Thus, a neural network classifier using a small set of extracted features as input may not fully exploit the power of distributive parallel processing. When a preprocessing stage is used, the higher-level ANN is at the mercy of the lower-level preprocessing stage. A crucial portion of the feature information may have been inadvertently excluded by the preprocessing even before it reaches the ANN classifier. To mimic the human perception of diagnostic medical images, it is important to apply the ANN to extracting features directly from the *raw* image data. Thus, the use of ANN for lower-level medical image processing should be a fruitful area that merits continuing research.

Another observation regarding ANN applications in medical imaging is that supervised learning has been the more dominant approach. The popularity of the feedforward back-propagation network may have contributed to this dominance. A supervised ANN classifier can also be trained on a continuing basis, hoping to improve upon the mistakes that the ANN has made on a retrospective basis. In contrast, the unsupervised neural network classifiers rely on their internal adaptation mechanisms to perform certain classification tasks. Although it is possible to improve their classification performance by optimizing the system parameters, such optimization is less intuitive and usually much more difficult to control.

B. Supervised versus Unsupervised Artificial Neural Network for Arteriogram Segmentation

In this study, we used the arteriogram segmentation problem as an example for lower-level processing of medical images. We developed a supervised ANN (the BP classifier) as well as an unsupervised ANN (the SA classifier) to classify pixels in arteriograms into either the vessel class or the background class. It was shown that both classifiers performed satisfactorily for arteriograms over a broad range of image quality. They also outperformed two other classifiers based on some more conventional approaches.

Although we ought to be prudent in generalizing our findings, the comparison of the supervised versus unsupervised classifier for this problem should provide a useful guideline for developing medical image processing systems. In Table II, we summarize the important features for the SA and BP classifiers. The main difficulty associated with the supervised BP classifier was the choice of its topology. The appropriate topology for our BP classifier was identified via a brute-force

Table II

Comparison between the SA and BP Classifiers

	SA classifier	BP classifier
Learning	Unsupervised	Supervised
Classification	Iterative	One-pass
	Converged fast	Feedforward
Mechanisms	Implemented internally	Learned from training
Preprocessing	No	No
Postprocessing	Yes	No
Empirical parameters	Input window size	Network topology
	Adaptation window size	Training set
	Learning rate	Learning rate
	Nonlinearity parameter	Momentum rate
		Training period

search. We experienced some very poor performance from BP neural networks with slightly different configurations. The performance of the BP classifier was also very sensitive to the choice of the training set, the learning rate, the momentum rate, and the training period. For clinical applications it is conceivable that a supervised ANN may not respond in a positive way to continuing training with new data; its performance may also degrade by overtraining.

On the other hand, the performance of the SA classifier was less sensitive to its parameters. There were also fewer parameters to be identified. For the arteriogram segmentation problem under investigation, the SA classifier stood out as the best performer when all things were considered. A drawback of the SA classifier is that the classification mechanisms must be studiously implemented into the adaptation algorithm, making it more difficult to generalize to other problems. The need for postprocessing is another minor drawback associated with the SA classifier.

C. Future Directions

Based on the results from this study, we attempt to identify some potentially fruitful directions for future research in applying ANNs to medical imaging. First, much can be learned about the distributive parallel processing of medical-image-based diagnostics by applying ANN models to lower-level processing tasks such as image enhancement, feature extraction, and segmentation. Second, the adaptation mechanisms in unsupervised ANNs should merit further studies for extracting various features in medical images such as malignant mass in mammogram, underperfused area in cardiac or brain SPECT, and lesion in brain MRI or CT. Third, general software tools especially for unsupervised classification and low-level processing should be developed to reduce the effort of adopting an ANN model for a specific clinical application in medical imaging.

Finally, in Fig. 19, we propose a generalized model for neural-network-based processing of medical images. Image features are extracted from pixel data and represented by designated hidden nodes in the network. Multimodality images can also be fused at an early stage of the neural network computing. Both unsupervised learning and supervised learning take place in the same system and interact with each other. The unsupervised learning is guided by adaptation schemes designed to extract specific features from the images. The supervised learning is based on retrospective data of known diagnostic outcomes. The system represents a unification among multimodality images, between lower-level processing and higher-level processing, and between supervised neural network computing and unsupervised neural network computing.

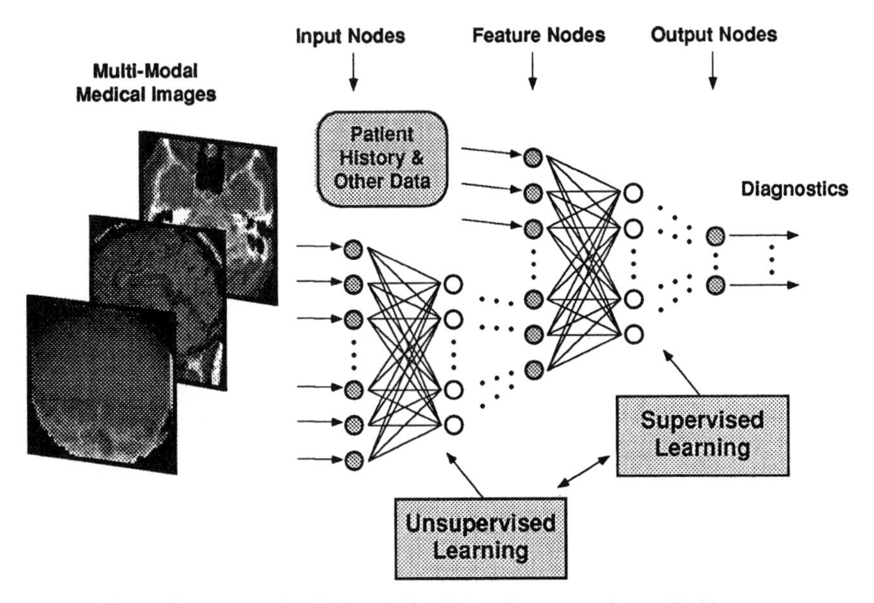

Figure 19 Proposed unified model for diagnostic system using medical images.

REFERENCES

[1] S. Webb. *The Physics of Medical Imaging*. Adam Hilger, New York, 1990.

[2] J. Radon. On the determination of functions from their integral values along certain manifolds. *IEEE Trans. Med. Imaging* MI-5:170–176, 1986. Translated by P. C. Parks.

[3] C. J. Thompson, Y. L. Yamamoto, and E. Meyer. Positome II: A high efficiency position imaging device for dynamic brain studies. *IEEE Trans. Nucl. Sci.* NS-26:583–389, 1979.

[4] A. S. Miller, B. H. Blott, and T. K. Hames. Review of neural network applications in medical imaging and signal processing. *Med. & Biol. Eng. & Comput.* 30:449–464, 1992.

[5] J. P. Kerr and E. B. Bartlett. A statistically tailored neural network approach to tomographic image reconstruction. *Med. Phys.* 22:601–610, 1995.

[6] M. T. Munley, C. E. Floyd, Jr., J. E. Bowsher, and R. E. Coleman. An artificial neural network approach to quantitative single photon emission computed tomographic reconstruction with collimator, attenuation, and scatter compensation. *Med. Phys.* 21:1889–1899, 1994.

[7] S. Cagnoni, D. Caramella, R. De Dominicis, and G. Valli. Neural network synthesis of spin echo multiecho sequences. *J. Digit. Imaging* 5:89–94, 1992.

[8] H. Yan and J. Mao. Data truncation artifact reduction in MR imaging using a multilayer neural network. *IEEE Trans. Med. Imaging* 12:73–77, 1993.

[9] Y. Hui and M. R. Smith. Comments on "Data truncation artifact reduction in MR imaging using a multilayer neural network." *IEEE Trans. Med. Imaging* 14:409–412, 1995.

[10] A. Adler and R. Guardo. A neural network image reconstruction technique for electrical impedance tomography. *IEEE Trans. Med. Imaging* 13:594–600, 1994.

[11] U. Raff, A. L. Scherzinger, R. F. Vargas, and J. H. Simon. Quantitation of grey matter, white matter, and cerebrospinal fluid from spin-echo magnetic resonance images using an artificial neural network technique. *Med. Phys.* 21:1933–1942, 1994.

[12] X. Li, S. Bhide, and M. R. Kabuka. Labeling of MR brain images using Boolean neural network. *IEEE Trans. Med. Imaging* 15:628–638, 1996.

[13] M. Özkan, B. M. Dawant, and R. J. Maciunas. Neural-network-based segmentation of multimodal medical images: A comparative and prospective study. *IEEE Trans. Med. Imaging* 12:534–544, 1993.

[14] K.-S. Cheng, J.-S. Lin, and C.-W. Mao. The applications of competitive Hopfield neural network to medical image segmentation. *IEEE Trans. Med. Imaging* 15:560–567, 1996.

[15] J.-S. Lin, K.-S. Cheng, and C.-W. Mao. A fuzzy Hopfield neural network for medical image segmentation. *IEEE Trans. Nucl. Sci.* 43:2389–2398, 1996.

[16] G. Coppini, R. Poli, and M. Rucci. A neural network architecture for understanding discrete three-dimensional scenes in medical imaging. *Comput. Biomed. Res.* 25:569–585, 1992.

[17] W. Qian, M. Kallergi, and L. P. Clarke. Order statistic-neural network hybrid filters for gamma camera-Bremsstrahlung image restoration. *IEEE Trans. Med. Imaging* 12:58–64, 1993.

[18] S.-C. B. Lo, S.-L. A. Lou, and S. K. Mun. Artificial convolution neural network techniques and applications for lung nodule detection. *IEEE Trans. Med. Imaging* 14:711–718, 1995.

[19] R. Nekovei and Y. Sun. Back-propagation network and its configuration for blood vessel detection in angiograms. *IEEE Trans. Neural Networks* 6:64–72, 1995.

[20] K. H. Chan, K. A. Johnson, J. A. Becker, A. Satlin, J. Mendelson, B. Garada, and B. L. Holman. A neural network classifier for cerebral perfusion imaging. *J. Nucl. Med.* 35:771–774, 1994.

[21] R. J. deFigueiredo, W. R. Shankle, A. Maccato, M. B. Dick, P. Mundkur, I. Mena, and C. W. Cotman. Neural-network-based classification of cognitively normal, demented, Alzheimer disease and vascular dementia from single photon emission with computed tomography image data from brain. *Proc. Nat. Acad. Sci. U.S.A.* 92:5530–5534, 1995.

[22] M. P. Page, R. J. Howard, J. T. O'Brien, M. S. Buxton-Thomas, and A. D. Pickering. Use of neural networks in brain SPECT to diagnose Alzheimer's disease. *J. Nucl. Med.* 37:195–200, 1996.

[23] H. Fujita, T. Katafuchi, T. Uehara, and T. Nishimura. Application of artificial neural network to computer-aided diagnosis of coronary artery disease in myocardial SPECT bull's-eye images. *J. Nucl. Med.* 33:272–276, 1992.

[24] D. Hamilton, P. J. Riley, U. J. Miola, and A. A. Amro. Identification of a hypoperfused segment in bull's-eye myocardial perfusion images using a feed forward neural network. *Br. J. Radiol.* 68:1208–1211, 1995.

[25] J. A. Scott and E. L. Palmer. Neural network analysis of ventilation-perfusion lung scans. *Radiology* 186:661–664, 1993.

[26] G. D. Tourassi, C. E. Floyd, H. D. Sostman, and R. E. Coleman. Artificial neural network for diagnosis of acute pulmonary embolism: Effect of case and observer selection. *Radiology* 194:889–893, 1995.

[27] R. E. Fisher, J. A. Scott, and E. L. Palmer. Neural networks in ventilation-perfusion imaging. *Radiology* 198:699–706, 1996.

[28] E. R. Kischell, N. Kehtarnavaz, G. R. Hillman, H. Levin, M. Lilly, and T. A. Kent. Classification of brain compartments and head injury lesions by neural networks applied to MRI. *Neuroradiology* 37:535–541, 1995.

[29] H. Azhari, S. Oliker, W. J. Rogers, J. L. Weiss, and E. P. Shapiro. Three-dimensional mapping of acute ischemic regions using artificial neural networks and tagged MRI. *IEEE Trans. Biomed. Eng.* 43:619–626, 1996.

[30] K. Suzuki, I. Horiba, and M. Nanki. Recognition of coronary arterial stenosis using neural network on DSA system. *Systems Computers Japan* 26:66–74, 1995.

[31] B. Zheng, W. Qian, and L. P. Clarke. Digital mammography: Mixed feature neural network with spectral entropy decision for detection of microcalcifications. *IEEE Trans. Med. Imaging* 15:589–597, 1996.
[32] B. Sahiner, H.-P. Chan, and M. M. Goodsitt. Classification of mass and normal breast tissue: A convolution neural network classifier with spatial domain and texture images. *IEEE Trans. Med. Imaging* 15:598–610, 1996.
[33] J. A. Baker, P. J. Kornguth, J. V. Lo, and C. E. Floyd, Jr. Artificial neural network: Improving the quality of breast biopsy recommendations. *Radiology* 198:131–135, 1996.
[34] V. Goldberg, A. Manduca, D. L. Ewert, J. J. Gisvold, and J. F. Greenleaf. Improvement in specificity of ultrasonography for diagnosis of breast tumors by means of artificial intelligence. *Med. Phys.* 19:1475–1481, 1992.
[35] B. G. Brown, E. Bolson, M. Frimer, and H. T. Dodge. Quantitative coronary arteriography: Estimation of dimensions, hemodynamic resistance, and atheroma mass of coronary artery lesions using the arteriogram and digital computation. *Circulation* 55:329–337, 1977.
[36] G. W. Vetrovec. Evolving applications of coronary angioplasty: Technical and angiographic considerations. *Amer. J. Cardiol.* 64:27E–32E, 1989.
[37] G. A. White, K. W. Taylor, and J. A. Rowlands. Noise in stenosis measurement using digital subtraction angiography. *Med. Phys.* 12:705–712, 1981.
[38] Y. Sun. Automated identification of vessel contours in coronary arteriograms by an adaptive tracking algorithm. *IEEE Trans. Med. Imaging* 8:78–88, 1989.
[39] I. Liu and Y. Sun. Recursive tracking of vascular networks in angiograms based on the detection-deletion scheme. *IEEE Trans. Med. Imaging* 12:334–341, 1993.
[40] I. Liu and Y. Sun. Fully automated reconstruction of 3-D vascular tree structures from two orthogonal views using computational algorithms and production rules. *Optical Eng.* 31:2197–2207, 1992.
[41] R. Kruger, C. Mistretta, and A. Crummy. Digital k-edge subtraction radiography. *Radiology* 125:243–245, 1977.
[42] D. Kottke and Y. Sun. Segmentation of coronary arteriograms by iterative ternary classification. *IEEE Trans. Biomed. Eng.* 37:778–785, 1990.
[43] D. E. Rumelhart, G. E. Hinton, and R. J. Williams. Learning internal representations by error propagation. *Nature* 323:533–536, 1986.
[44] F. Rosenblatt. The perceptron: A probabilistic model for information storage and organization in the brain. *Psychol. Rev.* 65:386–408, 1958.
[45] J. K. Kruschke. Improving generalization in back-propagation networks with distributed bottlenecks. In *IEEE–INNS International Joint Conference on Neural Networks*, pp. 443–447, 1989.
[46] B. Widrow and S. D. Stearns. *Adaptive Signal Processing*. Prentice-Hall, Englewood Cliffs, NJ, 1985.
[47] K. Steinbuch and U. A. W. Piske. Learning matrices and their applications. *IEEE Trans. Electron. Computers* 12:856–862, 1963.
[48] J. S. Koford and G. F. Groner. The use of an adaptive threshold element to design a linear optimal pattern classifier. *IEEE Trans. Inform. Theory* 12:42–50, 1966.
[49] I. P. Devyaterikov, A. I. Propoi, and Y. Z. Tsypkin. Iterative algorithms for pattern recognition. *Automation Remote Control* 28:108–117, 1967.
[50] D. Hush, B. Horne, and J. M. Salas. Error surface for multilayer perceptrons. *IEEE Trans. Systems, Man, Cybernet.* 22:1152–1161, 1992.
[51] R. Kohler. A segmentation system based on thresholding. *Comput. Graph. Image Proc.* 15:319–338, 1981.
[52] R. C. Gonzalez and P. Wintz. *Digital Image Processing*, pp. 351–450. Addison-Wesley, Reading, MA, 1987.

Paper Currency Recognition

Fumiaki Takeda
Technological Development Department
GLORY Ltd.
3-1, Shimoteno, 1-Chome, Himeji
Hyogo 670, Japan

Sigeru Omatu
Department of Computer and
Systems Sciences
College of Engineering
Osaka Prefecture University
Sakai, Osaka 593, Japan

I. INTRODUCTION

Up to now, we have proposed paper currency recognition methods by a neural network (NN) to aim at development for the new type of paper currency recognition machines [1–5]. Especially, we have proposed three core techniques using the NN. The first is the small-size neuro-recognition technique using masks [1–6]. The second is the mask determination technique using a genetic algorithm (GA) [7–12]. The third is the neuro-engine technique using a digital signal processor (DSP) [13–15]. In the first technique, we regard the sum of input pixels as a characteristic value. This is based on a slab-like architecture in Widrow's algorithm [6] which is invariant to various fluctuations of the input image.

Especially, in the neuro-paper currency recognition technique, we have adopted random masks in a preprocessor [1–5], which have masked some parts of the input image. The sum of nonmasked pixels by the mask is described as a slab value. This is the characteristic value of the input image. We input not pixel values but slab values to the NN. This technique enables us to realize a small-size neuro-recognition. However, in this technique, we must decide a masked area by the random numbers. So we cannot always get effective masks which reflect the difference between the patterns of input image to the slab values. We must optimize the masks and systematize their determination.

In the second technique, in order to determine the excellent masks which can generate the characteristic values of the input image effectively, we have adopted

Image Processing and Pattern Recognition
Copyright © 1998 by Academic Press. All rights of reproduction in any form reserved.

the GA [10–12] to the mask determination. This is a unique technique which is a searching procedure based on the mechanism of natural selection and natural genetics [11, 12]. The second technique on the mask determination can generate effective masks which satisfy the purposive generalization of the NN owing to the GA mechanism being like the evolutional process of a life [11, 12]. In this technique, we regard the position of a masked part as a gene. We operate "coding," "sampling," "crossover," "mutation," and "selection" for some genes. By repeating a series of these operations, we can get effective masks automatically.

In the third technique, we have developed a high-speed neuro-recognition board to realize the neuro-recognition machines [13–15]. In this neuro-recognition board, we have used a DSP, which has been widely used for image processing. The adopted DSP has the exponential function which is used in the sigmoid function $[f(x) = 1/(1+\exp(-x))]$ as a library. Furthermore, the DSP can execute various calculations of the floating-point variables. This matter enables us to implement the neuro-software, which is made on the EWS or the large computer, to this neuro-recognition board easily. Its computational speed is ten times faster compared with the current recognition machines.

In this chapter, we unify these three techniques for the paper currency recognition and describe this neuro-system technique. Then the possibility and effectiveness of this system technique in the experimental systems constructed by the various current banking machines are shown. The physical meaning of the unified system [14, 15] is made clear.

II. SMALL-SIZE NEURO-RECOGNITION TECHNIQUE USING THE MASKS

Here, we discuss the first technique. First of all, we describe its basic idea which comes from Widrow's algorithm [6]. Then we show the effectiveness for reduction of NN's scale with various experiments.

A. Basic Idea of the Mask Technique

We define a sum of input pixels as a characteristic value of the input image, which is described as a slab value [1–6]. We can get 31 as the slab value of pattern A in Fig. 1a. We can get 25 as the slab value of pattern B in Fig. 1b. In this case, the slab value is useful to the input of the NN. However, the slab value corresponding to pattern C in Fig. 2a is 23, while the slab value corresponding to pattern D in Fig. 2b is also 23. In this case, we cannot use the slab value as an input of the NN. We must reflect the difference of the input image to the slab value. This problem can be solved by adopting a mask [1–5] which covers some

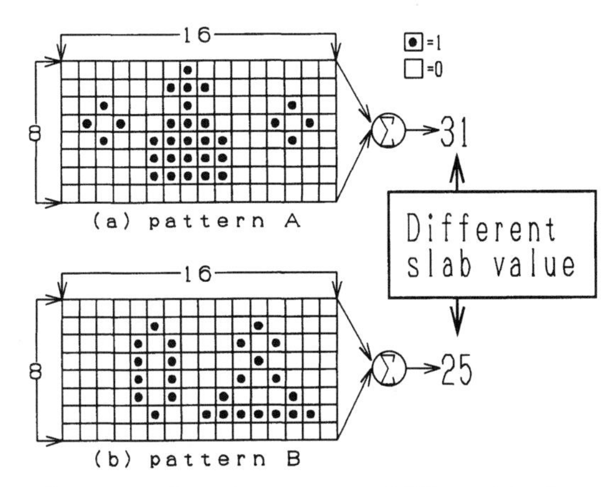

Figure 1 Different input patterns and different slab values.

parts of the input image in Fig. 3c. The slab value becomes 13 when the pattern C in Fig. 2a is covered by the mask. Otherwise, the slab value becomes 23 when the pattern D in Fig. 2b is covered by the mask. In this way, we can use the slab value as an input to the NN using the mask. Thus, the mask enables us to measure a two-dimensional image from various viewpoints as if we measured a three-dimensional object from various viewpoints. Furthermore, we use various masks and make some slab values from one input image since the probability to obtain effective slab values for pattern recognition becomes high [2, 3, 5].

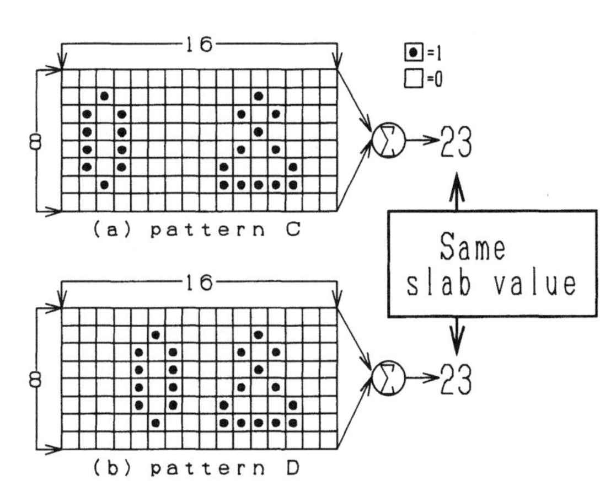

Figure 2 Different input patterns and same slab values.

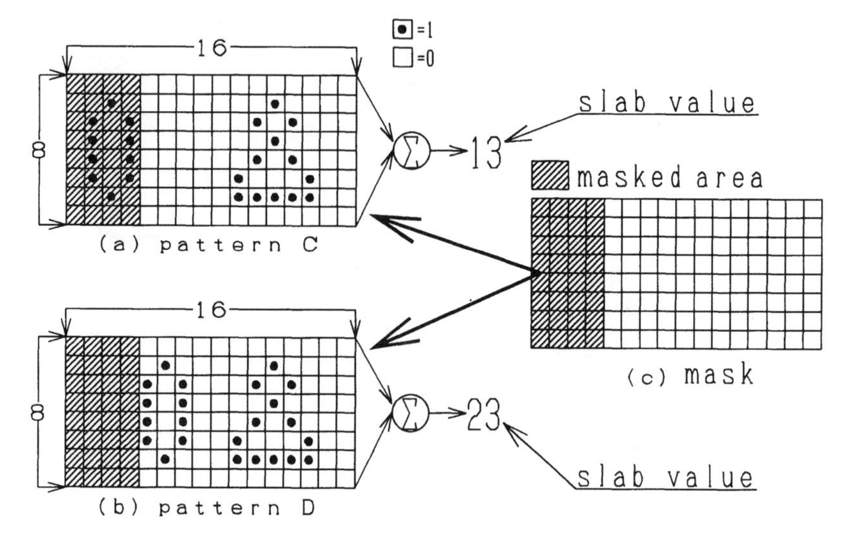

Figure 3 Mask and slab values.

As shown in Fig. 4, we show the construction of the mask processing for the NN. Some parts of the input image are covered with various masks in preprocessing. The sum of input pixels which are not covered becomes one slab value which is taken as an input of the NN.

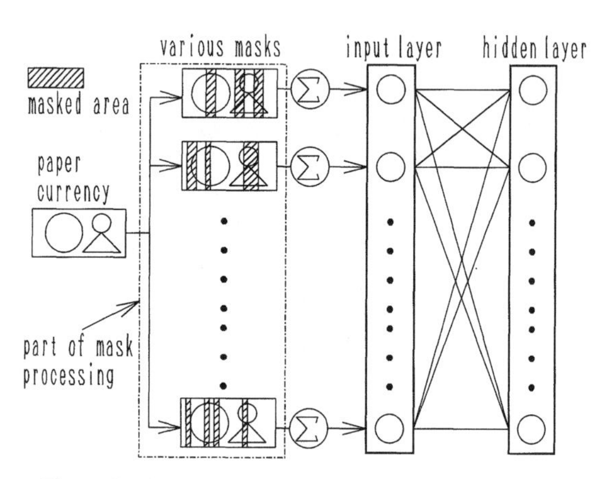

Figure 4 Construction of the mask processing for the NN.

B. Study of the Mask Parameters

We make some experiments for the mask parameters in order to standardize the mask technique. In the mask technique, we discuss mask number and its area with 12 alphabetical letters which are from "A" to "L" [2–5] and they are binary data written on an 8×8 matrix as shown in Fig. 5. We adopt the back propagation method with oscillation term [1–5] and this equation is given by

$$\begin{aligned}
\Delta W_{i,j}^{k-1\,k}(t) &= -\varepsilon d_j^k o_i^{k-1} + \alpha \Delta W_{i,j}^{k-1\,k}(t-1) + \beta \Delta W_{i,j}^{k-1\,k}(t-2), \\
d_j^m &= \left(o_j^m - y_j\right) f'(i_j^m), \qquad \text{for output layer,} \\
d_j^k &= \left(\Sigma W_{j,1}^{k\,k+1} d_1^{k+1}\right) f'(i_j^k), \qquad \text{for hidden layer,}
\end{aligned} \tag{1}$$

where $W_{i,j}(t)$ is the weight from unit i to j, $\Delta W_{i,j}(t)$ is the change of weight $W_{i,j}(t)$, d is the generalized error, o is the output unit value, t is the sample, i is the input unit value, y is the supervised value for the output unit, k is the layer number, ε is the positive learning coefficient, α is the proportional coefficient of inertia term, and β is the proportional coefficient of oscillation term. Especially, the β term has the role of escaping from a local minimum [1–5].

The neuro-weights are modified at the presentation of each alphabetical letter. We regard that convergence is completed when the summation of the squared error between the output unit value and the desired one for each pattern becomes

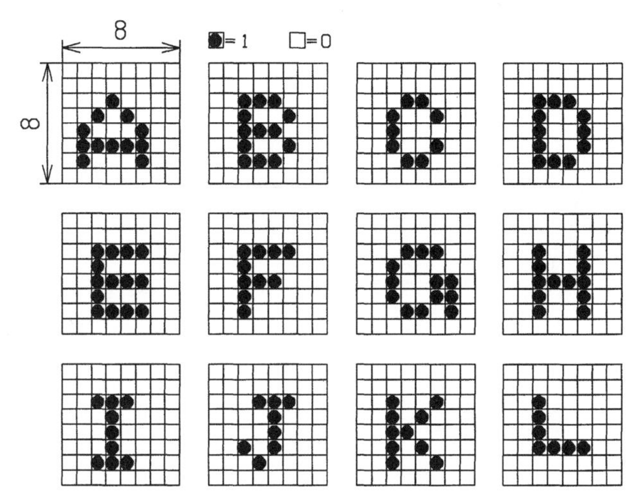

Figure 5 Alphabetical letters in 8×8 matrix.

less than a threshold value or its iteration number reaches a maximum number. This summation of the squared error is given by

$$E = \frac{1}{2} \sum_{p=1}^{N} \sum_{j=1}^{N} (o_{pj} - y_{pj}), \tag{2}$$

where N is pattern number. Here iteration number is defined as 1 in case of presenting from "A" to "L."

1. Mask Number

We discuss an effect of the mask number. First, we generate masks of the mask technique in the following way. We generate 64 ($= 8 \times 8$) random values among $[-1, 1]$ using random numbers and they are equal to the input pixels. We mask the pixels which correspond to minus values.

The mask numbers that we discuss are 2, 4, 8, 16, 24, and 32. Figure 6 shows the learning status for the six patterns of mask numbers until the iteration number reaches 30,000. The horizontal axis shows the iteration number and the vertical one shows the summation of the squared error which we have already described. From this figure, it is impossible to make the pattern recognition for the NN when mask numbers are 2 and 4. The learning can converge using more than eight masks. When we recognize alphabetical letters from "A" to "L" using these weights, we show every output unit value as shown in Fig. 7. From this figure, we can find that the recognition ability is almost the same when the mask number is more than 8. This matter shows that we can get enough output unit values for pattern recognition. Furthermore, we also recognize inputs with noise as shown in

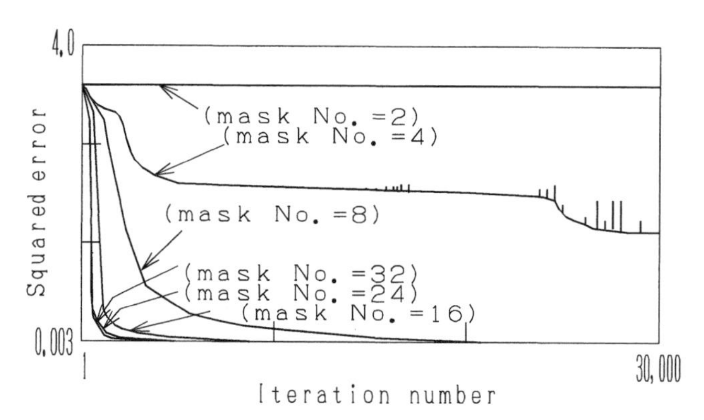

Figure 6 Relationship between the learning convergence and mask number.

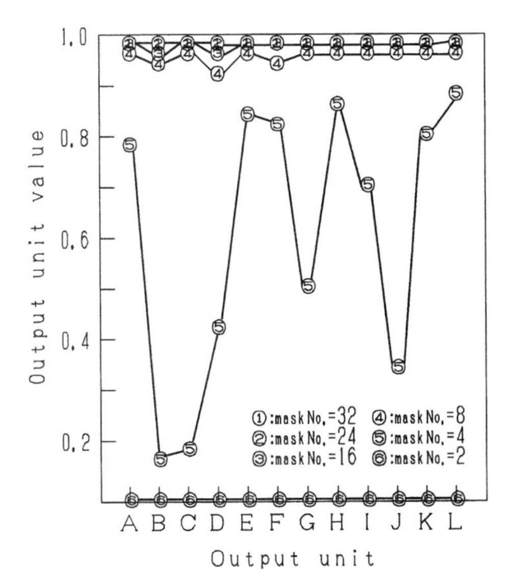

Figure 7 Output unit values for the various mask numbers.

Fig. 8 [4, 5]; its result is shown in Fig. 9. Here, "$*$" denotes the noise-added point where we change a 1 to a 0 or vice versa. From this result, the recognition ability depends on mask numbers. Thus, we select mask number 8 which is sufficient to get correct recognition from a series of experiments with the alphabetical letters [2–5].

2. Mask Area

We discuss a mask area in this section. First, we adjust a mask area with the alteration width of random numbers. Namely, the width of generating random numbers is $[-1, 1]$ as a basis. We change the width of generating random numbers

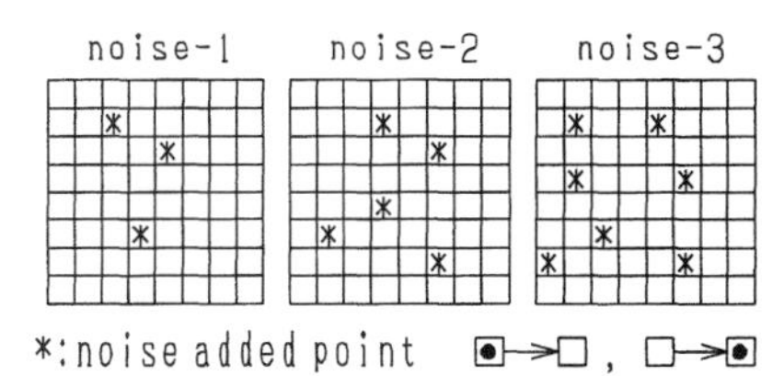

Figure 8 Various noises.

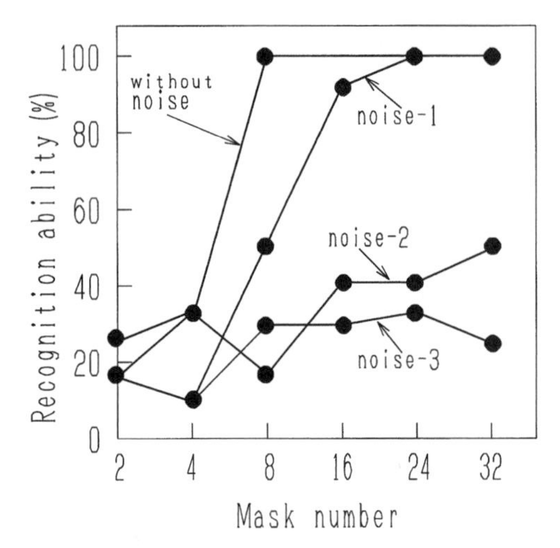

Figure 9 Result of robustness for the noisy input using the various mask numbers.

as [−2, 1], [−3, 1], or [−4, 1] according to increasing the mask area. Otherwise, [−1, 2], [−1, 3], or [−1, 4] is selected according to decreasing the mask area. Figure 10 shows the learning status until the iteration number reaches 30,000 when we alter the width of generating random numbers. From this figure, we can find that the learning convergence does not depend on the mask area. Still more, we show every output unit value as shown in Fig. 11. We can find that

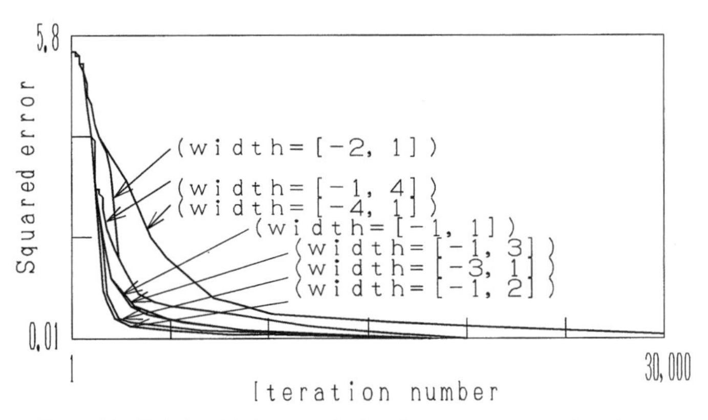

Figure 10 Relationship between the learning convergence and mask area.

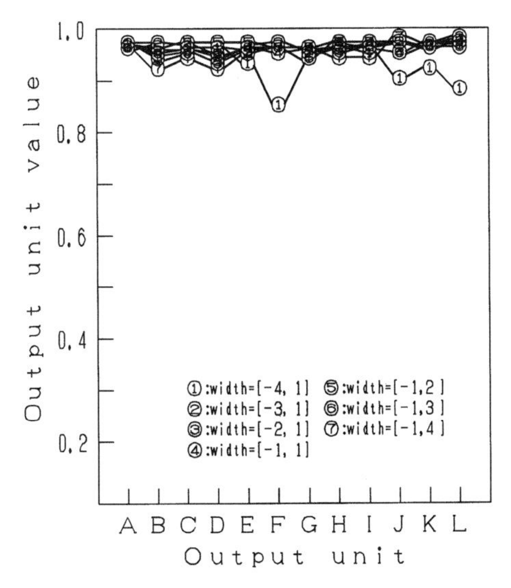

Figure 11 Output unit values for the various mask areas.

the recognition ability does not depend on the mask area from this result. We also recognize noisy inputs as in the discussion of the mask numbers [4, 5]. Its experimental result is shown in Fig. 12. From this result, the recognition ability does not depend on the mask area [4, 5].

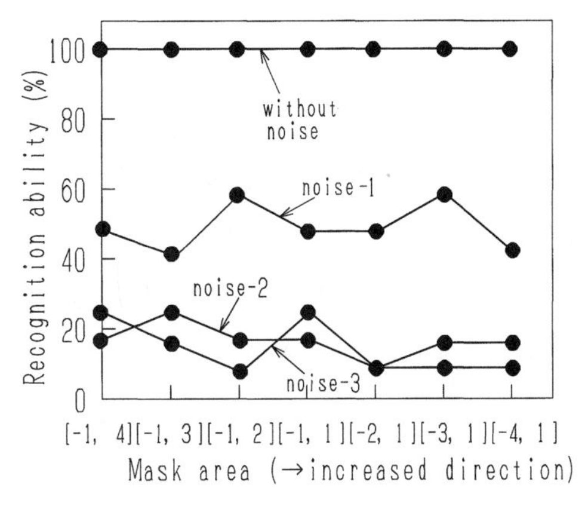

Figure 12 Result of robustness for the noisy input using the various mask areas.

C. Experiments of the Neural Network Scale Reduction Using the Masks

Here, we make some experiments to show the effectiveness of the scale reduction by the mask technique. The first is the case using the alphabetical letters and the second is the one using the paper currency.

1. Experiment Using the Alphabetical Letters

For comparison of the proposed technique with the conventional one, we consider the ordinary technique [1, 3–5] for the alphabetical letters. Figure 13 shows the NN constructions of both techniques for the alphabetical letters. In this or-

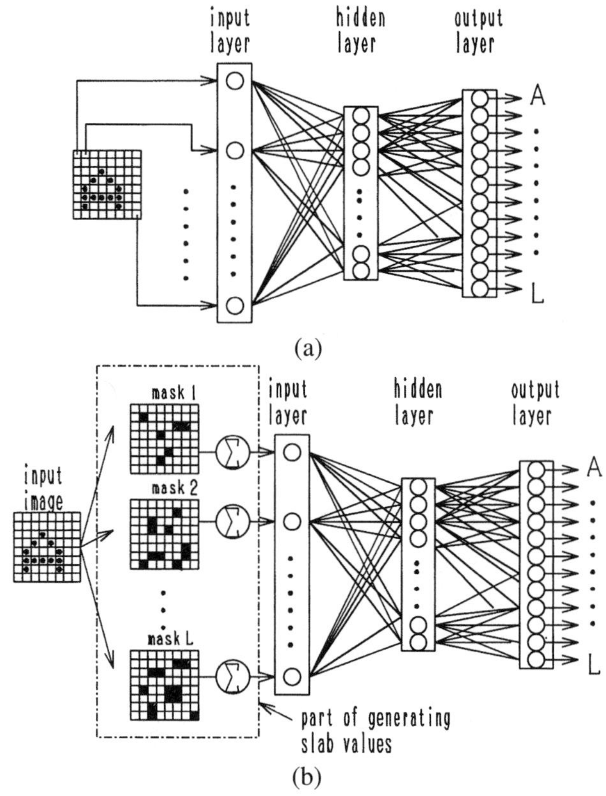

Figure 13 NN constructions of the ordinary technique and the proposed one for the alphabetical letters: (a) ordinary technique; (b) proposed technique.

dinary technique, we give directly the pixels to the input layer of the NN. However, this construction is three layers and the input unit number is 64 ($= 8 \times 8$). This is equal to pixel number. The hidden unit number is 32 and the output unit number is 12 which is equal to recognition patterns. Here, we have decided the hidden unit number of the ordinary technique through various experiments considering the recognition ability [1–3, 5]. The squared errors by the proposed technique and the ordinary one converged within 0.01. In both cases, recognition ratios are 100% by using the unknown data. Here, we regard the NN scale as the weight number which is (input unit number) $\times$ (hidden unit number) $+$ (hidden unit number) $\times$ (output unit number) [1–3, 5]. The weight number for each technique is the following:

- the number for the proposed technique is $8 \times 8 + 8 \times 12 = 160$,
- the number for the ordinary technique is $64 \times 32 + 32 \times 12 = 2432$.

In this way, we find that the NN scale can be reduced without spoiling recognition ability.

2. Experiment Using the Paper Currency

Here we use the Japanese paper currency data which are partly sensed to compare the scale of the proposed technique with that of the ordinary one [1–3, 5]. Figure 14 shows the NN constructions of both techniques for the paper currency. We directly input these sensed pixels to the NN. When we input the pixels to this ordinary technique, the input unit number is 128 ($= 32$ sample $\times$ 4 sensor) and this is equal to pixel number. The hidden unit number is 64. The output unit number is 12 and this is equal to the recognition pattern. Using another Japanese paper currency data which includes worn-out and defective ones, both of these recognition ratios are 100%. Still more, the weight number for each technique is the following:

- the proposed technique is $16 \times 16 + 16 \times 12 = 448$,
- the ordinary technique is $128 \times 64 + 64 \times 12 = 8960$.

Therefore, we find that the proposed technique is also effective for the paper currency data and does not spoil recognition ability.

III. MASK DETERMINATION USING THE GENETIC ALGORITHM

Here, we discuss the mask determination using the GA [11, 12]. First, we describe a few conventional mask determination methods and show the problem of each method on optimizing and systematizing the mask determination. Second, we show the basic idea of adopting the GA operations to the mask determination.

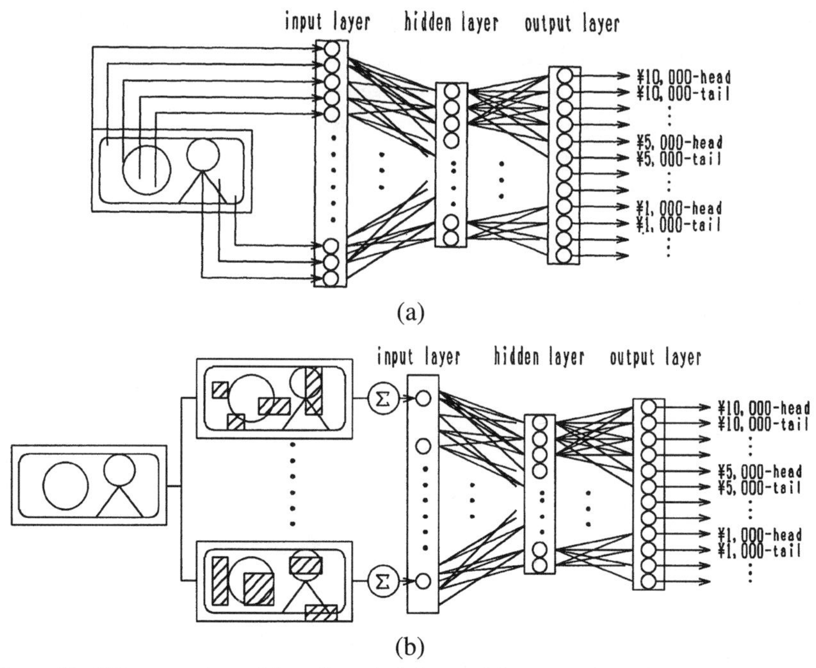

Figure 14 NN constructions of the ordinary technique and the proposed one for the paper currency: (a) ordinary technique; (b) proposed technique.

A. Conventional Mask Determination

1. Mask Determination by the Random Numbers

Initially, we determine the masks by the random numbers [1–5]. As shown in Fig. 15, we divide the paper currency by the least masked area (column) equally and each masked area is ordered. Here, the number of them is 16. We generate 16 random numbers among $[-1, 1]$ and they are equal to the column numbers. We mask the column whose number is equal to the ordered number of the random one which has a minus value. We repeat this procedure from the first random number to the sixteenth random one. So we can obtain one kind of mask. Second, we change the initial value which generates random numbers and repeat this procedure several times. Finally, we can obtain a series of plural nonduplicated masks. In the experiment, we decide 16 as the number of masks from the various kinds of simulation [4, 5]. Both the numbers of input units and the hidden ones are 16. The kinds of paper currency are US $1, $5, $10, $20, $50, and $100. Thus, the number of output units which corresponds to the kinds of paper currency becomes six.

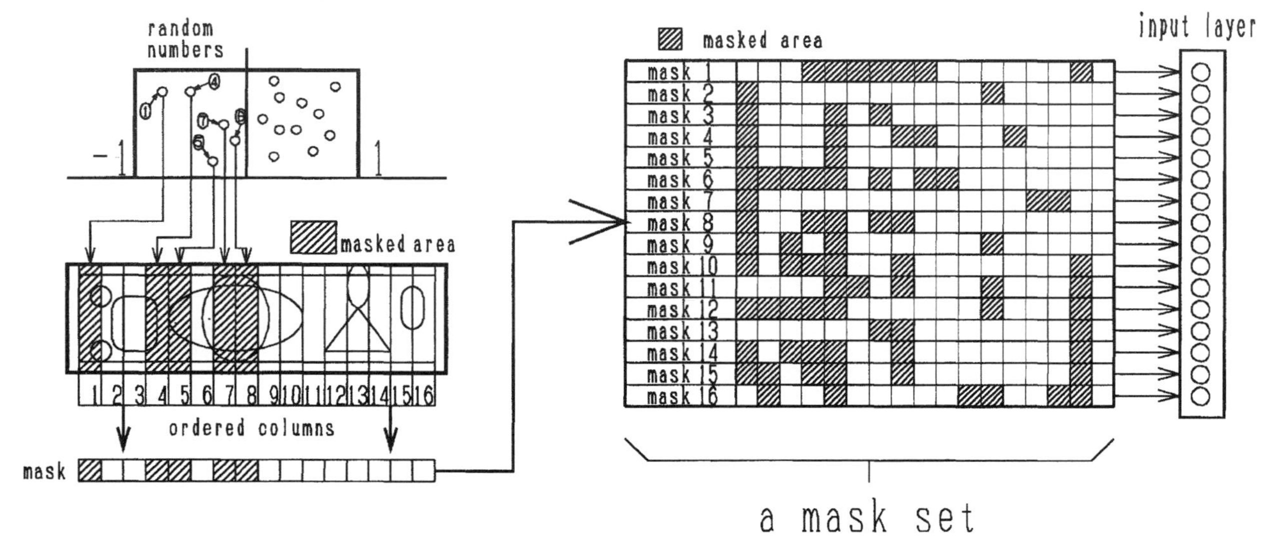

Figure 15 Determination procedure of random mask set.

Here, the NN needs plural masks and these are one treatment unit in the mask technique. After all, we describe these plural masks as a mask set and discriminate it from a mask. In the following, we decide 30 random mask sets and investigate their ability (generalization of the NN) using the unknown US dollars which include damaged paper currency and fluctuation error by conveyance.

In the experiment, we construct the experimental system using the current banking machine. It can sample image data such as 216×30 pixels which are represented by one byte gray level. Its conveyed speed is more than ten pieces per second. We adopt the back propagation method with oscillation term for learning [1–5]. We use ten pieces of paper currency for each kind as learning data. We define one iteration as learning from $1 to $100. We continue learning until the iteration number reaches 5000 times. To evaluate the method, 30 other pieces of paper currency for each kind are used.

From experimental results, the recognition abilities of the NN obtained by the 30 random mask sets are from 59% to 99% as shown in Fig. 16 [8–10]. In this way, generalization of the NN is largely influenced by mask sets. Furthermore, we cannot always have gotten the excellent mask sets by using the random numbers.

2. Every Mask Combination

It is supposed that we take a method to investigate every mask set [7–10]. Then every combination constructed by the least masked area can be considered as the mask. The inputs of every mask set can be generated. Learning should be executed by using each input. We have to investigate the ability with every mask set by using unknown data. In this case, we could choose the mask set which generates the input that shows the highest ability as the optimized one.

However, this method could be calculated in the case of a small number of mask set combinations. If the number of mask set combinations were to be increased, this method would no longer be effective and reasonable to determine the mask sets because the number of masks could be calculated as 2^M, where M denotes the number of the least masked area, and that is 16 in these experiments.

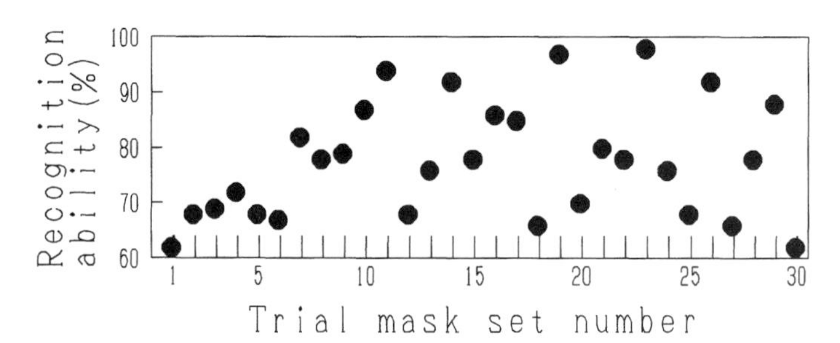

Figure 16 Abilities of the various random mask sets.

B. Basic Operations of the Genetic Algorithm

Under this background, we adopt the GA [11, 12] to the mask determination as shown in some figures. The basic operations are stated as follows.

Coding

First, we prepare some random mask sets which are candidates of the crossover. We represent the masked part as "1" and the nonmasked part as "0" in the mask as shown in Fig. 17a. This coding is easily understood and satisfies completeness, soundness, and nonredundancy, which are proposed as an evaluation standard of coding [12].

Sampling

Second, we regard the ability of the mask set as an evaluation value of the GA. We sample the mask sets which have the higher evaluation values as the parental mask sets [10, 14, 15]. As shown in Fig. 17b, this sampling method is similar to the roulette system which has the area in proportion to the evaluation value of the mask set [10, 12, 14, 15]. Furthermore, we scale the ability of the mask set to emphasize the superiority of the mask set [10, 12, 14, 15], which is given by

$$\text{evaluation value} = (\text{Ability of the mask set})^2. \tag{3}$$

Crossover

We crossover the half parts of genes in the two parental mask sets as shown in Fig. 17c. The crossover satisfies character preservation, which is proposed as an evaluation standard of crossover [12].

Mutation

Furthermore, as shown in Fig. 18a, to provide variety to the crossovered mask set, mutation, which reverses some bits of the genes in the mask, is randomly operated during the determination of the new mask. Learning is executed by using the inputs obtained by the mask sets. After that, using unknown data we investigate the generalization of the NN with mask sets, which means ability of the mask sets.

Selection

If we select only the descendant mask sets which satisfy purposive ability, there is some risk such that descendant mask sets will disappear in a few generations. We must maintain the number of descendant mask sets which are sampled for

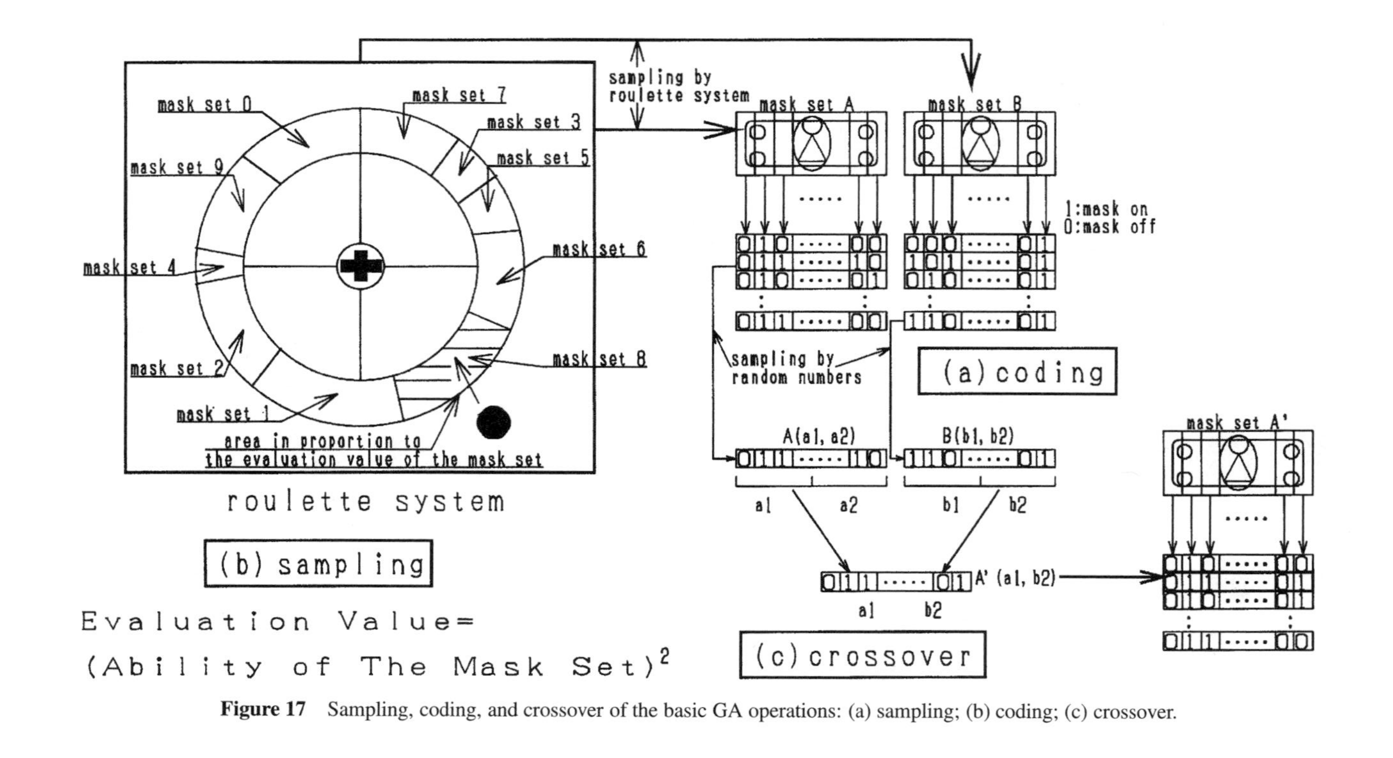

Figure 17 Sampling, coding, and crossover of the basic GA operations: (a) sampling; (b) coding; (c) crossover.

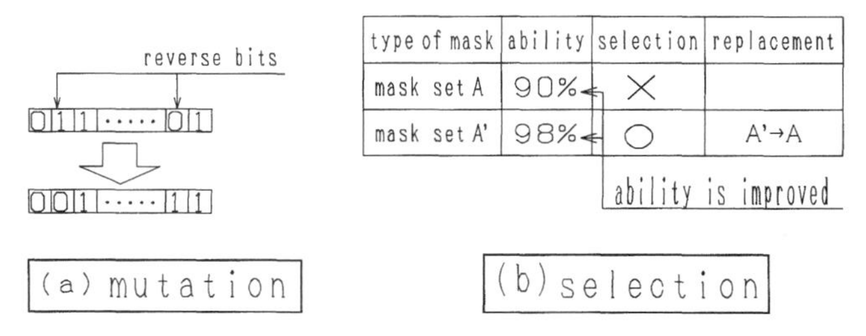

Figure 18 Mutation and selection of the basic GA operations: (a) mutation; (b) selection.

crossover in the next generation. As shown in Fig. 18b, we replace the parental mask set by the crossovered one when the ability of the crossovered mask set is better than that of the parental one [7–10, 14, 15]. Thus, the number of descendant mask sets is maintained. Finally, we show the flowchart of the GA operations in Fig. 19. By repeating a series of the GA operations, we can get excellent mask sets in a few number of generations. These mask sets enable us to shorten the learning time and to improve the generalization of the NN.

C. Experiments Using U.S. Dollars

The experimental condition is the same as in Section III.A.1. The number of mask sets is ten. We continue the GA operations until the purposive mask set whose ability is more than 95% is obtained. Figure 20a shows the transition of the ability of the mask set by the GA operations. We can obtain the purposive mask set (mask set 5) in the fifth generation. Furthermore, we change the initial mask sets and make an experiment one more time. Its result is shown in Fig. 20b. The purposive mask set (mask set 9) was obtained in the sixth generation. Its improvement rate is 20.3% [from 80.8% (initial) to 97.2% (final)]. For each experiment, the average ability of every ten mask sets (one point dotted line in the figure) is increased gradually.

From both experiments, we can obtain excellent mask sets in a few number of generations and automatically by the proposed GA operations. The possibility that the optimized mask set by the GA covers the area which have the picture similar to a watermark is supposed to decrease [10, 15]. We analyze the result of this mask determination of the experiment 2 in Fig. 20b. Figure 21 shows the changed genes of the initial mask set. In this result, the second, tenth, thirteenth, and fourteenth columns (arrow marks in the figure) have different pictures from each other for every kind of US dollars [14, 15]. Here, we regard a column as an important one when there are more than three bits which have "1." When the columns have

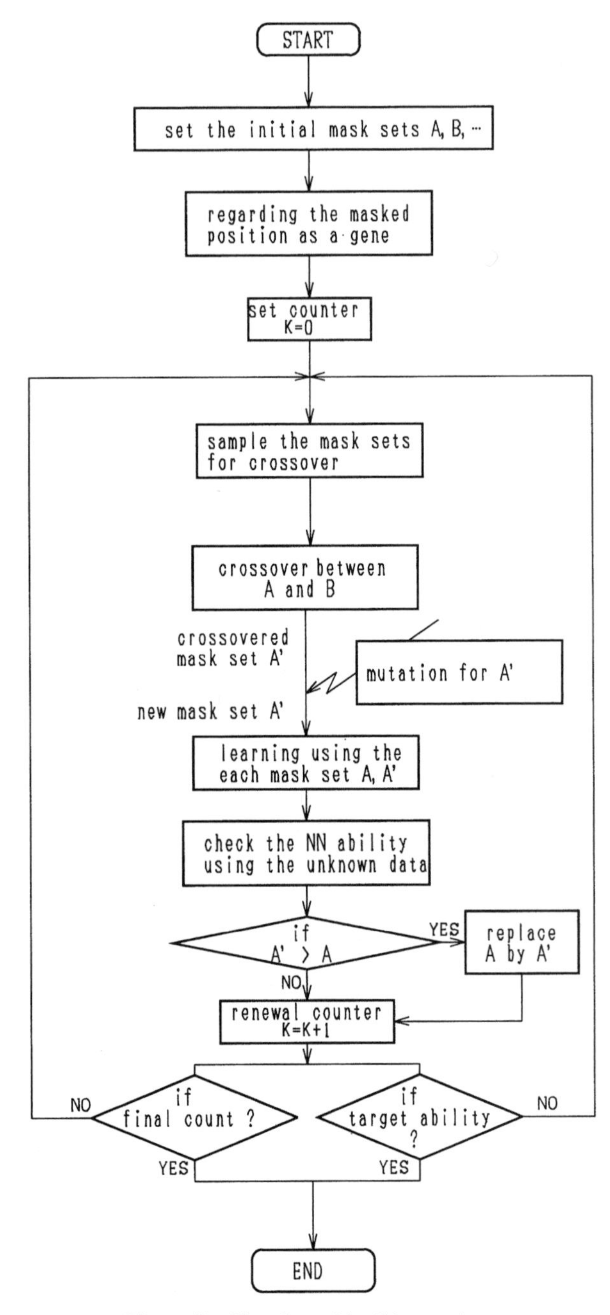

Figure 19 Flowchart of the GA operations.

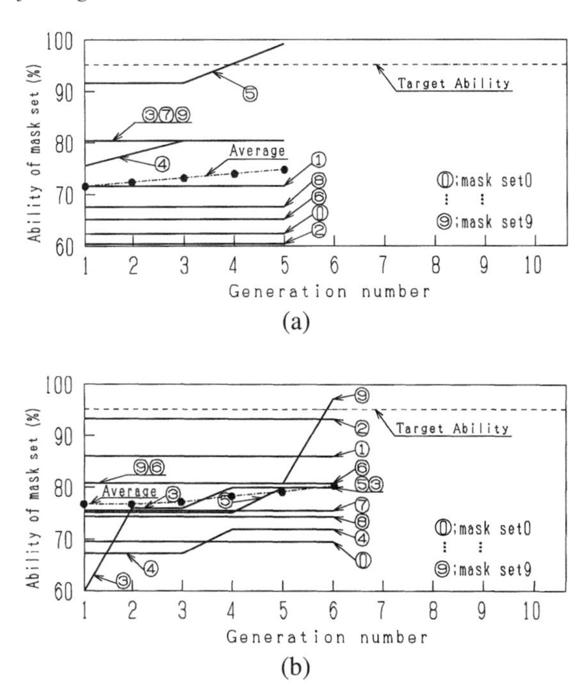

Figure 20 Transition for the ability of the mask sets by the GA: (a) experiment 1; (b) experiment 2.

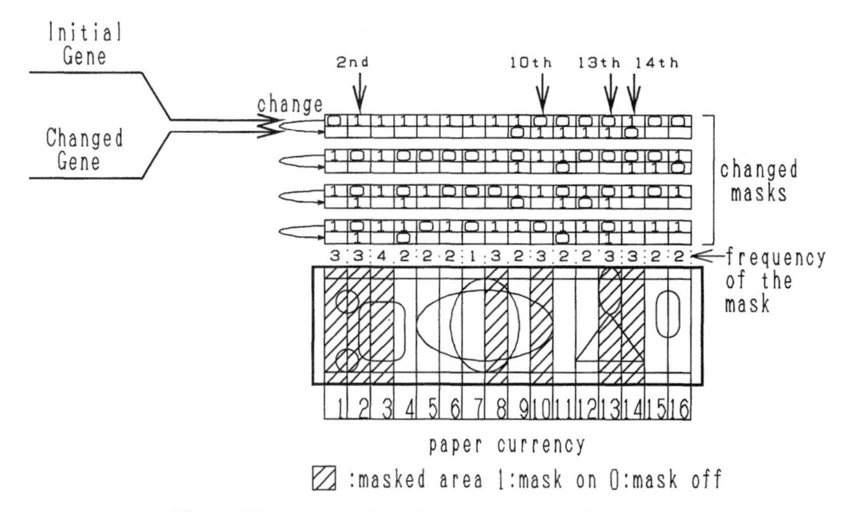

Figure 21 Analysis of the determined mask set by the GA.

a similar figure for every kind of US dollars, we have conventionally checked those columns and avoided them for the mask area manually [14, 15]. From this result, we suppose that the mask set is automatically optimized in some degree by the GA.

Thus, the proposed GA technique is effective to systematize the determination of the mask set. If the kind of paper currency is changed from US dollars to another kind, the better mask set to the paper currency can be easily and automatically determined in a short period by the proposed GA technique [14, 15].

IV. DEVELOPMENT OF THE NEURO-RECOGNITION BOARD USING THE DIGITAL SIGNAL PROCESSOR

A. Design Issue Using the Conventional Devices

We show the neuro-experimental systems which are developed using a single-board computer. Figure 22a is the original type and Fig. 22b is its portable one [5, 7, 13]. These experimental systems can recognize eight pieces of the paper currency per second. However, their recognition speed is not enough for real-time systems such as the banking machines, since we have to recognize one piece of the paper currency for several tens of seconds, which is the recognition interval of the paper currency. If we use the ordinary low-cost CPU (central processing unit) such as Intel's i80 series which is used in the current recognition machines, its calculation speed for the neuro-transaction is not enough to recognize the paper currency in real time [13].

Meanwhile, it has been reported that there are various neuro-devices such as a super parallel computer, a neuro-accelerator, and a special neuro-chip such as Intel's 80170NX as shown in Fig. 23 [13, 16–20]. Their calculation speeds are quite enough for the current banking machines. However, they are very expensive and are just on the way to development. Thus, we cannot adopt these neuro-devices to the design of the banking machines. We need another neuro-device which has low cost and whose calculation speed is enough for the real-time computation.

B. Basic Architecture of the Neuro-Recognition Board

To realize the neuro-paper currency recognition in the commercial products, we have developed a high-speed neuro-recognition board using the DSP as shown in Fig. 24 [7–10, 13–15]. Figure 24a shows the first type and Fig. 24b shows the second one. In Fig. 24a, the left side is the DSP circuit and the right side is the

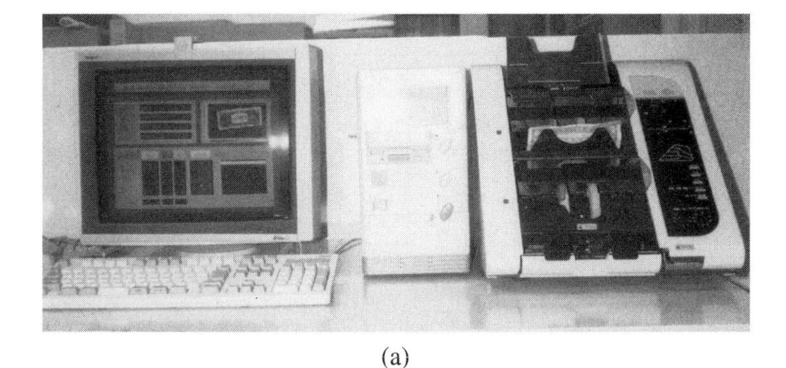

(a)

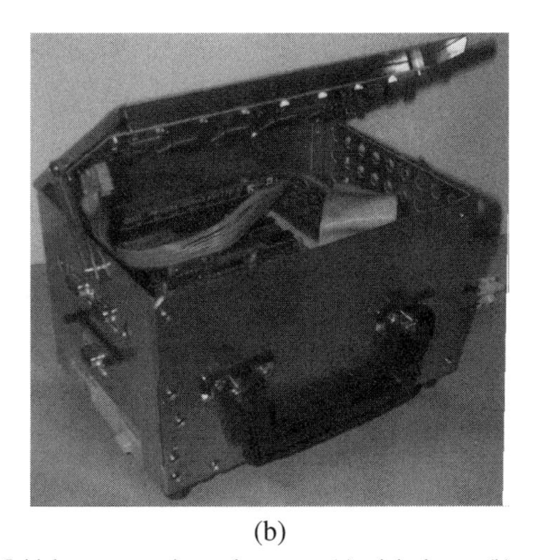

(b)

Figure 22 Initial neuro-experimental systems: (a) original type; (b) portable type.

interface circuit for the sensors. This DSP (TMS320C31) is produced by Texas Instruments. It runs under 33 MHz machine clock and its performance is 33.3 MFLOPS (million floating-point instructions per second) as shown in Fig. 23. Figure 25 shows the block diagram of the neuro-recognition board. The neuro-program boots up from EPROM (electrical programmable read only memory). The neuron's weights are saved in flash memory and they can be renewed by the connected extra-computer easily. Furthermore, the adopted DSP has the exponential function which is used in the sigmoid function $[f(x) = 1/(1 + \exp(-x))]$ as a library. This enables easy implementation of the neuro-algorithm from EWS (engineering work station) or another large computer to the real-time systems. Its

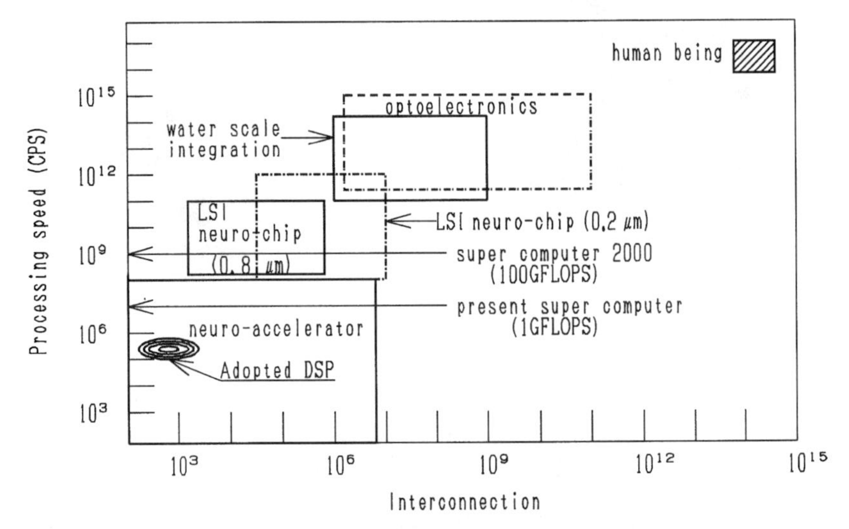

Figure 23 Comparison of the special neuro-devices.

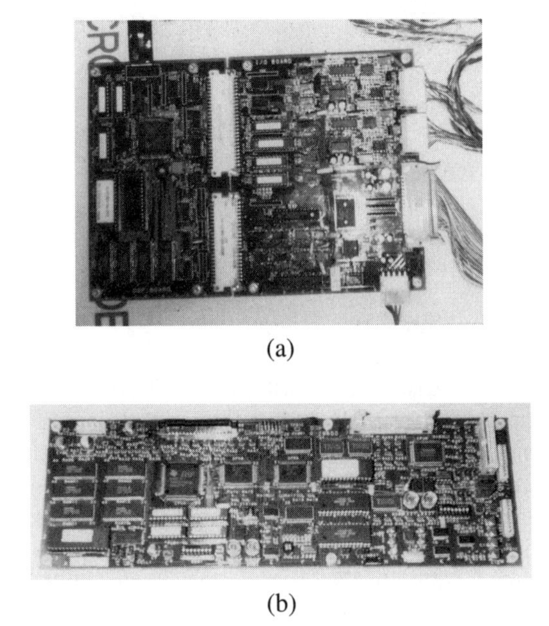

(a)

(b)

Figure 24 Feature of the neuro-recognition board: (a) first type; (b) second type.

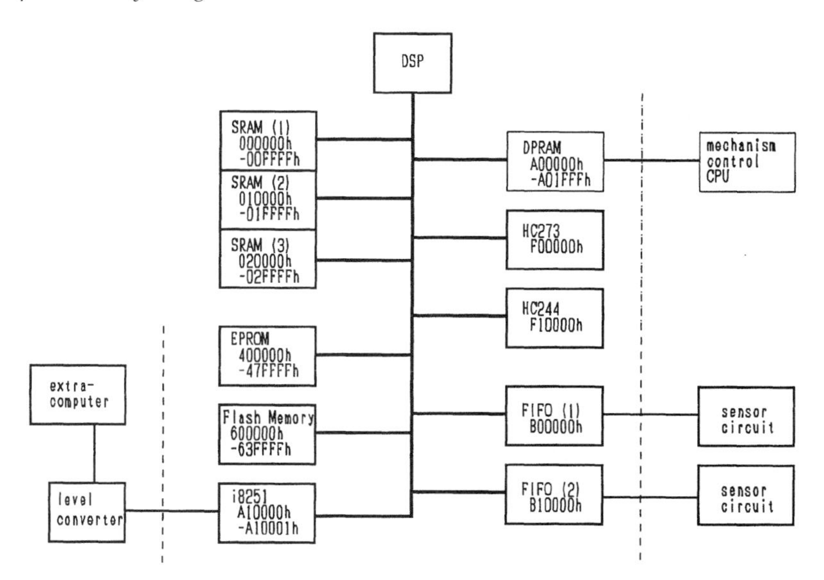

Figure 25 Block diagram of the neuro-recognition board.

computational speed is ten times faster compared with the current recognition machines [13–15]. Figure 26 shows the construction of the neuro-software modules. Core parts of this neuro-recognition algorithm are written in Assembly language and other parts are written in C language.

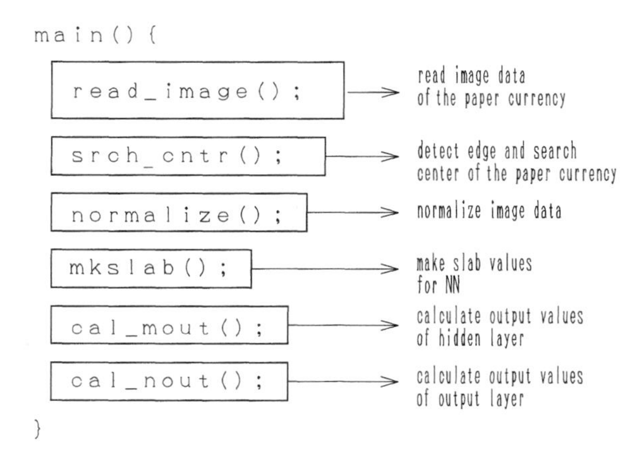

Figure 26 Construction of the neuro-software modules.

V. UNIFICATION OF THREE CORE TECHNIQUES

We unify the small-size neuro-recognition technique using masks, the mask determination technique by the GA, and the high-speed neuro-recognition board technique to realize the development of the worldwide paper currency recognition machine [14]. We have developed several business prototypes using the neuro-system technique as shown in Fig. 27. We have realized the neuro-banking machine which can transact the Japanese yen, the US dollar, the German mark,

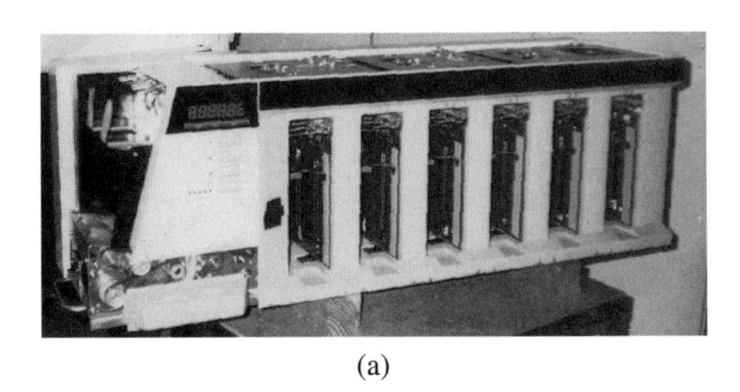

(a)

(b)

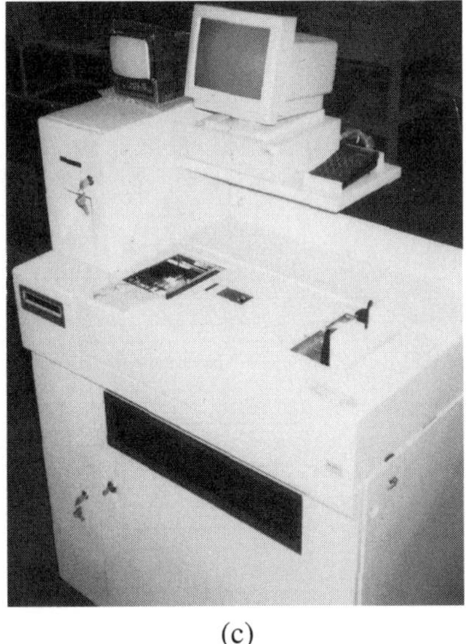

(c)

Figure 27 Business prototypes for currency recognition using the neuro-technique: (a) prototype 1; (b) prototype 2; (c) prototype 3.

the Belgian franc, the Korean won, the Australian dollar, and the British pound by only changing the neuro-weights and mask set. In these experiments, we use about 50 to 100 pieces of paper currency for each kind as learning data and evaluate more than about 20,000 pieces for each country's paper currency. Especially, we test the abilities for Japanese yen and US dollar using about 100,000 pieces of the paper currency which are sampled in the commercial market and involve worn-out, defective, and new paper currency. For every testing, recognition ability is more than 97%. There is no error recognition. Here, in these experiments, we regard a pattern according to the output unit which has the highest response value as a neuro-judged pattern. In order to increase the reliability of recognition, we check the highest value by a threshold level and check the difference between the highest response value and the second highest by another threshold level. Even if the neuro-judged pattern is correct because its unit has the highest response value, the paper currency will be rejected unless the above two checks are satisfied.

Furthermore, connecting the extra-computer to the neuro-recognition board, image data of the paper currency is transported to the extra-computer and learning is executed on it. After learning, the neuro-weights are easily downloaded to the flash memory. In this way, we can easily develop the paper currency recognition machines [14, 15]. Therefore, each development period for each country's paper currency needs less than one-fifth the work compared with the conventional developing style and its recognition method. We suppose that all calculation for recognition on one piece of the paper currency is 100; it needs 28, 35, and 37 for the detecting currency edge from the image frame, mask transaction, and neuro-calculation, respectively.

We illustrate the first construction of the NN for the US dollars as shown in Fig. 28. In this case, we use the random numbers to decide the mask sets. Since

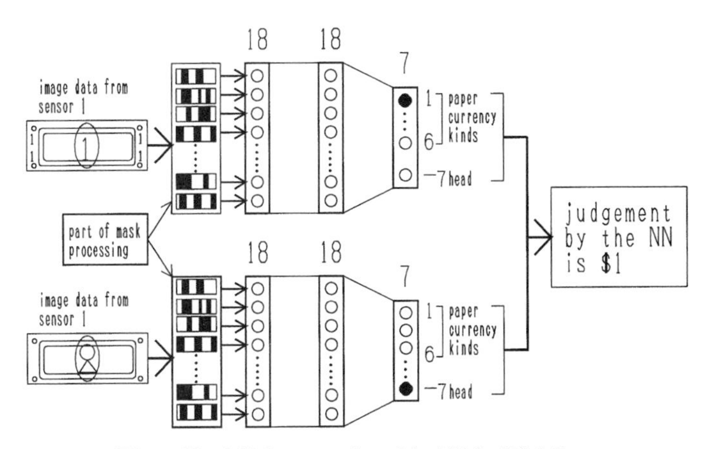

Figure 28 Initial construction of the NN for US dollars.

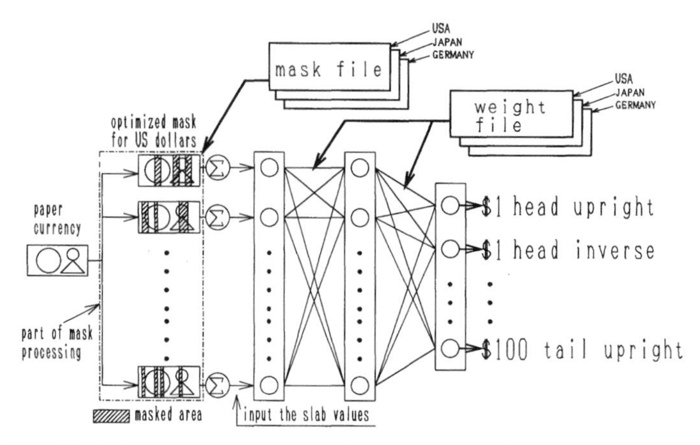

Figure 29 Universal construction of the NN using the optimized mask set by the GA.

all US dollars have similar patterns to each other and their basic color is green, recognition of US dollars is the most difficult problem [4, 9, 10]. In this figure, we use two NNs to recognize one piece of the paper currency. Namely, we sample the head and tail images [4, 5] of the paper currency at the same time using two sensors, up side one and down side one. One of the NNs obtains the tail images (landscape images). Then we decide the paper currency's kind by only that one NN which transacts the tail images, because the head images (figure images) of the paper currency are too similar to each other to recognize the currency's kind, while the tail images (landscape images) are not so similar to each other.

However, we optimize the mask set for US dollars using the proposed GA technique. Then we can also recognize kinds of US dollars by using the head images (figure images). We show the second construction of the NN for US dollars in Fig. 29. In this case, we need only one sensor's data to recognize the kind of US dollars owing to the excellent mask set [14, 15]. This construction can become a universal one for every kind of paper currency by changing the mask set and weights.

VI. CONCLUSIONS

We have proposed a paper currency recognition method using a NN. Especially, we have proposed three core techniques. The first is the small-size neuro-recognition technique using masks. The second is the mask determination technique using the GA. The third is the neuro-recognition board technique using the DSP. By unification of these three techniques, we confirmed realization of neuro-recognition machines which can transact various kinds of paper currency. The neuro-system technique enables us to accelerate the commercialization of a new type of banking machine in a short period and in a few trials.

Furthermore, this technique will be effective for various kinds of recognition applications owing to its high ability for recognition, high-speed transaction, short developing period, and reasonable cost. We suppose that it is effective enough to apply to not only paper currency and coins but also handwritten symbols such as election systems or questionnaires.

REFERENCES

[1] F. Takeda and S. Omatu. High speed paper currency recognition by neural networks. *IEEE Trans. Neural Networks* 6:73–77, 1995.

[2] F. Takeda, S. Omatu, T. Inoue, and S. Onami. A structure reduction of neural network with random masks and bill money recognition. In *Proceedings of the 2nd International Conference on Fuzzy Logic and Neural Networks IIZUKA'92*, Vol. 2, pp. 809–813. Iizuka, Japan, 1992.

[3] F. Takeda, S. Omatu, T. Inoue, and S. Onami. High speed conveyed bill money recognition with neural network. In *Proceedings of International Symposium on Robotics, Mechatronics and Manufacturing Systems'92*, Vol. 1, pp. 16–20. Kobe, Japan, 1992.

[4] F. Takeda and S. Omatu. Bank note recognition system using neural network with random masks. In *Proceedings of the World Congress on Neural Networks*, Vol.1, pp. 241–244. Portland, OR, 1993.

[5] F. Takeda and S. Omatu. Recognition system of US dollars using a neural network with random masks. In *Proceedings of the International Joint Conference on Neural Networks*, Vol. 2, pp. 2033–2036. Nagoya, Japan, 1993.

[6] B. Widrow, R. G. Winter, and R. A. Baxter. Layered neural nets for pattern recognition. *IEEE Trans. Acoust., Speech Signal Process.* 36:1109–1118, 1988.

[7] F. Takeda, S. Omatu, S. Onami, T. Kadono, and K. Terada. A paper currency recognition method by a small size neural network with optimized masks by GA. In *Proceedings of IEEE World Congress on Computational Intelligence*, Vol. 7, pp. 4243–4246. Orlando, FL, 1994.

[8] F. Takeda, S. Omatu, S. Onami, T. Kadono, and K. Terada. A paper currency recognition method by a neural network using masks and mask optimization by GA. In *Proceedings of World Wisemen/Women Workshop on Fuzzy Logic and Neural Networks/Genetic Algorithms of IEEE/Nagoya University*, Nagoya, Japan, 1994.

[9] F. Takeda and S. Omatu. A neuro-money recognition using optimized masks by GA. *Advances in Fuzzy Logic, Neural Networks and Genetic Algorithms, Lecture Notes in Artificial Intelligence* 1011, pp. 190–201. Springer-Verlag, Berlin/New York, 1995.

[10] F. Takeda and S. Omatu. A neuro-paper currency recognition method using optimized masks by genetic algorithm. In *Proceedings of IEEE International Conference on Systems, Man and Cybernetics*, Vol. 5, pp. 4367–4371. Vancouver, Canada, 1995.

[11] D. E. Goldberg. *Genetic Algorithms in Search, Optimization and Machine Learning*. Addison-Wesley, Reading, MA, 1989.

[12] Kitano. Genetic algorithm. *Sangyo Tosyo*, pp. 44–60, 1993. [in Japanese]

[13] F. Takeda and S. Omatu. Development of neuro-paper currency recognition board. *Trans. IEE Japan* 116-C:336–340, 1996. [in Japanese]

[14] F. Takeda and S. Omatu. A neuro-system technology for bank note recognition. In *Proceedings of Japan-USA Symposium*, Vol. 2, pp. 1511–1516. Boston, MA, 1996.

[15] F. Takeda and S. Omatu. A neuro-recognition technology for paper currency using optimized masks by GA and its hardware. In *Proceedings of International Conference on Information Systems Analysis and Synthesis*, pp. 147–152. Orlando, FL, 1996.

[16] J. Ghosh and K. Hwang. Critical issues in mapping neural networks on message-passing multicomputers. *Presented at the 15th International Symposium on Computer Architecture*, 1988.

[17] J. Alspector, R. B. Allen, V. Hu, and S. Satyanarayana. Stochastic learning networks and their electronic implementation. In *Proceedings of Neural Information Processing Systems—Neural and Synthetic*, Denver, CO, 1987.

[18] N. H. Farhat, D. Psalits, A. Prata, and E. Paek. Optical implementation of the Hopfield model. *Appl. Optics* 24:1469–1475, 1985.

[19] J. W. Goodman, F. I. Leonberger, S. Y. Kung, and R. A. Athale. Optical interconnection for VLSI systems. *Proc. IEEE* 72:850–866, 1984.

[20] R. Hect-Nielsen. Performance limits of optical, electro-optical and electronic neurocomputers. TRW Rancho Carmel AI Center Report, pp. 1–45, 1986.

Neural Network Classification Reliability: Problems and Applications

Luigi P. Cordella
Dipartimento di Informatica e Sistemistica
Università degli Studi di Napoli "Federico II"
Via Claudio, 21
I-80125 Napoli, Italy

Carlo Sansone
Dipartimento di Informatica e Sistemistica
Università degli Studi di Napoli "Federico II"
Via Claudio, 21
I-80125 Napoli, Italy

Francesco Tortorella
Dipartimento di Informatica e Sistemistica
Università degli Studi di Napoli "Federico II"
Via Claudio, 21
I-80125 Napoli, Italy

Mario Vento
Dipartimento di Informatica e Sistemistica
Università degli Studi di Napoli "Federico II"
Via Claudio, 21
I-80125 Napoli, Italy

Claudio De Stefano
Facoltà di Ingegneria di Benevento
Dipartimento di Ingegneria dell'Informazione ed Ingegneria Elettrica
Università degli Studi di Salerno
Piazza Roma, palazzo Bosco Lucarelli
I-82100 Benevento, Italy

I. INTRODUCTION

Classification is a process according to which an entity is attributed to one of a finite set of classes or, in other words, it is recognized as belonging to a set of equal or similar entities, possibly identified by a name. In the framework of signal and image analysis, this process is generally considered part of a more complex process referred to as pattern recognition [1]. In its simplest and still

Image Processing and Pattern Recognition
Copyright © 1998 by Academic Press. All rights of reproduction in any form reserved.

most commonly followed approach, a pattern recognition system is made of two distinct parts:

1. a description unit, whose input is the entity to be recognized, represented in a form depending on its nature, and whose output is a generally structured set of quantities, called features, which constitutes a description characterizing the input sample. A description unit implements a description scheme.
2. a classification unit, whose input is the output of the description unit and whose output is the assignment to a recognition class.

These two parts should not be considered perfectly decoupled, although this assumption is generally made for the sake of simplicity. In the following the term pattern recognition will be used in general to refer to the whole process culminating in classification, even if no hypotheses are made on the nature of the entities to be recognized. In fact, it is obvious that, in order to be classified, a sample entity has to be represented in terms that are suitable for the classifier. However, without affecting the generality of the treatment, examples will usually be taken from the field of image recognition.

Selecting the features to be used in the description phase is one of the most delicate aspects of the whole system design since general criteria for quantitatively evaluating the effects of the performed choices are not available. In the case of images, for instance, the main goal of the description phase is to transform a pictorial representation of the input sample, often obtained after a preliminary processing of the initial raw data, into an abstract representation made up of a structured set of numbers and/or symbols. Ideally, the whole process leading up to description should be able to select the most salient characteristics shared by all the samples of a same class so as to have identical descriptions for all of them, while keeping the descriptions of samples belonging to different classes quite separate. In practice, in application domains characterized by high variability among samples, it is extremely difficult to obtain descriptions near to the ideal ones. The aim of the classifier is to obtain the best possible results on the basis of the descriptions actually available.

Without losing generality, it can be said that a classifier operates in two subsequent steps:

(i) a training phase, during which it is provided with specific knowledge on the considered application domain, using information about a representative set of samples (training set) described according to the considered description scheme. If the samples of the training set are labeled, i.e., their identity is known, the training is said to be supervised; otherwise it is unsupervised.

(ii) an operative phase in which the description of a sample to be recognized is fed to the classifier that assigns it to a class on the basis of the experience acquired in the training phase. Equivalently, classification can be described as the matching between the description of a sample and a set of prototype descriptions generally defined during the training phase.

Different approaches to classification have been proposed in the past, from the purely statistical ones to those based on syntactic and structural descriptions, to hybrid schemes [2–4]. These will be briefly reviewed in Section II.

In recent years, artificial neural networks (ANN) [5] have come back into favor and their use for classification purposes has been widely explored [6]. In an international competition held in 1992 [7], about 40 character recognition systems were compared on a common data base of presegmented handwritten characters (NIST) [8], and the top ten used either neural classifiers or nearest neighbor methods [9]. There has been experimental evidence that the performance of neural network classifiers can be considered comparable to that obtained by using conventional statistical classifiers. Moreover, the ANN ability to learn automatically from examples makes them attractive and simple to use even in complex domains.

Regardless of the classification paradigm adopted, a problem of great practical interest lies in the evaluation of the reliability of the decisions taken by a classifier. Classification reliability can be expressed by associating a reliability parameter to every decision taken by the classifier. This is especially important whenever the classifier deals with input samples whose descriptions vary so much with respect to the prototypal ones that the risk of misclassifying them becomes high. In the framework of a recognition system, the knowledge of classification reliability can be exploited in different ways in order to define its action strategy. One possibility is to use it to identify unreliable classifications and thus to take a decision about the advantage of rejecting a sample (i.e., not assigning it to a class), instead of running the risk of misclassifying it. In practice, this advantage can only be evaluated by taking into account the requirements of the specific application domain. In fact, there are applications for which the cost of a misclassification is very high, so that a high reject rate is acceptable provided that the misclassification rate is kept as low as possible; a typical example could be the classification of medical images in the framework of a prescreening for early cancer detection. In other applications it may be desirable to assign every sample to a class even at the risk of a high misclassification rate; let us consider, for instance, the case of an optical character recognition (OCR) system used in applications in which a text has to be subjected to subsequent extensive editing by man. Between these extremes, a number of applications can be characterized by intermediate requirements. A wise choice of the reject rule thus allows the classifier behavior to be tuned to the given application.

Classification reliability also plays a crucial role in the realization of multi-classifier systems [10, 11]. It has been shown that suitably combining the results of a set of recognition systems according to a rule can give a better performance than that of any single system: it is claimed that the consensus of a set of systems based on different description and classification schemes may compensate for the weakness of the single system, while each single system preserves its own strength. The knowledge of classification reliability can be profitably used to define the combining criteria.

The main aspects of some of the most commonly used description and classification methods will be summarized in Section II. Neural networks and their use for classification will be discussed in Section III, while classification reliability in its different meanings will be reviewed in Section IV. Section V will be devoted to the problem of evaluating the classification reliability of neural classifiers; evaluation criteria for different neural network architectures will be proposed. Section VI discusses the problem of introducing a reject option and illustrates a method for selecting the reject threshold value in such a way as to obtain the best trade-off between recognition rate and reject rate, taking into account the specific requirements of the considered application. Finally, two applications of the illustrated techniques are discussed in Section VII: the first is in the field of automatic recognition of isolated handprinted characters, and the second refers to the automatic detection and identification of faults in electrical systems.

II. CLASSIFICATION PARADIGMS

One of the most widely followed approaches to classification, such that the term pattern recognition virtually identified with it until the late 1970s, is the statistical one [2]. According to it, a sample to be classified is characterized by a set of measures performed on it (feature vector) and then represented by a point in a feature hyperspace. Examples of widely used image features are moments of various order, transforms and series expansions, and local and geometric properties [12]. A potentially large set of easy-to-detect features can be initially extracted and then subjected to discriminant analysis methods in order to select a subset of features as much as possible uncorrelated.

According to the statistical approach, a training set made up of labeled samples is assumed to statistically represent the data set on which the classifier has to work. During the training phase, suitable algorithms exploiting knowledge about the training set make it possible to partition the feature hyperspace into regions (decision regions) and to associate each region to a class. Alternatively, the training phase results in the identification of the class prototypes. The aim of training algorithms is to perform the above tasks in such a way as to minimize the errors over the training set, and the representativeness of the training set is thus a necessary condition for the effectiveness of the method.

The problem can be formalized in the following terms: let $X = \{x_k\},\ k = 1, \ldots, r$, be the feature vector representing a generic sample and N the number of classes of interest. A canonical way of describing the functions of a classifier is as follows: the training phase leads to the definition of a set of functions $F_i(X),\ i = 1, \ldots, N$, such that $F_i(X) > F_j(X),\ i, j = 1, \ldots, N,\ i \neq j$, if X belongs to the ith class. In the feature space, the boundary between two classes C_i and C_j is given by the hypersurface for which $F(X) = F_i(X) - F_j(X) = 0$. Apart from the feature selection problem, the definition of the discriminating functions among classes is not at all trivial. Indeed, the simplest case is when the classes are linearly separable, i.e., the boundary surfaces between classes are hyperplanes of equation

$$F(X) = \sum_{k=1}^{r} a_k x_k + a_0 = 0. \tag{1}$$

Of course, it is possible to think of nonlinear classifiers where the discriminating function does not linearly depend on X.

From the operative point of view, different approaches are available according to the case in question. Assuming that X is a random vector, in order to assign it to the ith class, the Bayesian approach to classification entails first evaluating the *a posteriori* probability that, given the vector X, it belongs to the class C_i, i.e., $P(C_i|X)$. Such *a posteriori* probabilities can be evaluated once the *a priori* occurrence probability $P(C_i)$ of the class C_i has been given, together with the *a priori* probability $P(X|C_i)$ that the considered sample is X, after assuming that it belongs to C_i. X can be assigned to the class C_i, according to the Maximum Likelihood Decision Rule, if $P(C_i|X) > P(C_j|X)$, $i, j = 1, \ldots, N$, $i \neq j$.

If $P(X|C_i)$ is not known for the various classes, the parametric approach can be used. This assumes a functional form for the *a priori* probability density functions representing the distributions of the samples belonging to each class (e.g., the distributions may be Gaussian), and evaluates each $P(X|C_i)$ by computing a finite number of parameters characterizing the assumed distribution on the basis of the samples of the training set. The parametric approach can use well-known techniques of parameter estimation [2].

It may be desirable that, if the probability that a sample belongs to a certain class is not sufficiently higher than the probability that it belongs to any other class, the sample is rejected as not belonging to any of the defined classes. Note that misclassifications are generally less acceptable than rejects and that, for a properly defined reject rule, the curve representing the misclassification rate versus the reject rate for a given classifier is concave upward [13]. The problem of finding the best trade-off between the reject rate and the misclassification rate, so as to optimize the performance of the classifier, will be thoroughly discussed in Section VI.

Unfortunately, in most real applications a parametric form cannot be assumed for the probability density function, so that, in order to apply the likelihood rule, a nonparametric approach to the problem is the only possible one. One of the basic methods [14] estimates the unknown density by adding simple distributions weighted by suitable coefficients. Another aspect of the nonparametric approach is the design of a classifier without attempting to estimate the respective densities.

In this case the design of the classifier requires the definition of a function to be used as the classification rule. A classical example is the K-nearest neighbor rule [9]: when a sample is to be classified, its K nearest neighbors in the reference set are determined and the sample is assigned to the most frequent class among those of the nearest neighbors. The K-NN approach can be modified in order to allow sample rejection so that if at least K' (with $K' < K$) neighbors belong to the same class, the sample X is assigned to it; otherwise, it is rejected [15].

An alternative way of determining a set of decision regions in the feature space is to cluster the samples of the training set (whose identity, in this case, need not necessarily be known) according to some of the available techniques [16]. In order to achieve classification, the classes may have to be labeled after clustering.

Other available classification methods include the sequential method based on decision trees [17], which under certain hypotheses can be both fast and effective. According to this method, only a subset of the features chosen to characterize a sample is actually used to arrive at a decision about the class of the sample. The decision tree requires that the presence and/or value of a sequence of features is checked in the given sample. The first feature to be checked is suitably fixed and represents the root of the tree; the features to be considered in the subsequent steps of the decision process depend on the result of the check made at each previous step. The leaves of the tree may correspond to classes or rejects. The classification process implies that a path is followed from the root to one of the leaves. In the general case, more than one leaf may correspond to the same class. Feature selection and decision tree design may be particularly complex and can be carried out with either probabilistic or deterministic methods.

In the framework of image recognition, the structural approach to description and classification has also been followed since the 1970s [3, 4]. This approach attaches special importance to the feature selection problem: features representing image components that are meaningful from the geometric, morphological, or perceptive points of view are considered more reliable in order to obtain effective descriptions. The classification problem, which is central to the statistical approach, is here considered subordinate to the description problem; it is believed that an adequate description allows simple classification techniques to be used.

The main assumption of the structural approach is that every complex structure can be effectively subdivided into parts and described in terms of the component parts and their relationships. To be effective, a decomposition has to be stable with respect to the variations among samples belonging to the same class and such as not to destroy information needed to discriminate among classes. Although this is obviously not easy, the main reason why the approach is appealing is that, as the features are parts of the considered pattern, they can be perceptively appraised. This allows one to make some sort of *a priori* evaluation of the effectiveness of the descriptions.

The very nature of structural features allows them to give rise to descriptions outlining the structure of a pattern. Therefore, descriptions in terms of formal language sentences or attributed relational graphs can be conveniently employed. Accordingly, language parsers and graph inexact matching methods [18] can be used for classification. A more thorough discussion of the structural approach is beyond the scope of this chapter but the approach has been mentioned because some structural features will be used in one of the application cases illustrated in Section VII.

III. NEURAL NETWORK CLASSIFIERS

Artificial neural networks are an attempt to emulate the processing capability of biological neural systems. The basic idea is to realize systems capable of performing complex processing tasks by interconnecting a high number of very simple processing elements which may even work in parallel. The elements which substantially characterize the different types of ANNs that have been proposed so far are the network topology, the operation performed by the neurons, and the training algorithm. Almost all the network architectures share the simplicity of the elementary operations performed, the relatively straightforward use, and the short computation time in the operative phase, all of which make them particularly appealing for a number of applications and especially for classification. In this section we will assume that basic architectural and functional characteristics of neural networks are known, and we will point out some problems related to their use as classifiers.

The design of neural network classifiers implies a number of choices which can significantly influence the results during both the training phase (also referred to as learning phase) and the operative phase. According to the problem at hand, the network architecture can be chosen among the several ones proposed in the literature [19], and the network can be suitably sized by choosing the number of component neurons and the way they have to be interconnected. The number of neurons could be initially fixed and remain unchanged during the training phase or could be dynamically modified through appropriate techniques [20, 21]. In the former case, the most suitable number of neurons for the specific case has to be estimated before the learning phase on the basis of criteria depending on the sample distribution in the feature space [22]. In the latter case, the initial choice is less critical, but the methods that have so far been available for modifying the number of neurons cannot be considered generally applicable, since they have only been tested for simple case studies.

As for the training phase, possible choices regard the algorithm and modality of learning, selection of the training set, and determination of the optimal time to stop the learning procedure. Not only are different learning algorithms available for different network architectures, but initial conditions [5], training modality (supervised, unsupervised, or graded), and learning strategy [23] can be selected in a number of different ways. Criteria for selecting the optimal size of the training set have been proposed [5], as have suitable learning strategies for the case in which a slender number of samples is available for training [23]. Even the order in which the samples are fed into the net during training has to be controlled in order to avoid polarization effects which could lower performance in the operative phase [24].

The number of learning cycles performed affects the generalization capability of the network, i.e., its ability to correctly classify samples quite different from those present in the training set. In fact, if the number of learning cycles is too high, the network becomes too specialized on the training set, thus losing its generalization capability (overtraining phenomenon). A possible method to avoid the overtraining of the classifier is the one proposed in [23]. An additional set, called the training-test set and disjoined from the training set, is used to periodically check the error rate. While the error rate on the training set monotonically decreases with the number of learning cycles, the error rate on the training-test set first reaches a minimum and then increases. The minimum corresponds to the optimal number of learning cycles in order to avoid overtraining. Also the way the training set is chosen can influence the generalization capability of a neural classifier [25, 26].

In the following, some of the most frequently used neural network architectures will be illustrated, with reference to their use as classifiers and to the design problems discussed above, by outlining differences and common aspects. The considered neural network architectures (see Table I) are the multilayer perceptron (MLP) [27], the radial basis function network (RBF) [28], the learning vector quantization network (LVQ) [29], the self-organizing map (SOM) [29], the adaptive resonance theory network (ART) [30], the probabilistic neural network (PNN) [31], and the Hopfield network [32]. Networks can be subdivided into feedforward networks (Fig. 1a) where data flow one way from input to output, and recurrent networks for which the output values are fed back to input (Fig. 1b). Some networks (Fig. 1c) allow connections between neurons in the same layer (lateral connections). For feedforward networks a further subdivision can be made between feature-based and prototype-based networks: the former try to learn the functional mapping, normally nonlinear, existing between pairs of input–output vectors during the training phase, while the latter abstract the prototypes from the training set samples. Some relevant features of the classifiers which can be implemented with the considered network architectures will be summarized in the following.

The most frequently used feedforward network is the MLP [27] belonging to the feature-based category. In this case, learning can be seen as the process of fitting a function to a given set of data or, equivalently, of finding the hyperplanes separating the decision regions of the feature space.

The output layer is made of as many neurons as the number of classes. It would be expected that if, during training, a sample belonging to the kth class is presented to the network input, the kth output neuron will assume a value equal to

Table I
Some of the Most Frequently Used Neural Classifiers

Architecture	Connection scheme	Training modality	Learning rule	Learning algorithms
MLP	Feedforward	Supervised	Error-correction	Back-propagation
RBF	Feedforward	Supervised and unsupervised	Error-correction and competitive	RBF learning algorithm
LVQ	Lateral connection	Supervised or unsupervised	Competitive	LVQ1, RPCL, FSCL
SOM	Lateral connection	Unsupervised	Competitive	Kohonen's SOM
ART	Lateral connection	Unsupervised	Competitive	ART1, ART2
PNN	Feedforward	—	—	—
Hopfield	Recurrent	Unsupervised	Error-correction	Hebbian rule

1 while all the other outputs will assume a value equal to 0 (ideal output vector). In practice, the status of the output vector is generally different from the ideal one (i.e., the values of its elements may be numbers in the interval [0, 1]), and the input sample is attributed to a class according to some rule. The simplest rule is winner-takes-all, according to which the input sample is attributed to the class whose output neuron has the highest value.

As regards network sizing, the number of hidden layers and the number of neurons per layer influence the form of the decision regions in the feature space [22]. Too many neurons per layer may cause an overfitting of the data and the network risks becoming too specialized on the training samples. On the contrary,

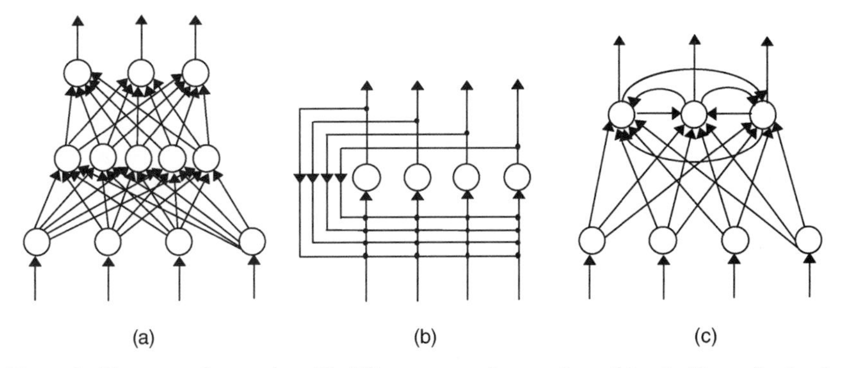

Figure 1 Three neural networks with different types of connections: (a) a feedforward network, (b) a network with lateral connections, and (c) a recurrent network.

with too few neurons, the network cannot reach a satisfactory recognition rate even on the training set. Different algorithms for the automatic sizing of an MLP net have been proposed [20, 33], in addition to methods for pruning oversized nets [21].

The MLP training modality is normally supervised; the most common algorithm is the back-propagation (BP), which is quite slow, but there are plenty of proposals aiming to make it faster, and there are also faster alternative training algorithms [34–36]. On the contrary, in the operative phase, the network is extremely fast. The presence of several local minima on the error surface may cause the training to stop in a minimum that could be very different from the absolute minimum. Suitable training algorithms are available to prevent the net getting trapped in a local minimum [37]. The presence of flat spots [38], i.e., net configurations such that the weight variations computed by the BP algorithm are very close to zero, implies that the training algorithm does not converge and is generally related to the problem of choosing the net initial conditions. In [5] some general criteria for a correct initialization, basically depending on the size of the input vector, are presented.

In the case of feature-based networks, the knowledge acquired during the training phase is completely spread over the net connection weights and therefore the single prototypes of the classes cannot be identified.

The LVQ classifier belongs to the so-called prototype-based category: during training it generates the prototypes of the classes by performing a clustering over the input space. In the operative phase, an input sample is assigned to the class characterized by the shortest distance from one of its prototypes. The Euclidean distance is commonly used. The set of weights associated with the connections between the neurons of the input layer and each of the output neurons represents a prototype generated by the network on the basis of the samples included in the training set. The learning algorithms can be either supervised or unsupervised and belong to the competitive learning category [23], i.e., they have the property that a competition among some or all of the neurons of the net always takes place before each learning step. At each step, the neuron winning the competition is allowed to modify its weight in a different way from that of the nonwinning units. In the supervised case, each output neuron is associated with one class before the training starts. In the unsupervised case, the output neurons must be labeled after training in order to allow classification; this can be done only after the whole training set has been examined, by associating a neuron to the class for which it obtained the highest winning frequency.

In both cases, identifying the prototypes gives rise to a Voronoi tessellation of the feature space. Each region of this partition is associated with a prototype and all the samples belonging to one region are attributed to the class of that prototype. One of the problems typical of this architecture is neuron underutilization: for particular configurations of the points representing the samples in the feature space, some neurons cannot modify their weights during training and remain unused. Algorithms substantially based on a modified distance calculation, which takes into account the number of times each neuron wins the competition, make it possible to overcome this problem [39–41].

Training is somewhat faster than for the MLP network and the overtraining problem is not nearly as important [23]. In order to guarantee the convergence of the training algorithm [29], the learning rate value has to be a decreasing function of the number of learning cycles (for instance, by varying the learning rate on the basis of the Robbins-Monro stochastic approximation [42]). This makes it possible to avoid the problem of choosing when the training has to be stopped. However, the results obtainable during training are significantly dependent on the initial value of the learning rate [43].

The SOM is another example of a prototype-based net. It works like the LVQ net, but since its training algorithm is unsupervised, its output neurons must eventually be labeled. In comparison with the LVQ net, the SOM has a bidimensional structure which makes it more suitable if it should be desirable to map the features of the input samples onto a bidimensional space. Moreover, its training algorithm makes it possible to update more than one neuron each time, in order to prevent the problem of neuron underutilization.

Another prototype-based architecture is the one founded on the adaptive resonance theory. This network has a complex training algorithm which tries to solve the so-called plasticity-stability dilemma [30], i.e., it aims to find a trade-off between the network ability to learn new samples (plasticity) and its ability to correctly classify the already seen samples (stability). The training algorithm generates new prototypes only when an input sample is sufficiently different from all the already generated prototypes. This eliminates the need to repeat the training if there are new samples to be learned. In contrast, if the MLP net is trained on a new set of samples, it forgets the previously learned set [24].

The RBF network represents a hybrid solution between feature-based and prototype-based architectures. In fact, it is made up of a hidden layer whose neurons are trained through competitive unsupervised algorithms and whose weight vectors represent the prototypes. These neurons are connected to an output layer of perceptrons. To work as a classifier, even in this case, the output layer has to be made up of as many neurons as there are recognition classes. The main difference between the RBF and the MLP nets is that the hidden neurons of the RBF have a Gaussian activation function instead of the sigmoidal function normally used for the MLP. The number of neurons of the hidden layer needed to solve a given problem may be significantly larger for the RBF than for the MLP. *Vice versa*, the duration of the training phase is significantly lower for the RBF than for the MLP, even if the latter is not trained with the BP algorithm [26]. Algorithms for optimally sizing the hidden layer have also been proposed for the RBF [41].

The basis for the PNN classifier is a probabilistic model. According to a nonparametric approach based on the Parzen method [14], the probability density functions of the samples of the training set are estimated and the *a posteriori* probabilities that a sample belongs to a given class are then computed. The input sample is assigned to the class with the highest *a posteriori* probability. Unlike the previous ones, this type of network does not have an explicit training phase, because it has as many neurons as the vectors of the training set. The main problems are the amount of memory needed and the amount of time necessary for classification. Methods for decreasing classification time by using a subset of the whole training set are proposed in [44, 45].

Finally, the Hopfield network is an example of a recurrent network classifier. During the training phase, the input samples are stored by associating them with the same number of stable states, i.e., states for which a suitable energy function associated with the net has reached a minimum. In the operative phase, an input sample is associated with the nearest stable state so that the net can work as a classifier once every possible state has been labeled. Nevertheless, the network may reach a stable state different from those reached in the training phase (spurious state); in this case, it is impossible to classify the sample. Moreover, unlike the other mentioned architectures, this net does not provide any output vector. The only output information, in the operative phase, is given by the stable state reached by the net.

IV. CLASSIFICATION RELIABILITY

In the field of classification, the term reliability can be (and sometimes has been) used with at least two different meanings. According to the first, classification reliability gives an estimate of the probability that the classifier assignment of an input sample to a class is correct. In this sense, it could be represented by a parameter associated with each decision taken by the classifier. The second meaning refers to a more global measure of the classifier performance in the specific task considered. In this case, a parameter associated with the classifier could measure its "effectiveness" in the context in which it is used. In the following, we will use the term classification reliability in the former meaning, whereas for the latter case we will use the term classifier effectiveness. A third way in which the term reliability might be used, with reference to the performance of a network when some of its internal connections are disabled, or more generally to the fault tolerance of the system, will not be considered in this chapter. The quantitative evaluation of both classification reliability and classifier effectiveness is of great practical importance, as will be shown below.

In the general frameworks of pattern recognition and statistical decision theory, the reliability problem has been tackled from various points of view. The Dempster–Shafer theory [46] and the fuzzy set theory [47] relate the problem of evaluating the reliability of a performed classification to that of the uncertainty measure. In the former case, a number in the range [0, 1] indicates belief in a hypothesis on the basis of a certain amount of evidence. In the fuzzy set theory, class membership is not binary, but is represented by the value of a function in the interval [0, 1].

In [10, 48], a reliability parameter, defined as the ratio of recognition rate to the sum of recognition rate and error rate, is used to measure the overall reliability of the considered classifier. The term reliability is used with a meaning similar to what we have here called classifier effectiveness, but the defined parameter does not take into account the peculiarities of the particular application domain.

Other approaches [15, 49] do not directly measure the reliability of a classification, but introduce costs to measure the risk of performing a classification and, using a probabilistic model, take the decision to classify or to reject on the basis of a minimum risk principle.

The reliability problem, in each of its possible meanings, has not often been considered in the literature regarding neural network classifiers. When it has been tackled, particular cases or specific frameworks have been considered. For instance, some authors have proposed criteria to improve classification reliability, intended as the ability of the classifier to reject significantly distorted samples [50, 51], but without giving a precise definition of classification reliability nor providing a method for measuring it. In [50], it is suggested using a neuron activation function different from the sigmoidal one with MLP classifiers, in order to obtain decision regions that are more strictly representative of the samples present in the training set and more sharply separated from each other. In this way, samples whose representative points fall outside these regions can be rejected and reliability can thus be improved. In [51], a system made up of two neural networks is illustrated: the first net works as a normal neural classifier, while the second is essentially devoted to detecting the errors made by the first one. This second network is trained with a training set containing the samples misclassified by the first network. Reliability is improved because some errors can be detected and the samples causing them rejected.

Other papers propose techniques to evaluate what we have called classification reliability, but they are based on criteria strictly depending on the architecture taken into account and cannot be immediately extended to other architectures. In [52], a new training algorithm for the MADALINE network architecture [24] is presented. A suitable function of the state of the output neurons is defined and a decision of the classifier is considered acceptable if the value of the function is higher than an upper reject threshold, unacceptable if it is below a lower threshold. Otherwise there are no elements for taking a decision. The thresholds are evaluated on the basis of the Dempster–Shafer theory [46], but without taking into account the requirements of the considered application domain. Moreover, the method is strictly dependent on the adopted network architecture and considers only nets with binary output vector. The system proposed in [53] integrates the fuzzy logic with a classic MLP network. Some fuzzy functions are used to identify unreliable classifications, but general criteria to define them are not given, the test data do not refer to a real problem, and the obtained results do not seem to be applicable outside the considered case.

The approach to the reliability problem presented below aims to be more general. A neural classifier is considered at a first level as a black box, accepting the description of a sample (e.g., a feature vector) as the input and supplying a vector of real numbers as the output. Nothing needs to be known about the network architecture nor about the way the learning process is carried out. After a formal definition of the parameters assumed to measure classification reliability and classifier effectiveness, a method for quantitatively evaluating them is illustrated. The situations in the feature space which can give rise to unreliable classifications are characterized and put in correspondence to the state of the classifier output vector. Therefore, the operative definition of the parameters allowing such situations to be recognized and enabling classification reliability to be quantified will depend on the considered neural architecture. In Section V, we will define the parameter measuring classification reliability for each of the different classifiers introduced in Section III. In the following sections, the parameter will be used in the context of a method implementing a reject option which can be regarded as optimal with respect to the considered application domain, and the results of the method applied in two complex classification problems will be presented.

V. EVALUATING NEURAL NETWORK CLASSIFICATION RELIABILITY

The low reliability of a classification is generally due to one of the following situations: (a) the considered sample is significantly different from those present in the training set, i.e., its representative point is located in a region of the feature space far from those occupied by the samples of the training set and associated to the various classes; (b) the point which represents the considered sample in the feature space lies where the regions pertaining to two or more classes overlap, i.e., where training set samples belonging to more than one class are present.

It may be convenient to distinguish between classifications which are unreliable because a sample is of type (a) or (b). To this end, let us define two parameters, say ψ_a and ψ_b, whose values vary in the interval [0, 1] and quantify the reliability of a classification from the two different points of view. Values near to 1 will characterize very reliable classifications, while low parameter values will be associated with classifications unreliable because the considered sample is of type (a) or (b). Note that in practice it is almost impossible to define two parameters such that each one identifies all and only the samples of one of the two types.

In the following the parameters ψ_a and ψ_b will be referred to as reliability parameters. With reference to neural classifiers, two parameters will be needed for each network architecture and each parameter shall be a function of the classifier output vector. A parameter ψ providing a comprehensive measure of the reliability of a classification can result from the combination of the values of ψ_a and ψ_b. A recent review of the several combination operators considered in the literature has been presented in [54]. For ψ we have chosen the form

$$\psi = \min\{\psi_a, \psi_b\}. \tag{2}$$

This is certainly a conservative choice because it implies that, for a classification to be unreliable, just one reliability parameter needs to take a low value, regardless of the value assumed by the other one. By definition, the ideal reliability parameter should assume a value equal to 1 for all the correctly classified samples and a value equal to 0 for all the misclassified samples. However, this will almost never happen in real cases. Let us suppose, for instance, that in the test set there is a sample belonging to the ith class, whose description is identical to that of some samples of the training set belonging to the jth class; this sample will certainly be misclassified, but the classifier will reach its decision with high reliability as it has no elements to judge it unreliable. An *a posteriori* evaluation of how good a reliability parameter actually is can be made by computing both the average of the parameter values associated with correct classifications and the average of the values associated with misclassifications. The nearer these values are to 1 and 0, respectively, the better the parameter works.

The operative definition of ψ requires the classifier to provide an output consisting of a vector the values of whose elements make it possible to establish the class a sample belongs to. Therefore a reliability parameter cannot be defined for the Hopfield network which, as it behaves like an associative memory, provides as output the stable state reached after minimizing its energy function, and thus only the information about the class attributed to the input sample.

The remaining neural classifiers can be grouped into three categories according to the meaning of their output vector. The MLP and RBF networks can be grouped together because for both of them the cardinality of the output vector is equal to the number of classes and, in the ideal case, only one vector element at a time can have a value equal to 1, while the remaining elements must have a value equal to 0. A second group can include the networks LVQ, SOM, and ART. Their output neurons provide a measure of the distance between the input sample and each of the class prototypes: the classification is performed by assigning the input sample to the class that has the shortest distance from it. The third group is made up of the PNN network only, whose output vector provides the probabilities that the input sample belongs to each class; the input sample is assigned to the class that maximizes this probability.

In the following, the reliability parameters will be defined for each of the above three groups of neural classifiers.

As for the classifiers of the first group, we saw that, in real cases, the output vector is generally different from the ideal one and an input sample is attributed to the class associated with the winner output neuron, i.e., the one having the highest output value. In practice, these networks use the value of the connection weights to obtain the hyperplanes defining the decision regions [22]. During the training phase, the network dynamically modifies the decision region boundaries in such a way as to provide, for each sample, an output vector as close as possible to the ideal one. Consequently, samples of type (a) may fall outside every decision region as they are very different from those which contributed to determining the hyperplanes separating the decision regions; in this case, all the output neurons should provide values near to 0. Therefore, an effective definition of the reliability parameter ψ_a can be the following:

$$\psi_a = O_{\text{win}}, \tag{3}$$

where O_{win} is the output of the winner neuron. In this way, the nearer to 0 the value of O_{win}, the more unreliable the classification is considered.

Samples of type (b), lying where two or more decision regions overlap, typically generate output vectors with two or more elements having similar values. Therefore, the classification reliability is higher when the difference between O_{win} and the output of the neuron having the highest value after the winner ($O_{2\text{win}}$) is greater. In this case a suitable reliability parameter is

$$\psi_b = O_{\text{win}} - O_{2\text{win}}. \tag{4}$$

Let us note that, since the values of the elements of the output vector are real numbers in the interval [0, 1], the reliability parameters ψ_a and ψ_b also assume values in the same interval, as required by their definition.

In conclusion, the classification reliability of classifiers from the first group can be measured by

$$\psi = \min\{\psi_a, \psi_b\} = \min\{O_{\text{win}}, O_{\text{win}} - O_{2\text{win}}\} = O_{\text{win}} - O_{2\text{win}} = \psi_b. \tag{5}$$

For classifiers of the second group, the values of the elements of the output vector give the distances of an input sample X from each of the prototypes W_i, $i = 1, \ldots, M$, with M generally greater than the number N of classes. Therefore, the winner neuron is the one having the minimum output value:

$$O_{\text{win}} = \min_i\{O_i\} = \min_i\{d(W_i, X)\}. \tag{6}$$

During successive learning cycles, as long as the samples of the training set are taken into account, some starting prototypes are updated and the feature space is partitioned in such a way that the final prototypes defined by the net are the centroids of the regions into which the feature space is partitioned. Obviously, since samples that are significantly different from those present in the training set have not contributed to generating the prototypes, their distance from the prototype associated to the winner neuron will be greater than that of the samples of the training set. Therefore, the reliability parameter ψ_a can be defined as

$$\psi_a = 1 - \frac{O_{\text{win}}}{O_{\text{max}}}, \tag{7}$$

where O_{max} is the highest value of O_{win} among those relative to all the samples of the training set. In this way, since it has to be expected that the value of O_{win} is high for samples of type (a), these will be classified with a low reliability (low values of ψ_a).

According to the above definition, the value of ψ_a ranges from 0 to 1 only for the samples belonging to the training set as the relation $O_{\text{win}} \leq O_{\text{max}}$ is certainly valid only for such samples. Therefore, to make the definition applicable when the classifier is in the operative phase, the previous expression has to become

$$\psi_a = \begin{cases} 1 - \dfrac{O_{\text{win}}}{O_{\text{max}}}, & \text{if } O_{\text{win}} \leq O_{\text{max}}, \\ 0, & \text{otherwise.} \end{cases} \tag{8}$$

On the other hand, samples of type (b) have comparable distances from at least two prototypes. Consequently, the reliability parameter ψ_b must be a function of both O_{win} and $O_{2\text{win}}$ (in this case the second winner neuron is the one having the second lowest distance from the input sample):

$$\psi_b = 1 - \frac{O_{\text{win}}}{O_{2\text{win}}}. \tag{9}$$

On the basis of this definition, ψ_b takes values ranging from 0 to 1, and the larger the difference $O_{2\text{win}} - O_{\text{win}}$ is, the higher the values of ψ_b are.

The classification reliability for the classifiers of the second group is given by

$$\psi = \min\{\psi_a, \psi_b\} = \min\left\{\max\left\{1 - \frac{O_{\text{win}}}{O_{\text{max}}}, 0\right\}, 1 - \frac{O_{\text{win}}}{O_{2\text{win}}}\right\}. \tag{10}$$

Finally, let us consider the case of the PNN classifier. In the classifier operative phase, the distances between the input sample X and every sample belonging to the training set are computed and consequently the probabilities P_i that X belongs to the ith class, for $i = 1, \ldots, N$, are evaluated. The input sample is assigned to the class associated to the winner neuron, for which the following relation holds:

$$O_{\text{win}} = \max_i \{h_i \cdot l_i \cdot P_i\}, \tag{11}$$

where h_i is the *a priori* occurrence probability of the ith class and l_i is the "loss" caused by a wrong assignment to the ith class. As the value of P_i depends on the whole training set, it is evident that samples of type (a), i.e., quite different from those of the training set, have a low probability of being attributed to any class. Therefore the parameter ψ_a can be defined as

$$\psi_a = \frac{O_{\text{win}}}{O_{\text{max}}}, \tag{12}$$

where O_{max} is the highest value of O_{win} on the samples belonging to the training set. Again, $\psi_a \leq 1$ only for the samples belonging to the training set since the relation $O_{\text{win}} \leq O_{\text{max}}$ is certainly valid only in this case, and the previous equation must become

$$\psi_a = \begin{cases} \dfrac{O_{\text{win}}}{O_{\text{max}}}, & \text{if } O_{\text{win}} \leq O_{\text{max}}, \\ 1, & \text{otherwise.} \end{cases} \tag{13}$$

This definition ensures that ψ_a assumes low values in the case of samples to be classified with a low reliability.

Samples of type (b) have similar probabilities of belonging to different classes so that ψ_b can be defined as

$$\psi_b = 1 - \frac{O_{\text{2win}}}{O_{\text{win}}}. \tag{14}$$

According to this definition, $0 \leq \psi_b \leq 1$ and the higher the probability is that the input sample belongs to the winner class rather than to the second winner class, the more reliable the classification will be.

The classification reliability for the PNN classifier is

$$\begin{aligned} \psi = \min\{\psi_a, \psi_b\} &= \min\left\{\min\left\{\frac{O_{\text{win}}}{O_{\text{max}}}, 1\right\}, 1 - \frac{O_{\text{2win}}}{O_{\text{win}}}\right\} \\ &= \min\left\{\frac{O_{\text{win}}}{O_{\text{max}}}, 1 - \frac{O_{\text{2win}}}{O_{\text{win}}}\right\}. \end{aligned} \tag{15}$$

VI. FINDING A REJECT RULE

A. METHOD

When the reliability of a classification is low, the question is: does one accept the decision of the classifier running the risk of an error, or reject it and consider the sample at hand as unrecognizable by the given classification system? In the former case, the risk of the decision being wrong increases as the classification reliability decreases. In the latter case, the sample has to be examined again with alternative techniques, generally by man. In both cases the choice implies a cost that has to be paid.

Finding a reject rule which achieves the best trade-off between error rate and reject rate is undoubtedly of practical interest in real applications. Nevertheless, very few results of general applicability are available in the literature. A significant contribution to the problem has been given by Chow in [13, 49, 55]. These papers describe an optimal reject rule and then derive a general relation between the error and reject rates. The basic assumption of the method is the knowledge, or the possibility of making a hypothesis, about the *a priori* probability distributions of the samples in the parametric space. In most recognition tasks, however, the underlying probability distributions are not known, nor can suitable hypotheses on their form be made, thus making Chow's approach not generally applicable [56].

The approach we propose is more general than the one mentioned above: an optimal reject rule is defined for a given classifier by taking into account only the value of the classification reliability ψ computed using information about the output vector of the 0-reject classifier, i.e., the classifier with no reject option. If the reliability is greater than a threshold σ, determined through a suitable algorithm, the classification is considered acceptable; otherwise, the input sample is rejected (Fig. 2). In this way, no *a priori* knowledge about the probability distributions is needed, and the classifier can be regarded up to a certain level as a black box, regardless of its architecture. It can be shown that our approach achieves, as its upper bound, the results that Chow's approach makes possible if the distributions are exactly known.

The introduction of a reject option gives rise to two opposite effects: on the one hand, the misclassified samples having a value of ψ less than σ are rejected, and this effect is undoubtedly desirable since the cost of a misclassification is always higher than the cost of a reject. On the other hand, also the correctly classified samples with values of ψ less than σ are rejected, and this is an undesirable side effect since it contributes to decrease the classification rate. This reduction partly reabsorbs the advantage obtained by introducing the reject option.

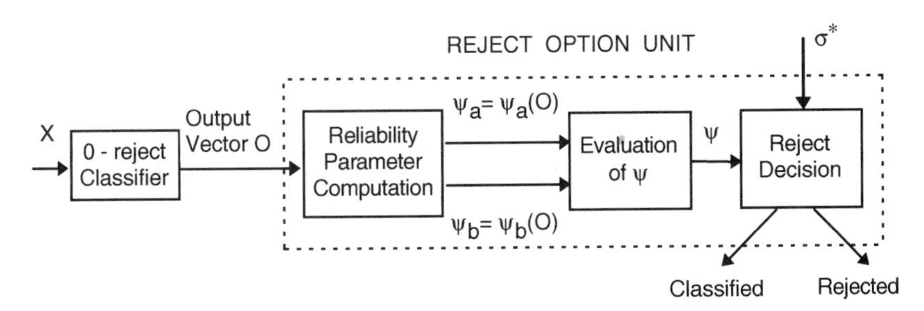

Figure 2 A block diagram of the proposed method: the reject option operates on the basis of the classification reliability ψ which is a function of the output vector of the neural classifier. σ^* is the optimal value of the reject threshold, established through a suitable algorithm.

In order to evaluate the real advantage of using the reject option and establish a criterion for fixing the reject threshold in such a way as to find the best trade-off between misclassifications and rejects, we introduce a function $\mathcal{P}$ that aims to quantify the classifier effectiveness in the considered application domain, while taking into account the two previous effects.

Let us call R_c the percentage of correctly classified samples (also referred to as recognition rate), R_e the misclassification rate (also called error rate), and R_r the reject rate. Since the effectiveness of a classifier depends on the results it produces, we can certainly write

$$\mathcal{P} = \mathcal{P}(R_c, R_e, R_r). \tag{16}$$

For $\mathcal{P}$ to actually measure the classifier effectiveness, it must satisfy at least two constraints:

(i) it must have a monotonic trend, increasing with respect to R_c and decreasing with R_r and R_e, that is:

$$\frac{\partial \mathcal{P}}{\partial R_c} > 0, \qquad \frac{\partial \mathcal{P}}{\partial R_r} < 0, \qquad \text{and} \qquad \frac{\partial \mathcal{P}}{\partial R_e} < 0, \tag{17}$$

(ii) it must be such that

$$\left|\frac{\partial \mathcal{P}}{\partial R_r}\right| < \left|\frac{\partial \mathcal{P}}{\partial R_e}\right|, \tag{18}$$

since it is expected that a misclassification negatively affects $\mathcal{P}$ more than a reject.

In principle, no further hypotheses are necessary on the form of $\mathcal{P}$.

Since we need a function measuring the actual effectiveness improvement when the reject option is adopted, independently of the absolute performance of the 0-reject classifier, it may be convenient to define the following function:

$$P = \mathcal{P}(R_c, R_e, R_r) - \mathcal{P}^0, \tag{19}$$

where $\mathcal{P}^0 = \mathcal{P}(R_c^0, R_e^0, 0)$ is the value of the function $\mathcal{P}$ when the classifier is used at 0-reject (i.e., when $R_r = 0$), and R_c^0 and R_e^0 are respectively the recognition rate and the error rate in the same case.

The functional dependence of P on the considered application can be expressed by attributing a cost to each error and to each rejection and a gain to each correct classification. For notation uniformity, let us denote these three quantities C_e, C_r, and C_c, respectively. Although such costs are, in general, functions of R_c, R_e, and R_r, for most of the applications they can be considered constant. In fact, the cost of a misclassification is generally attributed by considering the burden of locating and possibly correcting the error or, if this is impossible, by evaluating the consequent damage; the cost of a reject is that of a new classification using a different technique. It is reasonable to presume, although this may not be true in certain specific cases, that such a burden is generally independent of the relative number of correctly classified, misclassified, or rejected samples, i.e., C_c, C_r, and C_e are constant. Moreover, in the following, the function P will be assumed to be linearly dependent on R_c, R_e, and R_r since this is the most frequently occurring case and it simplifies the illustration of our method for determining the optimal reject threshold value for a given application. In [57], it is shown how the method can be extended to the case of a function P of generic form.

Taking all the above considerations into account, the function P can be written in the form

$$P = C_c\left(R_c - R_c^0\right) - C_e\left(R_e - R_e^0\right) - C_r R_r. \tag{20}$$

It can be noted that Eq. (20) satisfies the constraints of Eq. (17), and, as $C_e > C_r$, also the constraint of Eq. (18).

Since R_c, R_r, and R_e depend on the value of the reject threshold σ, P is also a function of σ. To highlight the dependence of P on σ, let us consider the occurrence density curves of correctly classified and misclassified samples as a function of the value of ψ. Let us call them $D_c(\psi)$ and $D_e(\psi)$, respectively. The trend of the curves (see Fig. 3) should be such that the majority of correctly classified sam-

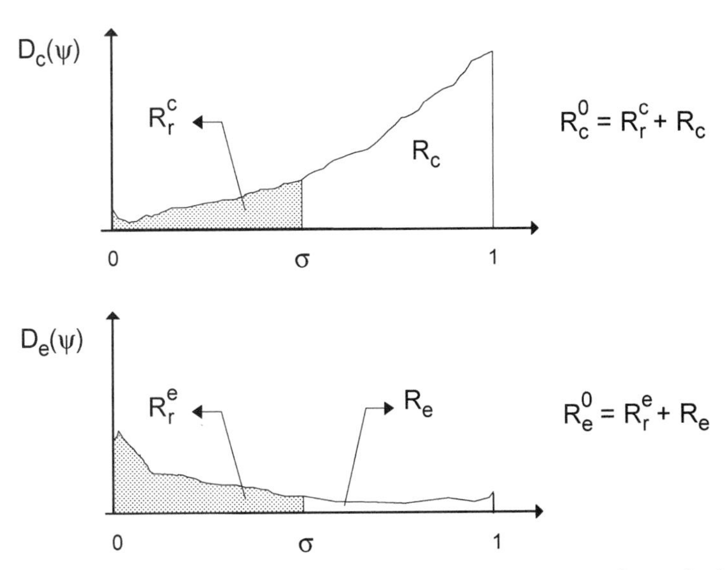

Figure 3 Qualitative trends of the curves $D_c(\psi)$ and $D_e(\psi)$. The percentages of correctly classified and misclassified samples which are rejected after the introduction of a reject threshold σ (denoted R_r^c and R_r^e respectively) are given by the gray areas. R_c (R_e) represents the percentage of samples which are correctly classified (misclassified) after the introduction of the reject option.

ples is found for high values of ψ, while misclassified samples are more frequent for low values of ψ.

For our purposes it is convenient to normalize the occurrence density curves so that their integrals extended to the interval $[\psi_1, \psi_2]$ respectively provide the percentage of correctly classified and misclassified samples having values of ψ ranging from ψ_1 to ψ_2. From this definition it follows that

$$R_c^0 = \int_0^1 D_c(\psi)\, d\psi, \tag{21}$$

$$R_e^0 = \int_0^1 D_e(\psi)\, d\psi. \tag{22}$$

The occurrence density curves $D_c(\psi)$ and $D_e(\psi)$ allow us to easily compute the reject rate R_r in case that a reject threshold σ is set on the value of ψ. In fact, R_r is a function of σ and is given by the sum of two terms: the percentage of correctly classified samples with a reliability ψ less than σ and the percentage of misclassified samples with a reliability ψ less than σ. With reference to Fig. 3, the two terms, denoted R_r^c and R_r^e, are given by the values of the gray areas in the two plots. Analytically,

$$R_r(\sigma) = R_r^c(\sigma) + R_r^e(\sigma) = \int_0^\sigma D_c(\psi)\, d\psi + \int_0^\sigma D_e(\psi)\, d\psi. \tag{23}$$

Analogously, when the reject option is activated, the percentages of correctly classified and misclassified samples are given by

$$R_c(\sigma) = \int_\sigma^1 D_c(\psi)\, d\psi, \tag{24}$$

$$R_e(\sigma) = \int_\sigma^1 D_e(\psi)\, d\psi. \tag{25}$$

Substituting Eqs. (21)–(25) into Eq. (20), it follows that

$$\begin{aligned} P(\sigma) = {} & C_c\left(\int_\sigma^1 D_c(\psi)\, d\psi - \int_0^1 D_c(\psi)\, d\psi\right) \\ & - C_e\left(\int_\sigma^1 D_e(\psi)\, d\psi - \int_0^1 D_e(\psi)\, d\psi\right) \\ & - C_r\left(\int_0^\sigma D_c(\psi)\, d\psi + \int_0^\sigma D_e(\psi)\, d\psi\right), \end{aligned} \tag{26}$$

and hence,

$$P(\sigma) = (C_e - C_r)\int_0^\sigma D_e(\psi)\, d\psi - (C_r + C_c)\int_0^\sigma D_c(\psi)\, d\psi. \tag{27}$$

From Eq. (27), it is evident that, since the two integral functions are monotonically increasing and $C_e > C_r$, an increase of σ implies that the first term contributes to increasing P, while the second decreases it.

The function P makes it possible to determine the optimal value σ^* of the reject threshold σ:

$$\sigma^*\colon\ P(\sigma^*) \geq P(\sigma), \qquad \forall\sigma \in [0, 1]. \tag{28}$$

In other words, σ^* is the threshold value for which the function P gets its maximum value. Once the cost coefficients have been fixed, the maximum value assumed by P obviously depends on the adopted reliability parameter.

In order to determine σ^*, let us assume we have a classifier operating at 0-reject, and a set S of labeled samples which are different from the training set. Under the hypothesis that the set S is adequately representative of the target domain, and once the cost coefficients characterizing the given application have been fixed, the optimal reject threshold value is obtained by finding the value of σ that satisfies Eq. (28) on the set S. For this purpose, let us calculate the derivative of Eq. (27) with respect to σ and make it equal to 0. We obtain

$$C_n D_e(\sigma) - D_c(\sigma) = 0 \quad \text{on } S, \tag{29}$$

where

$$C_n = \frac{C_e - C_r}{C_r + C_c}. \tag{30}$$

Henceforth, C_n will be referred to as the normalized cost.

In practice, the functions $D_c(\psi)$ and $D_e(\psi)$ are not available in their analytical form and therefore, in order to evaluate the solutions of Eq. (29), they should be experimentally determined in tabular form.

The process for determining σ^* is performed on the set S, once the cost coefficients for the given application domain have been fixed, and is described by the following algorithm:

1. The set S is submitted to the 0-reject classifier and then split into the subsets S_e of misclassified samples and S_c of correctly classified samples.
2. For each sample of the set S_c, the classification reliability value ψ is computed. The set of values of ψ obtained for S_c makes it possible to numerically determine the occurrence density function $D_c(\psi)$. In the same way, by using the set S_e, the function $D_e(\psi)$ can be determined.
3. The values of σ satisfying Eq. (29) can be determined with a numerical algorithm.
4. The value σ^* that corresponds to the absolute maximum of P is selected from the values computed at the previous step. It may happen that several values satisfy Eq. (29), because the density curves do not necessarily have

a monotonic trend. Thus, to be sure of obtaining the value σ^* that corresponds to the absolute maximum of P, it is necessary to determine first the values corresponding to all the relative maxima and then to select the value corresponding to the absolute maximum among them.

B. Discussion

The ideal behavior of a classifier with reject option should be such that the rejected samples are all and only those which, if not rejected, would be misclassified. In this case, no correctly classified samples would be rejected, the recognition rate would not decrease, and P would get its absolute maximum. For this to be possible, the nature and quality of the data in addition to the adopted reliability parameter should be such as to give rise to distributions such as those shown in Fig. 4. In this way, in fact, it will be possible to find two values for ψ, say ψ_1 and ψ_2, such that

$$\begin{aligned} D_c(\psi) &= 0, \qquad \forall \psi \leq \psi_2 \qquad \text{and} \\ D_e(\psi) &= 0, \qquad \forall \psi \geq \psi_1 \qquad \text{with } \psi_1 \leq \psi_2. \end{aligned} \tag{31}$$

The ideal value of P, indicated with P_{id}, that is the maximum allowed improvement of the classifier effectiveness, would therefore be obtained by choosing a

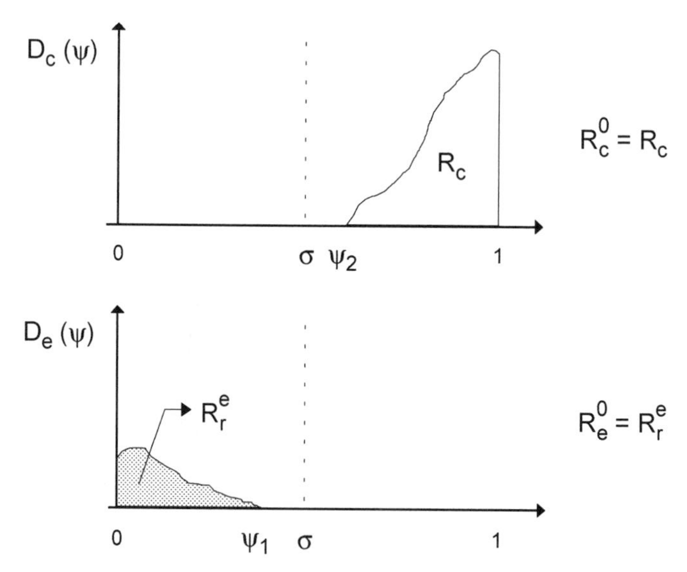

Figure 4 A case of distributions in which P_{id} can be achieved.

threshold value σ_{id} in the range $[\psi_1, \psi_2]$. In this case, Eq. (27) becomes

$$P_{\text{id}} = (C_e - C_r) \int_0^{\sigma} D_e(\psi)\, d\psi = (C_e - C_r) R_e^0. \tag{32}$$

As it is impossible to modify the data, the way of getting close to the ideal situation is to find a reliability parameter that, for the considered network architecture, makes it possible to influence the distributions $D_c(\psi)$ and $D_e(\psi)$ in such a way as to satisfy the constraints of Eq. (31). This aspect of the problem will not be further discussed here. However, in order to evaluate the degree of approximation to the ideal case achieved in a specific application, it is convenient to introduce the parameter

$$P_n = \frac{P}{P_{\text{id}}} \times 100. \tag{33}$$

In fact, P_n gives a measure of the percentage improvement of classifier effectiveness related to the maximum attainable improvement P_{id}. Moreover, the trend of P_n as a function of C_n can give useful information about the advantage achieved by introducing the reject option in a classification system as the requirements of the considered application domain vary.

One further consideration can be made on the basis of Eq. (29) which implies a relation between the optimal threshold value σ^* and the normalized cost C_n. In particular, it can be verified that all the triplets of cost coefficients $(C_c - k,\ C_r + k,\ C_e + k)$, obtained as k varies, provide the same value of C_n, and thus the same solution of Eq. (29). Moreover, approximate solutions of Eq. (29) can be obtained by neglecting C_c with respect to the other cost coefficients. Under this hypothesis, verified in many real applications, the normalized cost C_n assumes the form $C_n = C_e/C_r - 1$, and consequently the optimal reject threshold value depends on the ratio C_e/C_r.

In any case, when C_n is low (i.e., the difference $C_e - C_r$ is low), the advantage of introducing the reject option becomes negligible. In fact, from Eq. (27) it is evident that the percentage of samples turned from misclassified into rejected contributes to the increase of P proportionally to the difference $C_e - C_r$. Consequently, for $C_e \cong C_r$, the increment of P can become comparable to the decrement of P produced by the decrease of the classification rate.

VII. EXPERIMENTAL RESULTS

The performance of the method described in the previous section will be now illustrated with reference to two different applications: recognition of unconstrained handwritten characters and fault detection in electrical systems. The experimental results obtained in both cases will be discussed. The applications have been chosen because they represent critical recognition problems: both are characterized by a high variability among the samples belonging to a same class and by partial overlaps between the regions pertaining to different classes. In these conditions it is difficult to get high recognition rates and the availability of a reject option is particularly useful. Let us note that the emphasis here is not placed so much on the absolute performance of the description and classification techniques used, as on the improvement of classifier effectiveness achievable by introducing a reject option according to our method.

A. Case 1: Handwritten Character Recognition

Optical character recognition is one of the oldest application problems considered in pattern recognition. For its solution, a large number of statistical, structural, and hybrid methods have been proposed (reviews can be found in [58–60]). Although many OCR systems, suitable in a variety of applications, are today commercially available, the problem of recognizing unconstrained handprinted characters still remains unsolved and can be considered as a significant test bed for our method.

Recognition is made difficult by both a high degree of overlapping among classes and a high variability within each class; this is due partly to the quality of the original data, which comes from different writers with greatly varying drawing styles, and partly to the preprocessing and feature extraction phases, which can lose meaningful details of the character to be recognized. When significant character distortions occur, the uncertainty of the whole recognition process makes it essential to establish whether or not the decision of the classifier is acceptable or not, and therefore to estimate the classification reliability.

The characters used for the test are 7000 digits extracted from the ETL-1 character data base [61], containing 141,319 segmented characters produced by 1445 writers and belonging to 99 classes (digits plus latin, special, and katakana characters). In Fig. 5, some examples of digits are shown.

Characters are preliminarily submitted to a process that will represent them in terms of structural features [62]. The main steps of the process are briefly summarized in the following. Since the thickness of character strokes is generally not a significant feature for recognition purposes, character bit maps (Fig. 6a) are first thinned (Fig. 6b). Unfortunately, skeletonizing algorithms typically give rise to distorted representations of character shapes: the most significant shape distortions occur at joins and crossings of strokes. In order to better preserve the original shape information, a skeleton correction technique [63] is used; after this correction a character is represented by a set of polygonal lines (Fig. 6c). A further step consists of approximating pieces of polygonal lines with circular arcs (Fig. 6d) according to a method illustrated in [64].

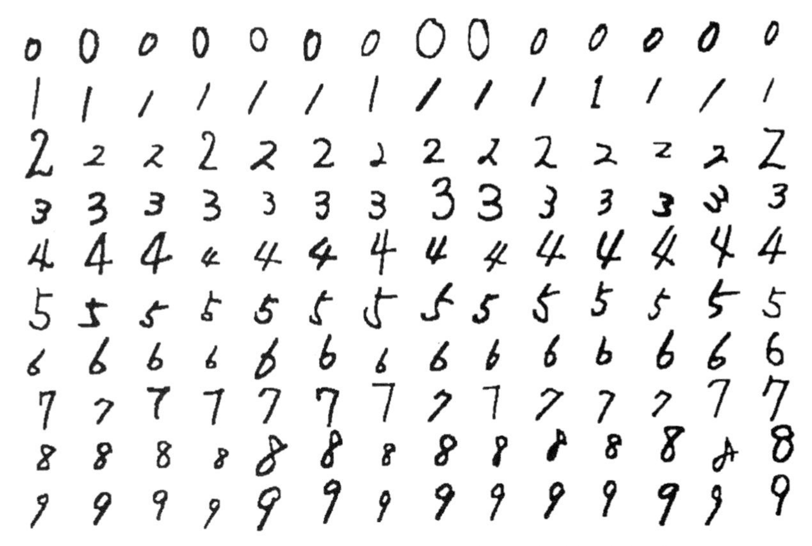

Figure 5 An example of the digits belonging to the ETL-1 data base.

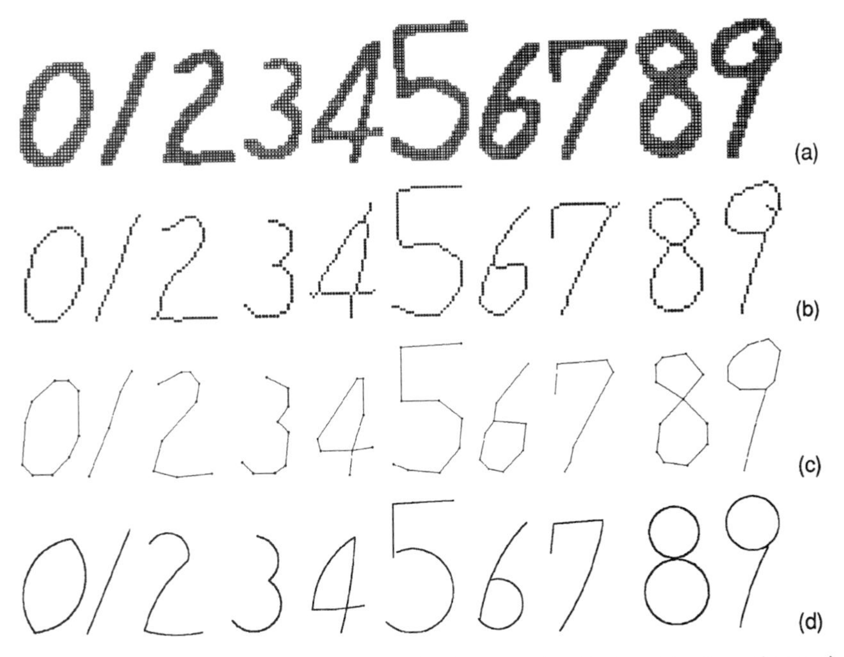

Figure 6 (a) The bit maps of some characters; (b) characters after thinning; (c) polygonal approximations after correcting shape distorsions; (d) representation in terms of circular arcs.

The aim of the above processing is to try to absorb as far as possible the large shape variability among different samples of the same class, singling out the most characteristic and invariant features for a recognition class. On this kind of character representation, the central moments up to the seventh order are evaluated. Moments of zero and the first order have been used to make the remaining moments invariant with respect to scale and translation, thus obtaining a feature vector made up of 33 components. Every component value is normalized so as to range between 0 and 1. For further details see [65].

The adopted classification system is an MLP made up of three fully connected layers, with 50 hidden neurons and a sigmoidal activation function. The standard BP algorithm was used for training with a constant learning rate equal to 0.5 and 10,000 learning cycles. A training set of 1000 samples and a test set of 5000 samples were used. The remaining 1000 samples were used for computing the reject threshold.

As regards the cost coefficients, we assumed $C_c = 1$, while the values for C_e and C_r were selected within the sets $\{6, 9, 12, 15, 18\}$ and $\{3, 4, 5\}$, respectively; under this assumption the normalized cost C_n ranges from 0.17 to 3.75. This choice seems adequate to include a bunch of real cases and makes it possible to verify the behavior of the method for taking up the reject option for a wide range of situations, from those for which $C_e \cong C_r$ (and thus $C_n \cong 0$) to those for which $C_e \gg C_r$ (and thus C_n is higher than 1). The former case refers to situations in which the occurrence of an error is not so detrimental as to induce to accept, in order to avoid it, a high reject rate which could imply a significant reduction of the classification rate. In the latter case, *vice versa*, the main requirement is to avoid as many errors as possible, even if classification rate significantly reduces.

Table II summarizes the results obtained with the 0-reject classifier and with the classifier with the reject option, showing, for each combination of the cost coefficients, the optimal threshold value σ^*, and the values of the parameters characterizing performance and effectiveness of the classifier. Another column lists the classifier effectiveness values in the ideal case (P_{id}). In order to properly interpret the data, it should remembered that by definition P represents the variation of classifier effectiveness with respect to $\mathcal{P}^0$.

The trend of the reject threshold σ^* as a function of C_n is plotted in Fig. 7.

In Fig. 8, R_c, R_e, and R_r are shown as a function of C_n. It is easy to verify that the trend of these curves is in agreement with the theoretical considerations made in Section VI. In particular, if the cost of a misclassification is low, the cost of a reject must be even lower and C_n is close to 0. In this situation, it is more convenient to accept a low reject rate so as to keep the recognition rate high, even at the risk of some misclassifications. The results are similar to those obtained at 0-reject. On the contrary, when the value of C_n increases, it is more convenient to reject an unreliably classified sample rather than run the risk of misclassifying it, even

Table II

Results at 0-Reject and with the Reject Option

			0-reject classifier[a]	Ideal case	Classifier with the reject option					
C_n	C_e	C_r	$\mathcal{P}^0$	P_{id}	σ^*	R_c	R_e	R_r	P	P_n
0.17	6	5	0.68	0.05	0.00	95.40	4.60	0.00	0.00	0.00
0.40	6	4	0.68	0.09	0.00	95.40	4.60	0.00	0.00	0.00
0.67	9	5	0.54	0.18	0.01	95.09	3.91	1.00	0.01	5.43
0.75	6	3	0.68	0.14	0.01	95.09	3.91	1.00	0.01	6.52
1.00	9	4	0.54	0.23	0.09	93.91	2.70	3.39	0.02	8.70
1.17	12	5	0.40	0.32	0.09	93.91	2.70	3.39	0.04	13.35
1.50	9	3	0.54	0.28	0.09	93.91	2.70	3.39	0.05	19.57
1.60	12	4	0.40	0.37	0.09	93.91	2.70	3.39	0.08	20.92
1.67	15	5	0.26	0.46	0.15	93.10	2.52	4.38	0.08	15.65
2.17	18	5	0.13	0.60	0.32	91.52	1.89	6.59	0.11	19.57
2.20	15	4	0.26	0.51	0.32	91.52	1.89	6.59	0.11	20.16
2.25	12	3	0.40	0.41	0.32	91.52	1.89	6.59	0.09	21.01
2.80	18	4	0.13	0.64	0.56	89.53	1.59	8.88	0.12	19.41
3.00	15	3	0.26	0.55	0.56	89.53	1.59	8.88	0.13	22.46
3.75	18	3	0.13	0.69	0.56	89.53	1.59	8.88	0.21	31.01

[a] The recognition rate at 0-reject is 95.4%.

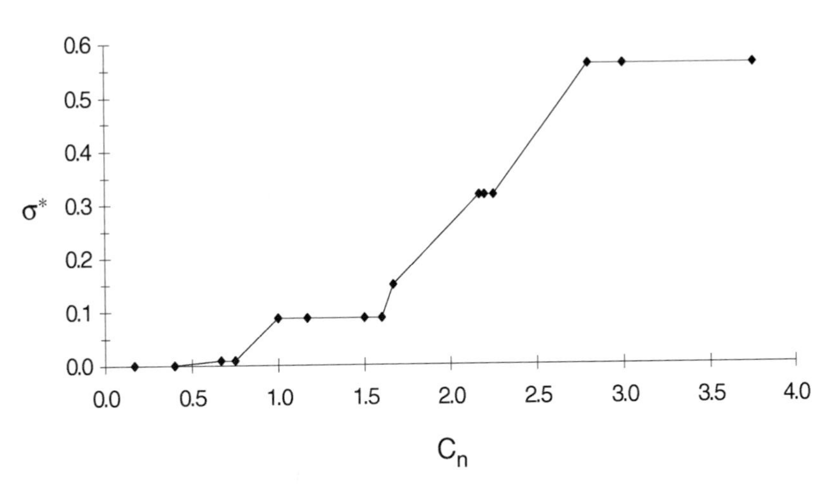

Figure 7 The trend of σ^* versus C_n for the test case 1.

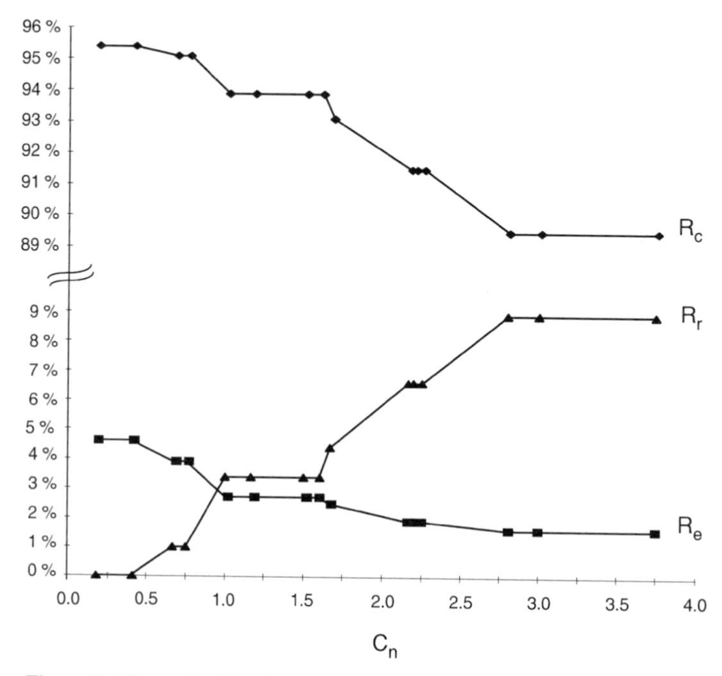

Figure 8 The trend of R_c, R_e, and R_r versus C_n when the reject option is used.

though this implies that some correctly classified samples could be rejected. Consequently, the value of the reject threshold rises with an overall decrease in both misclassified and correctly classified samples. Specifically, we observe a decrement of the recognition rate from 95.4%, for $C_n < 0.5$, to 89.5%, for $C_n = 3.75$, while the misclassification rate decreases from 4.6% to 1.6%, and the reject rate increases from 0.0% to 8.9%. The advantage attainable by exploiting the reject option can be made more evident by considering the relative variation of classification and misclassification rates, with respect to the 0-reject case, as a function of C_n (Fig. 9). It can be seen that, for high values of C_n, about 65% of the samples previously misclassified are now rejected, while the corresponding amount of correctly classified samples which are rejected is less than 6%.

As already said, a global evaluation of the advantage achieved when using the reject option can be obtained by considering the trend of P_n with respect to C_n. From the experimental results relative to this case (Fig. 10), it is evident that for high values of C_n, P_n reaches a maximum of more than 30%, demonstrating the convenience of using the reject option.

In conclusion, it is important to recall that the overall improvement of the classifier effectiveness is closely linked to the shape of the distributions D_c and D_e,

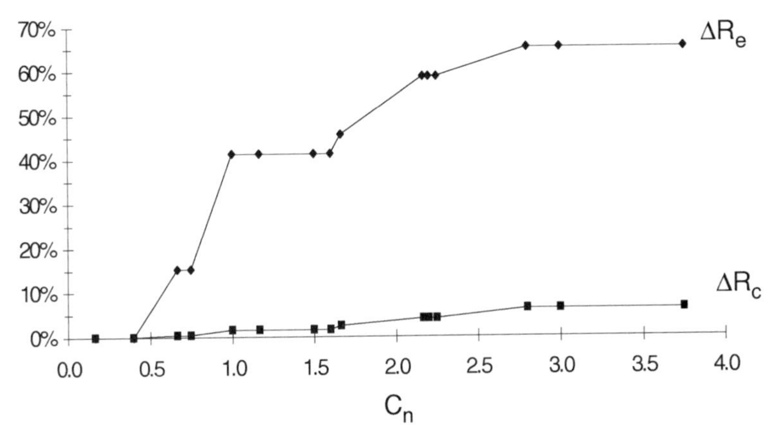

Figure 9 Percentage decrement of misclassification rate (ΔR_e) and recognition rate (ΔR_c) versus C_n obtained by using the reject option. The decrements are computed with respect to the corresponding rates at 0-reject.

which, in turn, depend not only on the data but also on the definition of ψ. In real situations, such as the one considered here, D_c and D_e are far from the ideal case since they overlap extensively (Figs. 11 and 12). This makes them not separable and thus the attainable improvement in classifier effectiveness, although valuable, is lower than in the ideal case.

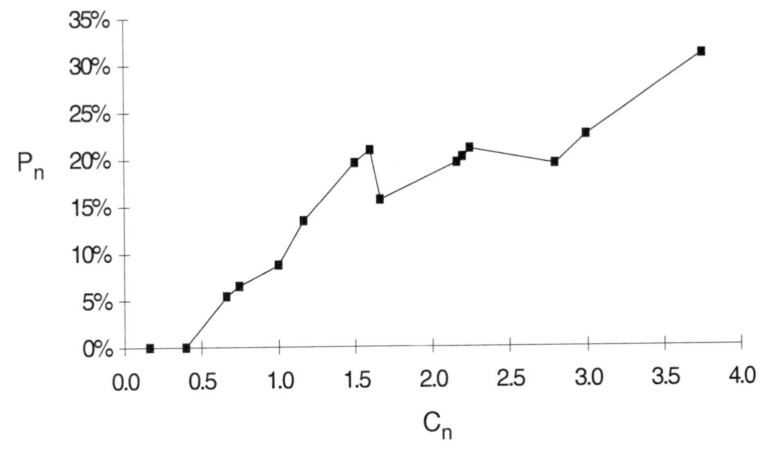

Figure 10 The trend of P_n versus C_n for the test case 1.

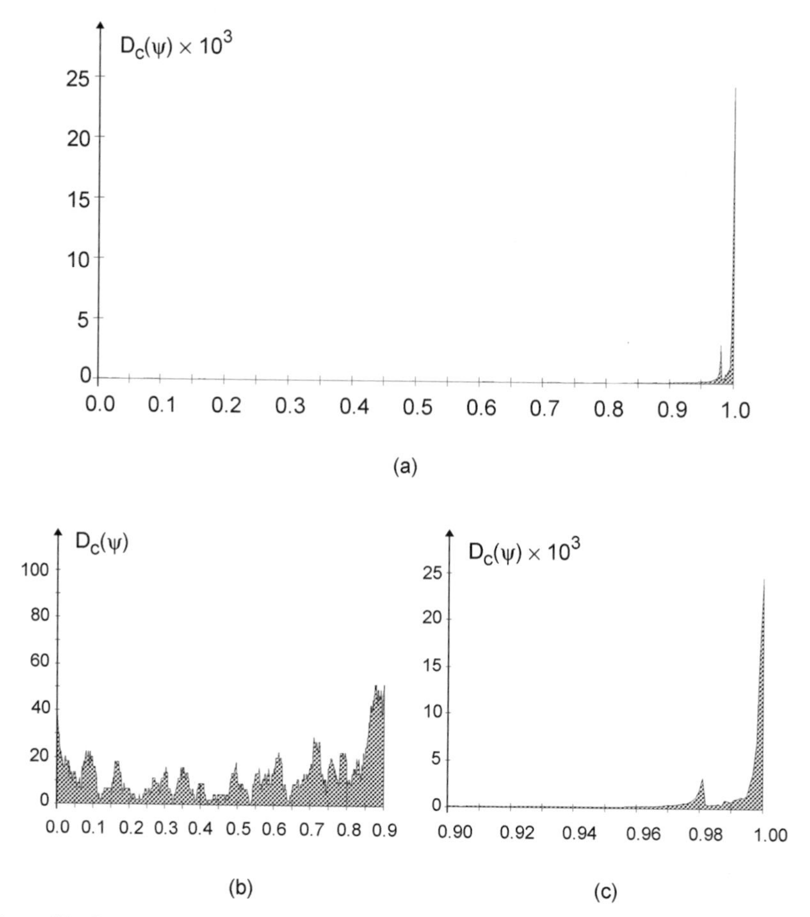

Figure 11 The distributions of correctly classified samples at 0-reject versus ψ. Since the values of D_C range over many orders of magnitude, the plots (b) and (c) show two parts of the plot (a) using different scales for the y axis.

B. Case 2: Fault Detection and Isolation

The second case deals with the detection and localization of faults in complex systems. This is a very crucial problem for the correct management of industrial plants, electrical networks and equipment, and many other systems.

Generally, a system S is monitored by means of a set of instruments (the measurement station) which measure the relevant parameters of S. The outputs of the instruments are fed to a fault detection and isolation unit (FDI) which, on the ba-

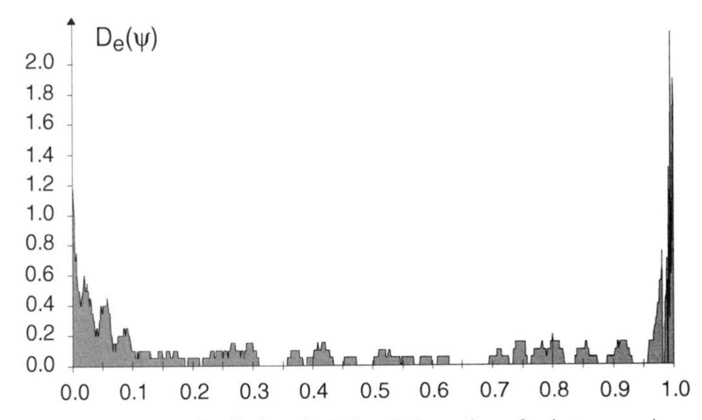

Figure 12 The distribution of misclassified samples at 0-reject versus ψ.

sis of the measurement values, should detect and localize the faults on the various components of S. Faults on the instruments should also be taken into account as atypical measurement values may indicate either a fault on S or a fault on the instruments. Consequently, to be effective, an FDI system should be able to classify an event as associated either to the presence of a fault on S or on some instrument, or to the absence of faults (Fig. 13).

To this end, many FDI techniques and architectures [66, 67] have been proposed in the literature; in particular, schemes based on "hardware redundancy" [68, 69] or analytical models of the expected measurement values [70, 71] have been widely investigated. Whatever approach is adopted, a primary issue to be considered is the evaluation of the classification reliability attainable with the FDI system. In other words, the matter in hand is to recognize whether a sample (de-

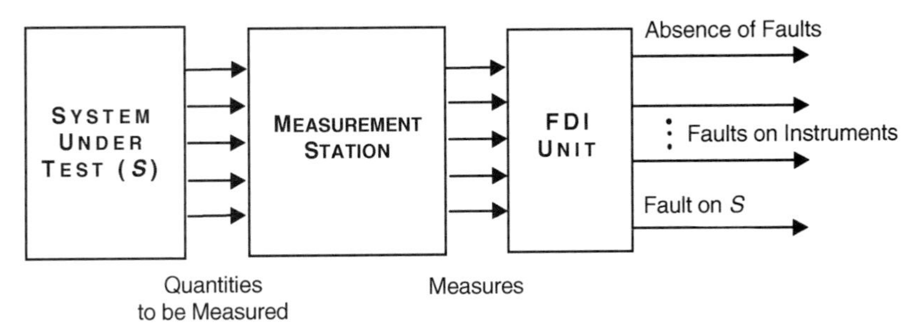

Figure 13 Block diagram of a generic FDI system.

scribed by a vector of measurements in this case) denotes the presence of a fault and, if one is indeed present, the nature of the fault. For this application, the sample description problem is relatively trivial, whereas classification is critical as serious consequences may occur if a fault remains undetected because of a recognition error. In these cases it is more advisable to reject the uncertain decision and ask for human intervention.

Neural network classifiers turn out to be particularly profitable [72] for this application because of their speed, which allows real-time fault localization, and because of the increased system effectiveness achievable by introducing the reject option. Moreover, the generalization capability of the neural classifier contributes, to some extent, to ensuring correct monitoring even when the system works outside its operating range, i.e., when the values of the parameters are different from the allowed ones, without causing yet a system malfunction. This property is referred to as the diagnostic robustness of the FDI system.

The neural-based FDI approach we considered has been applied to an automatic measurement station for induction motor testing (Fig. 14). The data acquisition board measured three phase voltages ($V1$, $V2$, $V3$), three line currents ($I1$, $I2$, $I3$), the motor angular speed (ω) and torque (T) and three phase powers ($P1$, $P2$, $P3$). A reference voltage (VR) was used to verify the correct operation of the data acquisition board. The mean values of the 12 instrument output signals, computed over 30 measurements, and the corresponding standard deviations have been assumed as input data for the neural network.

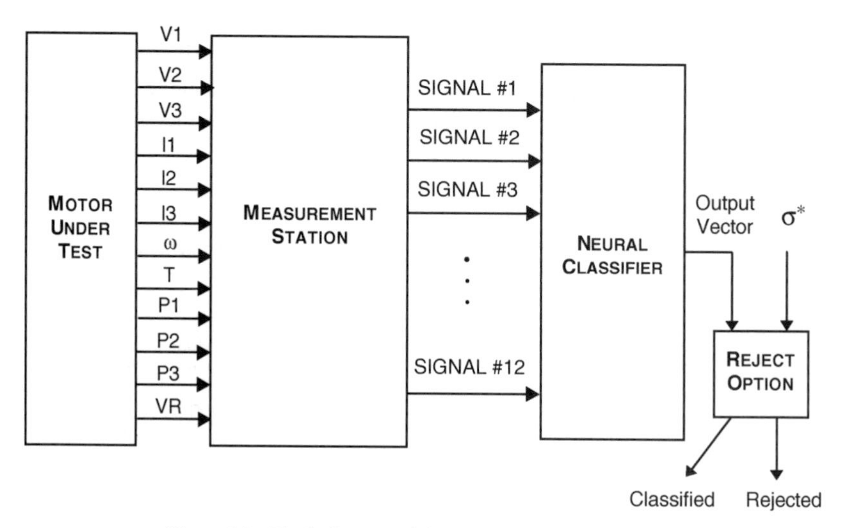

Figure 14 Block diagram of the realized neural FDI system.

As for the classifier architecture, two implementations of the MLP have been tested: the first with 12 input neurons, corresponding to the signal mean values, the second with 24 input neurons, 12 for the mean values and 12 for the standard deviations. After a preliminary test, the number of hidden layer neurons was fixed to 30 for both classifiers.

Besides one "unfaulty" class and one class for motor faults, 28 classes for instrument faults have been considered: 12 classes for short circuits on the data acquisition input channels, 12 classes for interruptions on the system wiring, and 4 classes for faults relative to the devices used to reduce the input currents to the transducers needed to measure the phase powers.

The training set was made up of 240 samples obtained from tests carried out by varying the motor current up to 110% of its maximum nominal value.

Both the set used for computing the reject threshold and the test set were made up of 72 samples, corresponding to 10 unfaulty conditions, 56 instrument faults (two cases for each considered fault) and 6 motor faults; these samples were obtained from tests carried out in operating conditions different from those of the training set. In order to evaluate the diagnostic robustness of the FDI system, a further test was conducted with 32 faults occurring outside the system operating range. Both the classifiers were trained using the standard BP algorithm, for 6000 learning cycles, with a learning rate of 0.5. The values of C_c, C_r, and C_e were chosen equal to 1, 2, and 6, respectively; therefore the reject threshold σ^* turned out to be equal to 0.29 for the 12-input classifier and to 0.44 for the 24-input classifier.

Table III shows the results obtained with the two networks on both training and test sets. The 24-input classifier performs well on the training set but gets significantly worse on the test set; this may be due to the excessive variability of the

Table III

Diagnostic Performance of the Neural FDI System[a]

	0-reject classifier		Classifier with the reject option			
	R_c^0	R_e^0	R_c	R_e	R_r	P_n
12 Input, TR	95.49	4.51	95.08	1.64	3.28	56.82
12 Input, TS	94.44	5.56	94.44	1.39	4.17	75.00
24 Input, TR	97.95	2.05	97.95	0.00	2.05	100.00
24 Input, TS	77.78	22.22	75.00	11.11	13.89	40.62
12 Input, TS1	93.75	6.25	90.63	3.12	6.25	12.64

[a] TR and TS mark results obtained on the training set and on the test set, respectively. TS1 refers to a case in which the motor current is external to the system operating range.

standard deviation values, showing that sample description using these features is not adequate. On the contrary, the performance of the 12-input classifier is quite good. In both cases it is evident that the introduction of the reject option makes it possible to reject a significant percentage of misclassified samples with a small reduction in the recognition rate.

However, the case of practical interest regards the 12-input classifier which achieves an effectiveness equal to 75% of the ideal one on the training set. The diagnostic robustness of this latter configuration was then evaluated. The last row of Table III shows the results obtained on a set of samples for a motor current equal to 120% of its nominal value and thus outside the system operating range: the recognition rate, although worse than that obtained within the nominal operating range, is still more than acceptable. Again, there is a significant decrease (about 50%) in the misclassification rate after the introduction of the reject option.

VIII. SUMMARY

In this chapter, the problem of assessing classification reliability, with special reference to the case of neural classifiers, has been addressed. After a review of the specific problem and the related topics, a method for using the information about classification reliability in order to find the best trade-off between reject rate and error rate has been illustrated. The method takes the following considerations as a starting point. When the reliability of a classification is low, the question is: does one accept the decision of the classifier running the risk of an error, or reject it and consider the sample at hand as not recognizable by the given classification system? In the former case, the risk of the decision being wrong increases as the classification reliability decreases. In the latter case, the sample has to be examined again with alternative techniques, generally by man. In both cases the choice implies a cost that has to be paid.

In practice, the definition of a parameter measuring classification reliability reflects the characteristics of the considered classifier; criteria for evaluating the classification reliability for a wide set of neural network classifiers have been proposed. They allow one to detect low reliable classifications in the case that the considered sample is either significantly different from those present in the training set, or similar to training set samples belonging to different classes. However, reliability alone is not sufficient to take a decision about the advantage of rejecting a sample instead of running the risk to misclassify it. This advantage can only be evaluated by taking into account the requirements of the specific application domain. In fact, for some applications the cost of a misclassification may be so high that a high reject rate becomes acceptable provided that the misclassification rate is kept as low as possible, while for other applications it may be desirable to assign every sample to a class even at the risk of a high misclassification rate.

Under hypotheses generally satisfied for a wide range of applications, these requirements can be expressed by attributing a cost to each misclassification, reject, and correct classification.

The method which is proposed computes the reject threshold value in such a way as to maximize the value of a function, which we have called "classifier effectiveness," taking into account the occurrence density distributions of correctly classified and misclassified samples, computed over a representative data set, as a function of classification reliability, together with the requirements of the application domain considered. The method does not require any *a priori* knowledge about the probability density distributions of the samples to be classified. Under these assumptions, the optimal reject threshold is the one for which the classifier effectiveness reaches its maximum value.

Assuming that in an ideal case it is possible to reject all and only those samples which, if not rejected, would be misclassified, it seemed convenient to compare the improvement P of the classifier effectiveness achieved when using the reject option with that achievable in the ideal case P_{id}.

The results of testing the method in two real applications, recognition of handwritten characters and fault detection and isolation in electrical systems, showed that the ratio P/P_{id} can be maintained relatively high even in recognition problems characterized by high variability among the samples of a same class and by partial overlap between the regions pertaining to different classes, so demonstrating the effectiveness of the approach.

REFERENCES

[1] R. O. Duda and P. E. Hart. *Pattern Classification and Scene Analysis*. Wiley, New York, 1973.

[2] K. Fukunaga. *Introduction to Statistical Pattern Recognition*, 2nd ed. Academic, New York, 1990.

[3] K. S. Fu. *Syntactic Methods in Pattern Recognition*. Academic, New York, 1974.

[4] T. Pavlidis. *Structural Pattern Recognition*. Springer-Verlag, Berlin, 1977.

[5] S. Haykin. *Neural Networks: A Comprehensive Foundation*. Macmillan College Publishing Co., New York, 1994.

[6] J. A. Anderson and E. Rosenfeld, Eds. *Neurocomputing: Foundations of Research*. MIT Press, Cambridge, MA, 1988.

[7] R. A. Wilkinson, J. Geist, S. Janet, P. Grother, C. Bruges, R. Creecy, B. Hammond, J. Hull, N. Larsen, T. Vogl, and C. Wilson. The first census optical recognition system conference. Technical Report NISTIR 4912, National Institute of Standards and Technology, Gaithersburg, 1992.

[8] M. D. Garris and R. A Wilkinson. *NIST Special Database 3. Handwritten Segmented Characters*. National Institute of Standards and Technology, Gaithersburg.

[9] T. M. Cover and P. E. Hart. *IEEE Trans. Inform. Theory* 13:21–27, 1967.

[10] L. Xu, A. Kryzak, and C. Y. Suen. *IEEE Trans. Systems Man Cybernet.* 22:418–435, 1992.

[11] R. Battiti and A. M. Colla. *Neural Networks* 7:691–707, 1994.

[12] A. Rosenfeld and A. C. Kak. *Digital Picture Processing*, Vol. II. Academic, Orlando, FL, 1982.

[13] C. K. Chow. *IEEE Trans. Inform. Theory* 16:41–46, 1970.

[14] E. Parzen. *Ann. Math. Statist.* 36:1065–1076, 1962.
[15] M. E. Hellman. *IEEE Trans. Systems Sci. Cybernet.* 6:179–185, 1970.
[16] M. R. Anderberg. *Cluster Analysis for Applications.* Academic, New York, 1973.
[17] J. R. Quinlan. *Mach. Learn.* 1:81–106, 1986.
[18] L. G. Shapiro and R. M. Haralick. *IEEE Trans. Pattern Anal. Mach. Intell.* 3:505–519, 1981.
[19] A. K. Jain, J. Mao, and K. M. Mohiuddin. *Computer* 29:31–44, 1996.
[20] Y. Hirose, K. Yamashita, and Y. Hijiya. *Neural Networks* 4:61–66, 1991.
[21] R. Reed. *IEEE Trans. Neural Networks* 4:740–747, 1993.
[22] R. P. Lippman. *IEEE ASSP Mag.* 4:4–22, 1987.
[23] R. Hecht-Nielsen. *Neurocomputing*. Addison-Wesley, Reading, MA, 1990.
[24] A. Freeman and D. M. Skapura. *Neural Networks: Algorithms, Applications and Programming Techniques*. Addison-Wesley, Reading, MA, 1992.
[25] J. de Villiers and E. Barnard. *IEEE Trans. Neural Networks* 4:136–141, 1993.
[26] M. T. Musavi, K. H. Chan, D. H. Hummels, and K. Kalantri. *IEEE Trans. Pattern Anal. Mach. Intell.* 16:659–663, 1994.
[27] D. E. Rumelhart, G. E. Hinton, and R. J. Williams. In *Parallel Distributed Processing* (D. E. Rumelhart and J. L. McClelland, Eds.), pp. 318–362. MIT Press, Cambridge, MA, 1986.
[28] D. S. Broomhead and D. Lowe. *Complex Syst.* 2:321–355, 1988.
[29] T. Kohonen. *Proc. IEEE* 78:1464–1480, 1990.
[30] S. Grossberg. *Cognitive Sci.* 11:23–63, 1987.
[31] D. F. Specht. *Neural Networks* 3:109–118, 1990.
[32] J. J. Hopfield. *Proc. Nat. Acad. Sci. USA* 79:2554–2558, 1982.
[33] T. Ash. Dynamic node creation in backpropagation networks. ICS Report 8901, Cognitive Science Dept., University of California, San Diego, 1989.
[34] R. P. Brent. *IEEE Trans. Neural Networks* 2:346–354, 1991.
[35] A. G. Parlos, B. Fernandez, A. F. Atiya, J. Muthusami, and W. K. Tsai. *IEEE Trans. Neural Networks* 5:493–497, 1994.
[36] S. Ergezinger and E. Thomsen. *IEEE Trans. Neural Networks* 6:31–42, 1995.
[37] T. P. Vogl, J. K. Mangis, A. K. Rigler, W. T. Zink, and D. L. Alkon. *Biolog. Cybernet.* 59:257–263, 1988.
[38] G. E. Hinton. Connectionist learning procedures. Technical Report CMU-CS-87-115, Computer Science Dept., Carnegie-Mellon University, Pittsburgh, 1987.
[39] D. DeSieno. *Proc. 2nd Annual IEEE Int. Conf. Neural Networks* 1:1117–1124, 1988.
[40] S. C. Ahalt, A. K. Krishnamurthy, P. Chen, and D. E. Melton. *Neural Networks* 3:277–290, 1990.
[41] L. Xu, A. Krzyzak, and E. Oja. *IEEE Trans. Neural Networks* 4:636–649, 1993.
[42] H. Robbins and S. Monro. *Ann. Math. Statist.* 22:400–407, 1951.
[43] C. De Stefano, C. Sansone, and M. Vento. In *Proceedings IEEE International Conference on Systems, Man and Cybernetics*, San Antonio, TX, pp. 759–764, 1994.
[44] P. Burrascano. *IEEE Trans. Neural Networks* 2:458–461, 1991.
[45] D. F. Specht. *IEEE Trans. Neural Networks* 1:111–121, 1990.
[46] G. Shafer. *A Mathematical Theory of Evidence.* Princeton University Press, New Jersey, 1976.
[47] L. H. Zadeh. *Inform. Contr.* 8:338–353, 1965.
[48] C. Y. Suen, C. Nadal, R. Legault, T. A. Mai, and L. Lam. *Proc. IEEE* 80:1162–1180, 1992.
[49] C. K. Chow. *IRE Trans. Electron. Computers* 6:247–254, 1957.
[50] G. C. Vasconcelos, M. C. Fairhust, and D. L. Bisset. *Pattern Recog. Lett.* 16:207–212, 1995.
[51] K. Kim and E. B. Bartlett. *Neural Comput.* 7:799–808, 1995.
[52] C. Tumuluri and P. K. Varsheny. *IEEE Trans. Neural Networks* 6:880–892, 1995.
[53] K. Archer and S. Wang. *IEEE Trans. Systems Man Cybernet.* 21:735–742, 1991.
[54] I. Bloch. *IEEE Trans. Systems Man Cybernet.—Part A: Systems and Humans* 26:52–67, 1996.
[55] C. K. Chow. In *Proceedings of the 3rd Annual Symposium on Document Analysis and Information Retrieval*, Las Vegas, pp. 1–8, 1994.

[56] K. Fukunaga and L. K. Kessel. *IEEE Trans. Inform. Theory* 18:814–817, 1972.
[57] L. P. Cordella, C. De Stefano, F. Tortorella, and M. Vento. *IEEE Trans. Neural Networks* 6:1140–1147, 1995.
[58] G. Nagy. In *Handbook of Statistics II* (L. Kanal and P. R. Krisnaiah, Eds.), pp. 621–649. North-Holland, Amsterdam, 1982.
[59] V. K. Govindan and A. P. Shivaprasad. *Pattern Recog.* 23:671–683, 1990.
[60] S. Mori, C. Y. Suen, and K. Yamamoto. *Proc. IEEE* 80:1029–1058, 1992.
[61] ETL-1 Character Data Base, collected by the Technical Committee for OCR at the Japan Electronic Industry Development Association and distributed by the Electrotechnical Laboratory.
[62] A. Chianese, L. P. Cordella, M. De Santo, A. Marcelli, and M. Vento. In *Recent Issues in Pattern Analysis and Recognition* (V. Cantoni, R. Oreutzburg, S. Levialdi, and G. Wolf, Eds.), *Lecture Notes in Computer Science*, Vol. 399, pp. 289–302. Springer-Verlag, New York, 1989.
[63] G. Boccignone, A. Chianese, L. P. Cordella, and A. Marcelli. In *Progress in Image Analysis and Processing* (V. Cantoni, L. P. Cordella, S. Levialdi, and G. Sanniti di Baja, Eds.), pp. 275–282. World Scientific, Singapore, 1990.
[64] A. Chianese, L. P. Cordella, M. De Santo, and M. Vento. In *Proceedings of the 6th Scandinavian Conference on Image Analysis*, Oulu, pp. 416–423, 1989.
[65] P. Foggia, C. Sansone, F. Tortorella, and M. Vento. Character recognition by geometrical moments on structural decompositions. Technical Report DIS-AV-96-12, Dipartimento di Informatica e Sistemistica, Università degli Studi di Napoli "Federico II," 1996.
[66] A. S. Willsky. *Automatica* 12:601–611, 1976.
[67] R. Patton, P. Frank, and R. Clark, Eds. *Fault Diagnosis in Dynamic Systems—Theory and Application.* Prentice-Hall International, Englewood Cliffs, NJ, 1989.
[68] R. N. Clark, D. C. Fosth, and V. M. Walton. *IEEE Trans. Aerosp. Electron. Syst.* 11:465–473, 1975.
[69] R. Isermann. In *Conference Record of IMEKO TC-10 Symposium*, Dresden, pp. 14–45, 1993.
[70] E. Y. Chow and A. S. Willsky. *IEEE Trans. Automat. Contr.* 29:603–614, 1984.
[71] P. M. Frank. *Automatica* 26:459–474, 1990.
[72] A. Bernieri, G. Betta, A. Pietrosanto, and C. Sansone. *IEEE Trans. Instrument. Measur.* 44:747–750, 1995.

Parallel Analog Image Processing: Solving Regularization Problems with Architecture Inspired by the Vertebrate Retinal Circuit

Tetsuya Yagi
Kyushu Institute of Technology
680-4 Kawazu, Iizuka-shi
Fukuoka Prefecture,
820 Japan

Haruo Kobayashi
Department of Electronic Engineering
Gumma University
1-5-1 Tenjin-cho
Kiryu, 376 Japan

Takashi Matsumoto
Department of Electrical, Electronics and Computer Engineering
Waseda University
Tokyo 169, Japan

I. INTRODUCTION

Almost all digital image processors employ the same architecture for the sensor interface and data processing. A camera reads out the sensed image in a raster scan-out of pixels, and the pixels are serially digitized and stored in a frame buffer. The digital processor then reads the buffer serially or as blocks to smooth the noise in the acquired image, enhance the edges, and perhaps normalize it in other ways for pattern matching and object recognition. There have been several attempts in recent years to implement these functions in the analog domain, to attain low-power dissipation and compact hardware, or simply to construct an electrical model of these functions as they are found in biological systems. For analog implementations, we must evaluate their performance in comparison with

Image Processing and Pattern Recognition
Copyright © 1998 by Academic Press. All rights of reproduction in any form reserved.

their digital counterparts and establish systematic techniques for their design and implementation. These considerations guided the work described here.

Single-chip analog image processor chips consist primarily of resistors and transistors in an array, and sometimes memory elements. A two-dimensional image is sensed by embedded photosensors at nodes in the array and converted to voltages or currents which drive the array. Computations are performed by physical laws underlying circuit behavior. These laws may be categorized as follows:

- Kirchhoff's laws together with Ohm's law imposed on resistor or transistor characteristics define the desired linear combination of signals. This contrasts with digital signal processors in which linear combinations are computed with multiply or add operations on binary words.
- The desired filtering is defined by a circuit (equilibrium) operating point. When subject to a stimulus, the circuit attains the operating point through dynamics defined by the parasitic capacitors. A clock is employed for memoryless filtering, and the circuit attains its equilibrium in real time.

This chapter describes how a class of parallel analog image processing algorithms is derived and how such algorithms can be implemented as parallel analog chips. The architectures for the chips described in this chapter are inspired by several physiological findings in lower vertebrates.

Section II explains several findings in lower vertebrate retinas in a manner which is accessible by engineers, while Section III describes algorithms, architectures, circuitry, and chip implementations. Section IV presents the circuit stability issues motivated by one of our vision chip implementations.

Until the early 1990s there had been only a small number of vision chips. Today, however, there are numerous vision chips (see the references). While some of the chip architecture are inspired by physiological facts, many others are based purely on engineering disciplines.

II. PHYSIOLOGICAL BACKGROUND

The retina is a part of the central nervous system in the vertebrate and plays important roles in early stages of visual information processing. The retina computes the image with a completely different algorithm or architecture from the one which most engineers are familiar with. Using this algorithm or architecture, the retina can perform real-time image processing with very low power dissipation. Inspired by such excellent performance and underlying network structure, vision chips have been designed using analog CMOS very large scale integrated circuit (VLSI) technology [1–8].

The retinas of lower verterbrates provide suggestive insights into designing the vision chips, since their visual functions are somewhat more sophisticated than those of mammal retinas. Therefore, the contents of this chapter are mainly inferred from physiological observations of the lower vertebrates. Most of these observations, however, are applicable to higher species including humans.

A. Structure of the Retina

The vertebrate retina is one of the few tissues of the nervous system in which electrical properties and structural organization of neurons are well correlated. Six principal cell types of neurons have been identified in the retina (see for review [9]). Figure 1 is a schematic illustration showing the gross structure of the vertebrate retina. Although the retina is transparent, the figure is colored for obvious reasons. Each of these principal cell types is classified into several subtypes, which are not shown in the figure to avoid complexities. In Fig. 1, the bottom side corresponds to the frontal surface of the retina from which the light comes through the optical apparatus of the eye. The light passes through the transparent retina to reach the photoreceptor array (gray), on which an image is projected. The light-sensitive pigment catches photons and a chemical reaction cascade tranduces it to a voltage response in the photoreceptor. The voltage signal is transmitted to the *second-order* neurons, which are the horizontal cell (blue) and the bipolar cell (red). The photoreceptors, horizontal cells, and bipolar cells interact with each other in the outer plexiform layer (OPL), which is a morphologically identifiable lamina seen in the cross section of the retina (indicated by arrow 1 in Fig. 1b). In this chapter, we refer to the neuronal circuit consisting of these three types of neurons as the outer retinal circuit.

The bipolar cell transmits the output of the outer retinal circuit to the amacrine cell (white) and the ganglion cell (yellow). The bipolar cells, amacrine cells, and ganglion cells interact in the inner plexiform layer (IPL), which is another morphologically identifiable lamina seen in the cross section of the retina (indicated by arrow 2 in Fig. 1b). The neuronal circuit consisting of these three types of neurons is referred to as the inner retinal circuit in this chapter. There are two distinguishable information channels in the inner retinal circuit. One channel is sensitive to static stimuli, and the other, to moving stimuli. The interplexiform cell (light green) is a unique neuron which provides a feedback pathway from the IPL to the OPL. The function of the interplexiform cell (IP cell) will be discussed in Section II.D. The outer retinal circuit and the inner retinal circuit are important portions to study how visual information is processed in the retinal circuit.

The retina consists of several layers of neuronal networks. The neurons belonging to the same types are arranged in two-dimensional arrays to aggregate in separate layers. The photoreceptors are arranged in a two-dimensional array. The photoreceptor mosaic is relevant to visual resolution under optimal viewing conditions. Other types of neurons are also arranged in two-dimensional arrays.

Figure 1 Schematic illustration of the vertebrate retina. (a) Overview of the retina. The six major types of neurons are distinguished by different colors. Gray, photoreceptors; red, bipolar cells; blue, horizontal cells; white, amacrine cells; orange, ganglion cells; light gray, interplexiform cells. (b) Cross section of the retina. The arrows indicate the outer plexiform layer (1) and the inner plexiform layer (2). Reprinted with permission from T. Matsumoto, H. Kobayashi, and T. Yagi, *J. IEICE* 76:783–791, 1993.

The layered structure is conserved in all vertebrate retinas, from fish to mammals. The visual information is processed in successive stages, from one layer to the next layer, with convergences and divergences of the wiring in the retina. The interaction between these layers includes feedback connections from proximal layers to distal layers as well as feedforward connections. It is noteworthy that these interconnections are made only between nearby neurons. This effective wiring is achieved because of the layered architecture. Since the voltage signal is transmitted only to neighboring neurons, the signal distortions are minimal and the calculation in the circuit is carried out with analog voltage signals (except in the ganglion cells). The ganglion cells give rise to action potentials to send the outputs to the brain.

B. Circuit Elements

1. Chemical Synapse

The interaction between the neurons takes place with two types of mechanisms, the chemical synapse and the electrical synapse. At the chemical synapse, the signal is transmitted by a chemical substance, the neurotransmitter. Figure 2a illustrates the signal transmission at the chemical synapse. When the voltage signal reaches the nerve terminal of the presynaptic neuron, which sends the signal to another neuron, the transmitter is secreted from the terminal. The transmitter reaches the receptor molecule of the membrane of the postsynaptic neuron, which receives the signal from the presynaptic neuron and opens channels of specific ions. As a consequence, the currents carried by the ions change the membrane

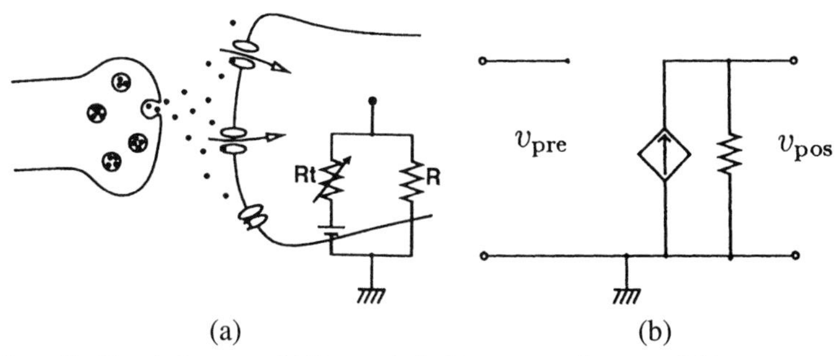

Figure 2 Chemical synapse. (a) The terminal of the presynaptic neuron (left) secretes neurotransmitters. The neurotransmitters open ionic channels in the membrane of the postsynaptic neuron (right). (b) The signal transmission at the chemical synapse is modeled by the voltage-controlled current source. The postsynaptic currents are generated by the neurotransmitters which are controlled by the membrane voltage of the presynaptic terminal, v_{pre}.

potential, which is the voltage inside the cell in reference to the outside space. In other words, the membrane current of the postsynaptic neuron is controlled by the voltage of the presynaptic neuron at the chemical synapse. When the signal is weak and the change of membrane conductance at the postsynaptic site is small compared with the input conducatance of the postsynaptic neuron, the transmission at the chemical synapse is modeled by the voltage-controlled current source as shown in Fig. 2b. Because the signal transmission takes place through several intermediate processes, there exists an inherent time delay in the transmission at the chemical synapse.

Several neurotransmitter molecules have been identified in the retina. The membrane potential of the postsynaptic neuron shifts in either positive (depolarization) or negative direction (hyperpolarization) depending on the neurotransmitter molecules. Glutamic acid is one of the major excitatory neurotransmitters which depolarize the membrane potential. γ-aminobutyric acid (GABA) is one of the major inhibitory neurotransmitters which hyperpolarize the membrane potential. A particular transmitter activates corresponding receptor molecules and generates currents carried by specific ions. Therefore, the temporal properties of signal transmission depend on transmitter molecules.

2. Electrical Synapse

The currents spread directly into neighboring neurons through the electrical synapse. Gap-junctions are a typical electrical synapse occasionally found in the retina (Fig. 3a). The gap-junctions provide a low-resistance pathway between neighboring neurons. The currents flowing through the gap-junctions are ordinarily bidirectional, and therefore the gap-junctions are modeled by resistors. The voltage signals pass to neighboring cells without time delay through the gap-junctions. The electrical synapse plays important roles in visual information processing in the retina, especially in the outer retina as shown later. As was explained in Section II.A, homogeneous types of neurons are arranged in two-dimensional arrays. When neighboring neurons of a homogeneous type are connected by electrical synapse over the lamina, the neurons constitute an electrically continual network. This network is called a neuronal syncytium, and the voltage signals conducting in such a syncytium are described by the analog network model as shown in Fig. 3b [10, 11]. In this figure, the arrangement of neurons is treated as one-dimensional. We treat the two-dimensional arrangement of neurons as a one-dimensional array in this section, for simplicity. The one-dimensional model can be directly applied to analyze the responses to illuminations which induce homogeneous voltage change in one direction. For example, a long slit of light induces the voltage change which is homogeneous along the long axis, and the current spreads only perpendicular to the slit [12, 13]. Although the applications of this one-dimensional model are limited to such stimuli, it is still useful to gain qualitative insights for the properties of voltage responses generated two-dimensionally.

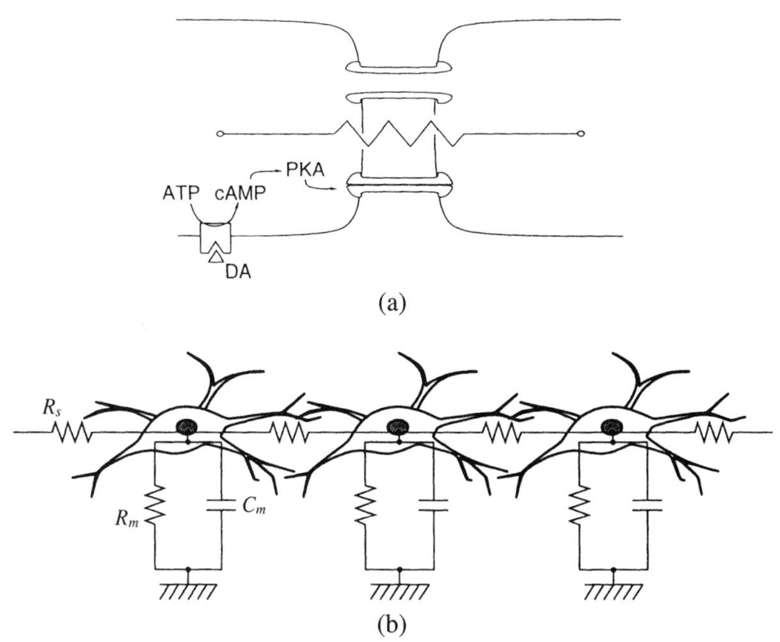

Figure 3 Electrical synapse. (a) The gap-junction provides a low-resistance pathway between cells and thus is described by a resistor. In some cases, the gap-junction is controlled by intracellular messenger machineries (see Section II.D). (b) Neuronal syncytium. Neighboring neurons are connected by gap-junctions and neurons constitute an electrically continual network.

Each neuron is represented by a parallel RC circuit in Fig. 3b. R_m and C_m correpond to the total membrane resistance and capacitance of a single neuron. The resistance of the electrical synapse connecting neighboring neurons is represented by R_s. The spatio-temporal properties of the voltage signal can be described by analytical solutions derived from the model. The solutions are obtained in the frequency domain, so the time course of voltage responses is calculated with the aid of the inverse fast fourier transform.

Let $v_k(\omega j)$, $j = \sqrt{-1}$, be the membrane voltage and $u_k(\omega j)$ be the synaptic current generated in the kth neuron of the syncytium. We assume that the distribution of the current is symmetrical, that is, $u_k(\omega j) = u_{-k}(\omega j)$. Applying Kirchhoff's current law in the frequency domain at each node, we obtain

$$\frac{(v_k(\omega j) - v_{k-1}(\omega j))}{R_s} + \frac{(v_k(\omega j) - v_{k+1}(\omega j))}{R_s} + \frac{v_k(\omega j)}{Z_m(\omega j)} = u_k(\omega j),$$

$$k = 1, \ldots, n.$$

Here,

$$Z_m(\omega j) = \frac{R_m}{1 + R_m C_m \omega j}$$

is the membrane impedance of each neuron. When the current $u_0(\omega j)$ is generated only at the neuron numbered 0, the voltage response of the kth neuron becomes [14, 15]

$$v_k(\omega j) = R_s \frac{u_0(\omega j)}{\sqrt{c^2(\omega j) - 4}} \rho(\omega j)^k, \tag{1}$$

where

$$\rho(\omega j) = \frac{-c(\omega j) - \sqrt{c(\omega j)^2 - 4}}{2}.$$

$c(\omega j)$ is a function of membrane impedance and coupling resistance expressed as

$$c(\omega j) = -\left(2 + \frac{R_s}{Z_m(\omega j)}\right).$$

Equation (1) is the line spreading function, expressed in the frequency domain, of the spatio-temporal properties of the voltage response in the single-layer syncytium. Specifically, the spatial distribution of an arbitrary frequency component of the voltage response is described by $\rho(\omega j)$. The modulus of $\rho(\omega j)$ gives the rate of response decay while the voltage conducts to the neighboring cell. The argument of $\rho(\omega j)$ gives a phase shift of the response during the conduction.

When the retina is illuminated with image which induces currents, $I_l(\omega j), l = 1, \ldots, n$, in the lth cell, the voltage response of the kth cell is expressed by the convolution of Eq. (1) with the light-induced current, i.e.,

$$V_k(\omega j) = \frac{R_s}{\sqrt{c^2(\omega j) - 4}} \sum_{l=1}^{\infty} \rho(\omega j)^{|k-l|} I_l(\omega j).$$

The response waveform can be obtained by transforming the above equation into the time domain with the aid of the inverse fast fourier transform (FFT) algorithm.

In some cases, the resistance of the electrical synapse is modulated through intracellular chemical reactions. An example of such modulation found in the horizontal cell is shown in Fig. 3a. A modulatory signal transmitted by dopamine activates the receptor on the horizontal cell membrane [16]. The activated receptor increases the concentration of cyclic AMP (cAMP). cAMP is one of the potent intracellular messenger substances influencing a wide variety of neuronal functions, and in this case cAMP closes the gap-junctions to increase the resistance connecting the horizontal cells [17]. This modulation is considered to be relevant to the adaptive control of the spatial filtering properties of the outer retinal circuit [2, 15]. The architecture of adaptive silicon retina discussed in Section III is inspired by this modulatory mechanism.

3. Properties of a Single Cell

The retinal network filters the input image with its dynamics. Since a single neuron is a basic component of the circuit, the filtering properties of the retinal circuit are highly relevant to the electrical properties of a single cell. The conductance of a single cell can be directly measured by electrophysiological experiments [18]. To measure the conductance, the single cell has to be isolated from the retinal tissue to curtail interactions with other neurons. Such isolation is obtained by treating the retina with an enzyme, e.g., papain [19]. Figure 4a shows the membrane current of a single bipolar cell induced by steps of voltage. The membrane voltage was *clamped* at −30 mV, where the membrane current was almost 0 pA, and then stepped to different levels. The physiological range of the membrane voltage of outer retinal neurons is between −60 mV and −10 mV. The current is as small as several picoamperes and almost linear (but in a biological sense) in the physiological voltage range (Fig. 4b). When the step of voltage deviates from the range, prominent nonlinear responses are seen. It is noteworthy that the conductance of the membrane increases significantly out of the physiological range. The increment of conductance prevents the membrane voltage from abnormal excursions by the shunting effect. The current responses to the voltage

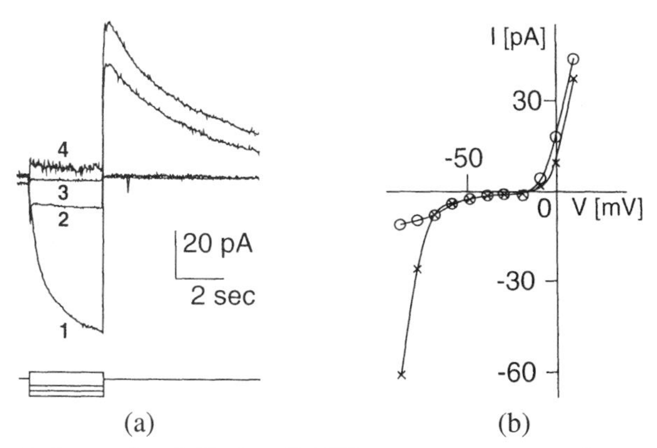

(a) (b)

Figure 4 Electrical properties of a bipolar cell. (a) The current responses to voltage steps of different levels are superimposed (upper trace). The lower trace illustrates the voltage steps. (b) The current-to-voltage relation replotted from (a). The amplitudes of current responses are measured at 0.5 sec (x) and 3 sec (o) after the onset of voltage steps. (Hayashida and Yagi, unpublished data.)

pulses are usually less than 50 pA in the physiological range in the outer retinal neurons. These observations indicate that the retinal nuerons compute the visual information with extremely small voltages and currents.

C. Outer Retinal Circuit

Most neurons have a membrane potential of about −70 mV at rest and respond to stimuli with action potentials. In contrast, the outer retinal neurons have a membrane potential of about −30 mV in the dark, indicating that they are in an excited state when the stimuli are absent. These neurons respond to light with slow voltage changes which are referred to as the graded voltage response. The slow time course of the response is mainly due to the chemical reaction cascade of the phototransduction process in photoreceptors.

1. Photoreceptors

There are two types of photoreceptors, the rod and the cone. The rod has a high sensitivity to light and operates at night. The cone is much less sensitive to light and operates in the daytime. The cone is classified into three subtypes corresponding to different spectral sensitivities to light wavelenghts. These subtypes are the red cone, green cone, and blue cone. Figure 5a shows a set of light-induced responses of a cone obtained with intracellular recordings [20]. In this figure, responses to different intensities of flash are superimposed. The flash duration is 10 msec. The membrane voltage of the cone is about −30 mV in the dark. The flashes of light generate hyperpolarizing voltage responses. The amplitude of response increases as the illumination becomes brighter. The response amplitude of the cone reaches its peak 50 to 100 msec after the onset of the flash, depending on the light-adaptation. The chemical reaction of the phototransduction process is much slower than operation of CMOS transistors. However, the cascade of chemical reactions can achieve an extremely high amplification of the signal.

The graded potential also has an advantage in integrating and averaging the photon signal over time. The time course of rods is much slower than that of cones. It takes several hundreds of milliseconds for the rod response to reach its peak amplitude. The high sensitivity and the slow time course of rod response are suitable for detecting a small number of photons.

Figure 5b shows the maximum amplitudes of the responses to different intensities of light plotted as a function of log light intensity (intensity response plot) [21]. The left curve is the intensity response plot for rods, and the right one is that for cones. In each case, the response amplitude increases with a shallow gradient when the intensity of light is low. The gradient of the response increase becomes the highest near half of the saturation amplitude where the resolution of the light

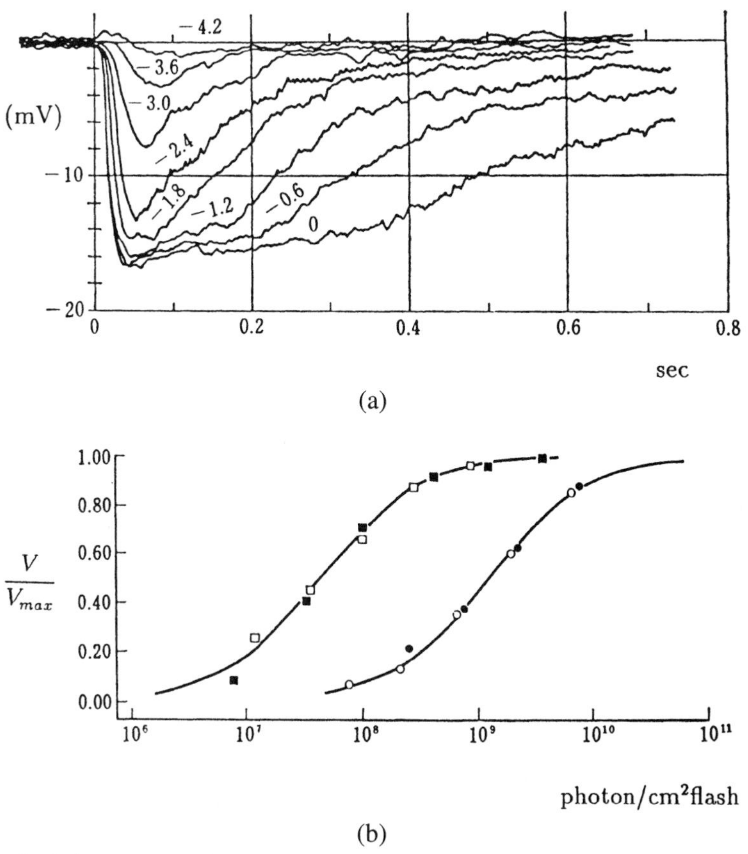

Figure 5 Voltage response of photoreceptors. (a) Responses of the cone to flash of different intensities. Responses are superimposed. From Baylor and Fuortes (1979), reprinted with permission. (b) The intensity response plot of cone (right) and rod (left) responses. Reprinted with permission from G. L. Fain and J. E. Dowling, *Science* 180:1178–1181, 1973 (Copyright 1973 American Association for the Advancement of Science).

intensity is highest. Around this intensity region, the amplitude of the voltage response is proportional to the log light intensity. The response amplitude reaches its saturation voltage with a shallow gradient. As shown in the figure, each of the photoreceptors detects the intensity difference in 3 log units. Therefore, the rod and cone together cover the light intensity of 4 to 5 log units. The relation between the maximum amplitude (V) and light intensity (I) is described by the Michaelis-Menten equation [20]:

$$V = V_{\max}\left(\frac{I}{I+\sigma}\right).$$

Here V_{max} is the amplitude at saturation. σ is the intensity which gives a response of $V_{max}/2$ and relates to the degree of light sensitivity.

Neighboring photoreceptors are coupled electrically by gap-junctions to constitute a syncytium [23, 24], even though the coupling is not as tight as the horizontal cell (see Section II.C.2). The significance of electrical coupling is thought to be the reduction of noise occurring in the retinal neuronal circuit. When cells are electrically coupled, the current generated in a single cell spreads into neighboring cells. Since the intrinsic noise in each cell is not correlated, the signal-to-noise ratio can be improved when the image has an appropriate size [10, 25]. It is also likely that the electrical coupling masks random variations of electrical properties of each cell. The electrical coupling, however, blurs the image. Thus, the coupling strength between neighboring cells is a critical parameter to be determined by the trade-off between these conflicting factors. The analog CMOS VLSI encounters a similar problem, i.e., random variations of transistor offsets.

2. Horizontal Cell

The horizontal cells also give rise to graded potential responses. It is well known that the horizontal cells exhibit large receptive fields which sometimes cover almost the entire retina [26, 27]. This is because neighboring horizontal cells are tightly coupled electrically by gap-junctions. The electrical coupling between neighboring cells is found in the photoreceptors as well as bipolar cells [26], but the coupling is not as intensive as in the horizontal cells.

Figure 6 shows a piece of evidence demonstrating the electrical coupling between horizontal cells. The schematic drawing in the figure explains the recording method. The response of a horizontal cell to a brief flash of light is recorded with a microelectrode (indicated by the thin triangle) connected to the operational amplifier (a). In this experiment, a narrow brief flash was first placed above the recorded horizontal cell (A), then displaced by 0.2-mm steps from the recorded cell (B and C). The experimentally recorded responses are shown in (b). The responses to the flash placed at A, B, and C are superimposed in the figure. The response was clearly observed for the flashes B and C, in which the distance from the recorded cell far exceeds the dimension of the horizontal cell. This indicates that the response to the flash in the recorded cell is propagated through the electrical synapse.

3. Bipolar Cell

The bipolar cell is the first neuron which exhibits a $\nabla^2 G$-like receptive field in the vertebrate visual system [29]. In the bipolar cell, the response to an illumination placed above the center region of the receptive field antagonizes the one placed above the surrounded region. Figure 7a presents the response of a bipolar

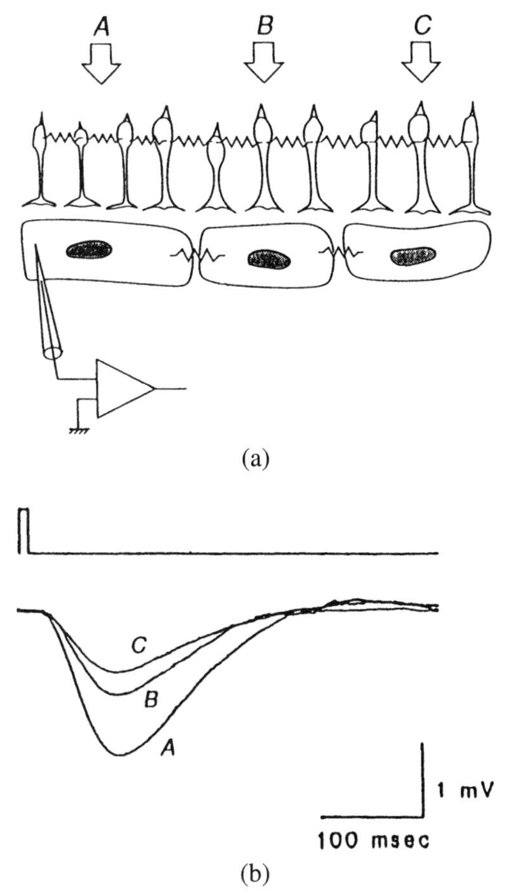

Figure 6 Spatial properties of horizontal cell response. (a) Schematic illustration of the experimental procedure. Intracellular voltages are recorded with a microelectrode (thin triangle) connected to an amplifier. The retina is illuminated by a narrow slit of flash as it is displaced at A, B, and C. (b) Voltage responses to the flash. The responses to the slit at A, B, and C are superimposed. Upper trace indicates the timing of flash.

cell showing the antagonistic receptive field (Sakakibara and Yagi, unpublished data). In this experiment, the responses of a bipolar cell to spots of light with different diameters were recorded. As the diameter of the spot of light centered above the recorded bipolar cell increased up to 0.3 mm, the response amplitude increased. However, the response amplitude decreased and finally the polarity of the response reversed when the diameter was further increased. This indicates that there exists a receptive field surround which antagonizes the receptive field center.

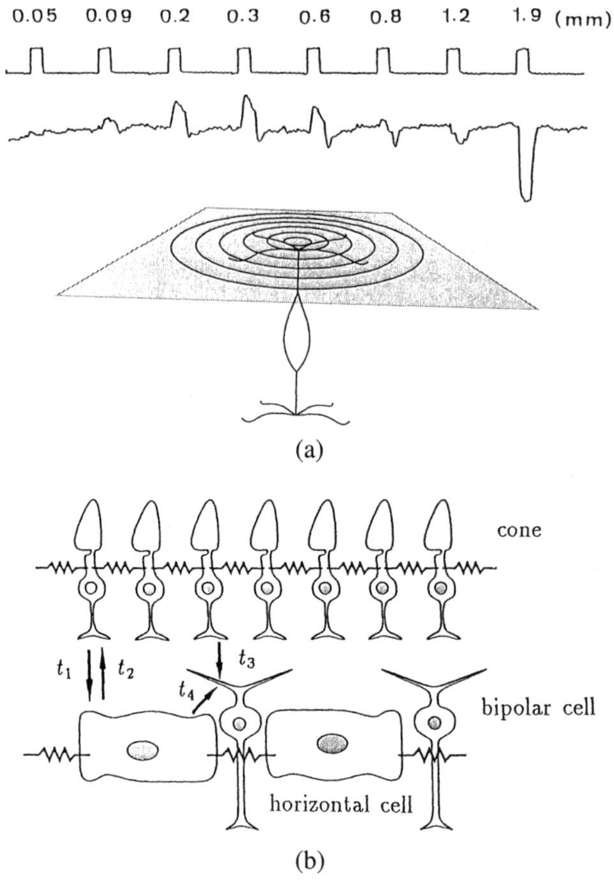

Figure 7 Receptive field of the bipolar cell. (a) A spot of light was centered on the recorded bipolar cell and voltage responses were obtained as its diameter was increased. The voltage responses are shown in lower trace. (b) The interaction among the photoreceptors, horizontal cells, and bipolar cells. The chemical synapses are indicated by arrows. The electrical synapses are indicated by resistors. (Sakakibara and Yagi, unpublished data.)

It is widely believed that the center response is mediated by the direct input from the photoreceptor and the antagonistic surround response is mediated by the horizontal cell. The neuronal circuit which generates the bipolar cell receptive field is illustrated in Fig. 7b. The bipolar cell receives inputs from the photoreceptor and the horizontal cell. These two inputs antagonize to produce the $\nabla^2 G$-like receptive field. The narrow center reflects the input from the photoreceptor and the wide surround reflects that of the horizontal cell.

D. Neuronal Adaptation

The $\nabla^2 G$-like receptive field has two effects on image processing. It smooths a noisy image and enhances the image contrast [28]. Since these two requirements, i.e., smoothing and contrast-enhancement, contradict each other, the receptive field size of the bipolar cell should be adjusted depending on the signal-to-noise ratio of the input image. Several pieces of evidence indicate that the outer retinal circuit adaptively modulates spatial filtering properties under a different visual environment. Previous physiological experiments revealed that the receptive field of the horizontal cell is controlled by the IP cell. The IP cell (light green in Fig. 1) is believed to be a centrifugal neuron innervating to the horizontal cell. Its cell body is located near the IPL (arrow 2 in Fig. 1b) with ascending axons to horizontal cells [16]. It was found that a physiologically active substance, dopamine, is released from the IP cell and reduces the receptive field size of the horizontal cell by decreasing the conductance of electrical synapses connecting neighboring cells [17]. Since the receptive field surround of the bipolar cell is mediated by the horizontal cell, it is natural that the receptive field of the bipolar cell is controlled by the IP cell. More recently, it was demonstrated that the effect of dopamine on the horizontal cell is mimicked by exposing the retina in the light-adapted state [31]. These observations indicate that the receptive field size of the horizontal cell is reduced in the light-adapted state, and consequently the receptive field of the bipolar cell is sharpened.

We hypothesize that the IP cell adaptively controls the receptive field size of the horizontal cell according to the signal-to-noise ratio. If we assume that the intrinsic noise is constant regardless of the adaptation level of the retina, the relative magnitude of noise to signal is small in the daytime since the light intensity of the signal image is high. In that situation, the bipolar cell receptive field is to be sharpened to gain spatial resolution. The spatial filtering properties of the outer retinal circuit are described in terms of this adaptive hypothesis in the following section.

E. Analog Network Model of Outer Retina

Based on the physiological bakcground described previously, the outer retinal circuit is expressed by the analog network model (Fig. 8). Each photoreceptor is represented by a membrane impedance $Z_{m1}(\omega j)$ and each horizontal cell by $Z_{m2}(\omega j)$. The resistance R_{s1} represents the electrical coupling between photoreceptors. The resistance R_{s2} represents the electrical coupling between horizontal cells. The resistance connecting neighboring horizontal cells, R_{s2}, is variable, since it is modulated by the IP cell. In this model, the coupling between bipolar cells is neglected for simplicity. We denote the light-induced current of cones with

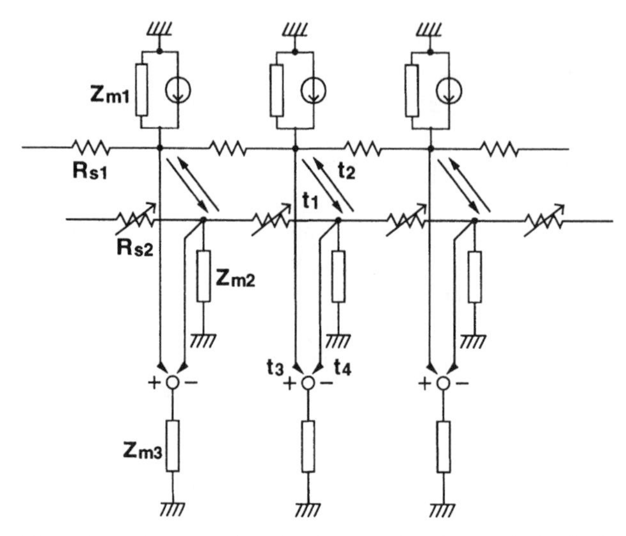

Figure 8 Analog network model of outer retina. See text for the detail.

u and the light-induced voltage responses of cone, horizontal cell, and bipolar cell with $\mathbf{v}^1$, $\mathbf{v}^2$, and $\mathbf{x}$, respectively, i.e.,

$$\mathbf{u} = \begin{pmatrix} u_0(\omega j) \\ u_1(\omega j) \\ \vdots \\ u_n(\omega j) \end{pmatrix}, \qquad \mathbf{v}^1 = \begin{pmatrix} v_1^1(\omega j) \\ v_2^1(\omega j) \\ \vdots \\ v_n^1(\omega j) \end{pmatrix},$$

$$\mathbf{v}^2 = \begin{pmatrix} v_0^2(\omega j) \\ v_1^2(\omega j) \\ \vdots \\ v_n^2(\omega j) \end{pmatrix}, \qquad \mathbf{x} = \begin{pmatrix} x_1(\omega j) \\ x_2(\omega j) \\ \vdots \\ x_n(\omega j) \end{pmatrix}.$$

Here, $u_k(\omega j)$ $(k = 1, \ldots, n)$ is the light-induced current of the kth cone expressed in the frequency domain. $v_k^1(\omega j)$ and $v_k^2(\omega j)$ are the voltage responses of the kth cone and horizontal cell. $x_k(\omega j)$ is the kth bipolar cell response. The strength of the synaptic input from the photoreceptor to the horizontal cell is expressed by $t_1(\omega j)$ (siemens). The strength of the feedback synaptic input from the horizontal cell to the photoreceptor is expressed by $t_2(\omega j)$ (siemens). The synaptic inputs to the bipolar cell from the photoreceptor and horizontal cell are expressed by

$t_3(\omega j)$ and $t_4(\omega j)$, respectively. These synaptic strengths are defined by the ratio of postsynaptic current induced by neurotransmitters to the presynaptic voltage. The synaptic strengths are also expressed in the frequency domain.

Applying Kirchhoff's current law at each node representing the cones and horizontal cells, we obtain a set of matrix equations for cone and horizontal cell responses:

$$\mathbf{C}_1\mathbf{v}^1 + t_2(\omega j)R_{s1}\mathbf{v}^2 = -R_{s1}\mathbf{u}(\omega j), \tag{2}$$

$$t_1(\omega j)R_{s2}\mathbf{v}^1 + \mathbf{C}_2\mathbf{v}^2 = \mathbf{0}. \tag{3}$$

Here $\mathbf{C}_1$ and $\mathbf{C}_2$ are

$$\mathbf{C}_1 = \begin{pmatrix} c_1 & 2 & 0 & \dots & 0 & 0 \\ 1 & c_1 & 1 & 0 & \dots & 0 \\ 0 & 1 & c_1 & 1 & \ddots & \vdots \\ \vdots & \ddots & \ddots & \ddots & \ddots & 0 \\ 0 & \dots & 0 & 1 & c_1 & 1 \\ 0 & 0 & \dots & 0 & 1 & c_1+1 \end{pmatrix},$$

$$\mathbf{C}_2 = \begin{pmatrix} c_2 & 2 & 0 & \dots & 0 & 0 \\ 1 & c_2 & 1 & 0 & \dots & 0 \\ 0 & 1 & c_2 & 1 & \ddots & \vdots \\ \vdots & \ddots & \ddots & \ddots & \ddots & 0 \\ 0 & \dots & 0 & 1 & c_2 & 1 \\ 0 & 0 & \dots & 0 & 1 & c_2+1 \end{pmatrix},$$

and

$$c_1 = -\left(2 + \frac{R_{s1}}{Z_{m1}(\omega j)}\right),$$

$$c_2 = -\left(2 + \frac{R_{s2}}{Z_{m2}(\omega j)}\right).$$

On combining Eqs. (2) and (3), we find matrix equations,

$$(\mathbf{C}_2\mathbf{C}_1 - t_1t_2R_{s1}R_{s2}\mathbf{E})\mathbf{v}^1 = -R_{s1}\mathbf{C}_2\mathbf{u}, \tag{4}$$

$$(\mathbf{C}_1\mathbf{C}_2 - t_1t_2R_{s1}R_{s2}\mathbf{E})\mathbf{v}^2 = t_1R_{s2}R_{s1}\mathbf{u}. \tag{5}$$

Here, $\mathbf{E}$ is the identity matrix.

When only the cone numbered 0 is stimulated and the current $u_0(\omega j)$ is induced, the solutions of Eqs. (4) and (5) become [15]

$$v_k^1(\omega j) = A_1(\omega j)\rho_1(\omega j)^k + A_2(\omega j)\rho_2(\omega j)^k, \quad (6)$$

$$v_k^2(\omega j) = B_1(\omega j)\rho_1(\omega j)^k + B_2(\omega j)\rho_2(\omega j)^k. \quad (7)$$

Here,

$$\rho_1(\omega j) = -\alpha(\omega j) - \sqrt{\alpha(\omega j)^2 - 1},$$

$$\rho_2(\omega j) = -\beta(\omega j) - \sqrt{\beta(\omega j)^2 - 1},$$

and

$$\alpha(\omega j) = \frac{c_1(\omega j) + c_2(\omega j)}{4} + \frac{1}{4}\sqrt{(c_1(\omega j) - c_2(\omega j))^2 + 4t_1(\omega j)t_2(\omega j)R_{s1}R_{s2}},$$

$$\beta(\omega j) = \frac{c_1(\omega j) + c_2(\omega j)}{4} - \frac{1}{4}\sqrt{(c_1(\omega j) - c_2(\omega j))^2 + 4t_1(\omega j)t_2(\omega j)R_{s1}R_{s2}}.$$

$A_1(\omega j)$, $A_2(\omega j)$, $B_1(\omega j)$, and $B_2(\omega j)$ are found from the boundary conditions. Solutions (6) and (7) indicate that the amplitude of an arbitrary frequency component of voltage response decays with two coefficients, $\rho_1(\omega j)$ and $\rho_2(\omega j)$. When the retina is illuminated with an arbitrary image and the current $u_k(\omega j)$ is induced, the voltage responses of the cone and horizontal cell are obtained by convolution of Eqs. (6) and (7) with $u_k(\omega j)$, respectively.

The voltage distribution of the bipolar cell response, $\mathbf{x}$, is expressed simply by the difference between the cone response and the horizontal cell response [30], i.e.,

$$\mathbf{x} = Z_{m3}t_3\mathbf{v}^1 + Z_{m3}t_4\mathbf{v}^2. \quad (8)$$

Here Z_{m3} is the membrane impedance of the bipolar cell.

If we focus on the spatial distribution of the bipolar cell response in the steady state (equilibrium point), the spatial filtering properties of the bipolar cell are found to be characterized in terms of the standard regularization theory, with which some early vision problems are solved as minimization of quadratic cost functions [33]. We demonstrate how the outer retinal circuit naturally solves regularization problems with the cost function derived from the model. Since we consider the voltage distribution of the RC circuit at equilibrium, the membrane can be replaced by pure resistors instead of impedances. Combining Eqs. (4), (5), and (8), we obtain the equation for the voltage distribution of the bipolar cell response:

$$(\mathbf{C}_1\mathbf{C}_2 - t_1t_2R_{s1}R_{s2}\mathbf{E})\mathbf{x} = -t_3R_{m3}R_{s1}\mathbf{C}_2\mathbf{u} + t_1t_4R_{m3}R_{s1}R_{s2}\mathbf{u}. \quad (9)$$

If we ignore the boundary effect, which is often feasible when the network is stable (see Section IV), and substitute $\mathbf{C}_1$ and $\mathbf{C}_2$ of Eq. (9) by $(\mathbf{L} - R_{s1}/R_{m1}\mathbf{E})$ and $(\mathbf{L} - R_{s2}/R_{m2}\mathbf{E})$, using

$$\mathbf{L} = \begin{pmatrix} -2 & 1 & 0 & \dots & 0 \\ 1 & -2 & 1 & 0 & \dots \\ 0 & 1 & -2 & 1 & \dots \\ \vdots & \ddots & \ddots & \ddots & 1 \\ 0 & \dots & 0 & 1 & -2 \end{pmatrix},$$

then we find that the response of the bipolar cell is described by an equation similar to the Euler equation ([2], see also Section III for details). Specifically,

$$\mathbf{x} - \mathbf{d} - \lambda_1 \mathbf{L}\mathbf{x} + \lambda_2 \mathbf{L}^2 \mathbf{x} = \mathbf{0}. \tag{10}$$

Here

$$\lambda_1 = \frac{R_{m1}/R_{s1} + R_{m2}/R_{s2}}{1 - t_1 t_2 R_{m1} R_{m2}}, \qquad \lambda_2 = \frac{R_{m1} R_{m2}/(R_{s1} R_{s2})}{1 - t_1 t_2 R_{m1} R_{m2}},$$

and

$$\mathbf{d} = \nu R_0 \mathbf{u} - R_0 \mathbf{L}\mathbf{u},$$

where

$$R_0 = \frac{R_{m1} R_{m2} R_{m3} t_3}{(1 - t_1 t_2 R_{m1} R_{m2}) R_{s2}}, \qquad \nu = R_{s2}\left(\frac{1}{R_{m2}} + \frac{t_1 t_4}{t_3}\right).$$

As was defined, $\mathbf{u}$ is the light-induced current of cones. R_0 has the dimension of resistance (ohm) and ν is a dimensionless constant. Therefore, $\nu R_0 \mathbf{u}$ designates the spatial voltage distribution proportional to the input image, provided that the light-induced current is proportional to the illumination intensity. Similarly, $R_0 \mathbf{L}\mathbf{u}$ designates the spatial voltage distribution which is proportional to the second-order difference of the input image. Note that the second-order difference operation enhances the contrast of image as well as noise. As will be shown in Section III, Eq. (10) gives the solution which minimizes the cost function,

$$J(\mathbf{x}) = (\mathbf{x} - \mathbf{d})^T (\mathbf{x} - \mathbf{d}) + \lambda_1 (\mathbf{D}\mathbf{x})^T (\mathbf{D}\mathbf{x}) + \lambda_2 (\mathbf{L}\mathbf{x})^T (\mathbf{L}\mathbf{x}). \tag{11}$$

In other words, the bipolar cell responses distribute themselves so as to minimize the function (11). The first term of the cost function (11) decreases as $\mathbf{x}$ becomes closer to $\mathbf{d}$, which is composed of the raw input image, $\nu R_0 \mathbf{u}$, and the contrast-enhanced image, $R_0 \mathbf{L}\mathbf{u}$. The second and the third terms of the cost function (11) decrease as $\mathbf{x}$ is smooth. Thus, the distribution of the bipolar cell response is determined by the trade-off between contrast-enhancement and smoothing operations. It is interesting to study how the resistance of gap-junctions between horizontal cells, R_{s2}, affects the spatial filtering properties of bipolar cells. The effect of R_{s2} is unambiguously predicted from the cost function (11). When R_{s2} decreases, R_0 increases and the contrast-enhanced image is emphasized. The weight on the raw image, νR_0, does not change since R_{s2} is canceled. This is an important feature since the response to the background illumination is not modulated even though R_{s2} changes. λ_1 and λ_2 also increase as R_{s2} decreases, and thus the filtered image becomes smoother. In the following, these effects are demonstrated by simulations.

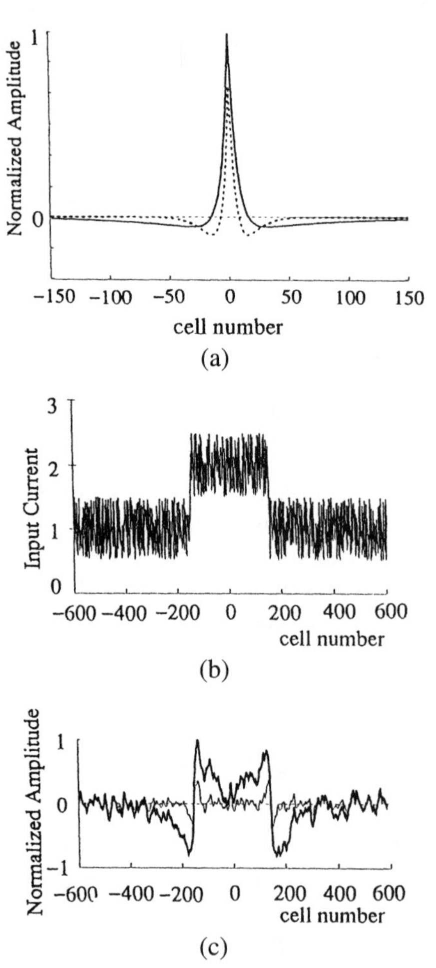

Figure 9 Simulation of neuronal adaptation. (a) The receptive field of the bipolar cell calculated with different values of R_{s2}. (b) Noisy input image. (c) The spatial distribution of bipolar cell response calculated with high R_{s2} (thin line) and with low R_{s2} (thick line).

The receptive field of the bipolar cell was calculated from the model with different values of R_{s2} (Fig. 9a). In this calculation, ν was taken to be zero for simplicity. Figure 9c illustrates the spatial distribution of the bipolar cell response to a noisy input image. The noisy input image is composed of a square object spreading from cell number -150 to 150 and random noise (Fig. 9b). As shown in Fig. 9c, the image filtered with the high R_{s2} is still noisy, and the contrast enhancement seen as the "Mach Band"-like phenomenon is hardly distinguished from the noise (thin line). The high R_{s2} enhanced the amplitude of noise as well as the contrast of the image. However, it is easier to specify the contrast-enhancement effect when R_{s2} is low (thick line). It is also evident that the gain of the bipolar cell response increases as R_{s2} decreases. Note that the response to background illumination does not change even though R_{s2} changes.

The vision chip with light-adaptive architecture inspired by these physiological and computational backgrounds is explained in the next section.

III. REGULARIZATION VISION CHIPS

A. INTRODUCTION

This section first explains how regularization problems which naturally arise in early vision problems can be solved in a completely parallel manner using layered analog resistive networks. The second part of this section presents complete details of the smoothing contrast-enhancement (SCE) vision chip which has a double-layer architecture inspired by the physiological results discussed in the previous section. The chip solves first-order and second-order regularization problems simultaneously and outputs their difference. Since the chip is equipped with a photosensor array and an analog processing array, the execution is extremely fast, typically several microseconds. Implementation of the chip requires no special technology; it uses a standard 2 μm CMOS process. The third part of this section described light-adaptive architecture and its CMOS circuitry. This adaptation mechnism enables automatic adjustement of the SCE filter scale in accordance with the intensity of input images. This is inspired by the horizontal cell adaptation mechanism explained in the previous section.

B. TIKHONOV REGULARIZATION

When a solution to an operator equation (not necessarily linear)

$$Av = d, \qquad v \in X, \quad d \in Y, \tag{12}$$

loses existence or uniqueness or continuity in d, Eq. (12) is called ill-posed. Ill-posedness typically arises when "data" d is noisy while the solution v sought should be reasonably smooth. It can also result from the nature of A. The Tikhonov regularization [34–36] converts Eq. (12) into a family of minimization problems:

$$G(v, d, \lambda) = \|Av - d\|^2 + \lambda\Omega(v), \tag{13}$$

where $\|\cdot\|$ denotes a norm (on Y), Ω: $X \to \mathcal{R}$ is continuous and strictly convex, $\lambda > 0$. If $Av^* = d^*$, then under reasonable conditions, Eq. (13) regularizes Eq. (12) in the sense that for any ϵ-neighborhood $N_\epsilon(v^*)$ of v^* (with respect to an appropriate topology), there is a δ-neighborhood $N_\delta(d^*)$ of d^* such that if $d \in N_\delta(d^*)$, and if $\lambda(\delta) > 0$ is appropriate, then there is a unique $v(d, \lambda(\delta)) \in N_\epsilon(v^*)$ which minimizes Eq. (13). It should be noted, however, than when d is noisy, choosing the best λ is another interesting as well as difficult problem because one needs to take into account the statistics of d [37–39], and its is outside the scope of this paper.

Now a typical "stabilizer" $\Omega(v)$ in Eq. (13) is of the form

$$\Omega(v) = \sum_{r=1}^{P} \int_D C_r(x) \left(\frac{d^r v(x)}{dx^r} \right)^2 dx, \tag{14}$$

where $C_r(x) \geq 0$ and $D = [a, b]$ is the domain of the problem. If Eq. (13) with Eq. (14) can be written as

$$G(v, d, \lambda) = \int_D F(v(x), v^{(1)}(x), \dots, v^{(P)}(x), x, d(x), \lambda)\, dx, \qquad v^{(r)} = \frac{d^r v}{dx^r}, \tag{15}$$

where F is "well-behaved," then the variational principle gives the Euler equation

$$\sum_{r=0}^{P} (-1)^r \frac{d^r}{dx^r} \frac{\partial}{\partial v^{(r)}} F(v(x), v^{(1)}(x), \dots, v^{(P)}(x), x, d(x), \lambda) = 0, \tag{16}$$

with natural boundary conditions

$$\sum_{r=0}^{q} (-1)^r \frac{d^r}{dx^r} \frac{\partial}{\partial v^{(P-q-r)}} F(v(x), v^{(1)}(x), \dots, v^{(P)}(x), x, d(x), \lambda) = 0,$$
$$\text{at } x = a, b \quad \text{for} \quad q = 0, 1, \dots, P. \tag{17}$$

It should be observed that because of the particular form of Eq. (14), the Euler equation Eq. (16) necessarily contains terms of the form

$$\left(\frac{d^{2r} v}{dx^{2r}} \right)(x), \qquad r = 1, \dots, P. \tag{18}$$

Namely, if the stabilizer Eq. (14) contains the rth-order derivative, one needs to implement the $2r$th-order derivative operation for solving the regularization problem.

For the sake of simplicity, the independent variable x has been one-dimensional. Two-dimensional problems will be discussed in the next subsection.

In the following, we will formulate the regularization problem as a minimization problem on a finite-dimensional space instead of approximating the Euler equation because

(a) in a chip implementation the space variable x takes finite discrete values,
(b) the formulation naturally leads to our layered architecture,
(c) discrete approximation of the Euler equation Eq. (16) together with the natural boundary condition Eq. (17) in a consistent manner is not straightforward. Boundary conditions are important since inadequate boundary conditions even lead to instability [40, 41]. Our discrete formulation given below naturally incorporates Eqs. (16) and (17),
(d) most of the vision chips fabricated/proposed so far, including the filter described in Section III.D, are on a hexagonal grid instead of a square grid (see Section III.C for reasons). A rigorous approximation result on a hexagonal grid will be rather involved.

Thus let $\mathbf{v} = (v_1, \ldots, v_n)^T \in \mathcal{R}^n$. Then the derivatives in Eq. (14) should be replaced by the differences, e.g.,

$$\left(\frac{dv}{dx}\right)(x) \to v_k - v_{k-1}, \qquad \left(\frac{d^2v}{dx^2}\right)(x) \to v_{k-1} + v_{k+1} - 2v_k. \tag{19}$$

These operations are conveniently expressed by

$$\left(\frac{dv}{dx}\right)(\cdot) \to \mathbf{Dv}, \qquad \left(\frac{d^2v}{dx^2}\right)(\cdot) \to \mathbf{Lv}, \tag{20}$$

where

$$\mathbf{D} = \begin{bmatrix} 1 & 0 & 0 & \ldots & . & 0 \\ -1 & 1 & 0 & \ldots & . & 0 \\ 0 & -1 & 1 & \ldots & . & 0 \\ . & . & . & \ldots & . & . \\ . & . & . & \ldots & . & . \\ 0 & 0 & 0 & \ldots & 1 & 0 \\ 0 & 0 & 0 & \ldots & -1 & 1 \end{bmatrix}, \qquad \mathbf{L} = \begin{bmatrix} -2 & 1 & 0 & \ldots & . & 0 \\ 1 & -2 & 1 & \ldots & . & 0 \\ 0 & 1 & -2 & \ldots & . & 0 \\ . & . & . & \ldots & . & . \\ . & . & . & \ldots & . & . \\ 0 & 0 & 0 & \ldots & -2 & 1 \\ 0 & 0 & 0 & \ldots & 1 & -2 \end{bmatrix}. \tag{21}$$

Note that although $\mathbf{D}$ is not symmetric, $\mathbf{D}^T\mathbf{D}$ is symmetric and

$$\mathbf{D}^T\mathbf{D} = -\mathbf{L}, \tag{22}$$

where T denotes transpose of a matrix. Therefore, the regularization problem for the finite-dimensional space case calls for minimizing

$$G(\mathbf{v}, \mathbf{d}, \lambda) = \|\mathbf{A}\mathbf{v} - \mathbf{d}\|^2 + \lambda \sum_{r=1}^{P} \begin{cases} \sum_k C_r(k)(\mathbf{L}^{r/2}\mathbf{v})_k^2, & r\text{: even,} \\ \sum_k C_r(k)(\mathbf{D}\mathbf{L}^{(r-1)/2}\mathbf{v})_k^2, & r\text{: odd,} \end{cases} \tag{23}$$

where $\mathbf{d} = (d_1, \ldots, d_n)^T \in \mathcal{R}^n$, $C_r(k) \geq 0$, and $(\mathbf{L}^{r/2}\mathbf{v})_k$ [respectively $(\mathbf{D}\mathbf{L}^{(r-1)/2}\mathbf{v})_k$] is the kth component of $\mathbf{L}^{r/2}\mathbf{v}$ [respectively $\mathbf{D}\mathbf{L}^{(r-1)/2}\mathbf{v}$]. Differentiating Eq. (23) with respect to $\mathbf{v}$ and setting it to zero, one has

$$\frac{1}{2}\frac{\partial G}{\partial \mathbf{v}} = \mathbf{A}^T(\mathbf{A}\mathbf{v} - \mathbf{d}) + \sum_{r=1}^{P}(-1)^r \lambda_r \mathbf{L}^r \mathbf{v} = \mathbf{0}, \tag{24}$$

where

$$\lambda_r := \lambda C_r$$

are called the *hyperparameters*.

Consider, for instance, $P = 2$, $\lambda_2 \neq 0$, $\lambda_1 = 0$, which amounts to

$$\mathbf{v} - \mathbf{d} + \lambda_2 \mathbf{L}^2 \mathbf{v} = \mathbf{0}. \tag{25}$$

Note that

$$\mathbf{L}^2 = \begin{bmatrix} -2 & 1 & 0 & 0 & . & . & . & 0 \\ 1 & -2 & 1 & 0 & . & . & . & 0 \\ 0 & 1 & -2 & 1 & . & . & . & 0 \\ . & . & . & . & . & . & . & . \\ . & . & . & . & . & . & . & . \\ . & . & . & . & . & . & . & . \\ 0 & 0 & 0 & . & . & 1 & -2 & 1 \\ 0 & 0 & 0 & . & . & 0 & 1 & -2 \end{bmatrix} \begin{bmatrix} -2 & 1 & 0 & 0 & . & . & . & 0 \\ 1 & -2 & 1 & 0 & . & . & . & 0 \\ 0 & 1 & -2 & 1 & . & . & . & 0 \\ . & . & . & . & . & . & . & . \\ . & . & . & . & . & . & . & . \\ . & . & . & . & . & . & . & . \\ 0 & 0 & 0 & . & . & 1 & -2 & 1 \\ 0 & 0 & 0 & . & . & 0 & 1 & -2 \end{bmatrix}$$

$$= \begin{bmatrix} * & -4 & 1 & 0 & . & . & . & 0 \\ -4 & 6 & -4 & 1 & . & . & . & 0 \\ 1 & -4 & 6 & -4 & . & . & . & 0 \\ . & . & . & . & . & . & . & . \\ . & . & . & . & . & . & . & . \\ . & . & . & . & . & . & . & . \\ 0 & 0 & 0 & . & . & -4 & 6 & -4 \\ 0 & 0 & 0 & . & . & 1 & -4 & * \end{bmatrix}, \tag{26}$$

where $* = 5$ due to the "boundary effect." One sees that the kth component of Eq. (25) in the "interior" reads

$$v_k - d_k + \lambda_2\left[6v_k - 4(v_{k-1} + v_{k+1}) + v_{k-2} + v_{k+2}\right] = 0. \tag{27}$$

A direct implementation of Eq. (27) is given by Fig. 10 where

$$g_0, g_1 > 0, \qquad g_2 < 0, \qquad g_1 = 4|g_2|, \tag{28}$$

because the Kirchhoff current law (KCL) gives

$$-(g_0 + 2g_1 + 2g_2)v_k + g_1(v_{k-1} + v_{k+1}) + g_2(v_{k-2} + v_{k+2}) + u_k = 0. \tag{29}$$

Therefore, $\lambda_2 = g_0/|g_2|$, $d_k = \lambda_2 u_k$. This is what has been done in [40–42]. For a general r, matrix $\mathbf{L}^r$ is of the form

$$\mathbf{L}^r = \begin{bmatrix} a_0 & a_1 & a_2 & . & a_r & 0 & 0 & . & . & 0 \\ a_1 & a_0 & a_1 & a_2 & . & a_r & 0 & . & . & 0 \\ a_2 & a_1 & a_0 & a_1 & . & . & . & . & . & 0 \\ . & a_2 & a_1 & a_0 & . & . & . & . & 0 & 0 \\ a_r & . & . & . & . & . & . & . & a_r & 0 \\ 0 & a_r & . & . & . & . & . & . & . & a_r \\ 0 & 0 & . & . & . & . & a_0 & a_1 & a_2 & . \\ . & . & . & . & . & . & a_1 & a_0 & a_1 & a_2 \\ 0 & . & . & 0 & a_r & . & a_2 & a_1 & a_0 & a_1 \\ 0 & . & . & 0 & 0 & a_r & . & a_2 & a_1 & a_0 \end{bmatrix}, \tag{30}$$

where the boundary effects are not explicitly shown in order to save the space. Equation (30) shows that direct implementation requires connections between every pair of the **kth nearest nodes** for **all** $k \leq r$ with possibly negative conductance. As will be shown in Section III.F, $r = 2$ is already very difficult to implement due to wiring complexity.

The architecture given below solves the Pth-order regularization prblem with only wiring between nearest nodes and without negative conductance. The following fact shows that the network given by Fig. 11 (in one-dimension) solves the Pth-order regularization problem for all P, $1 \leq P \leq N$, simultaneously, where $\mathbf{A} = \mathbf{1}$ and $C_r(k)$ is independent of k. Proof is found in [2].

Fact 1. Consider the network given by Fig. 11a (in one-dimension) where the symbol given in Fig. 11b stands for a voltage-controlled current source, and $g_{m_i}, g_{s_i} > 0, i = 1, \ldots, N$. Gain T_i is assumed to be constant unlike in Section II where T_i can depend on ω.

(i) The network is temporally stable in the sense that for any symmetric positive definite (not necessarily diagonal) parasitic capacitance matrix,

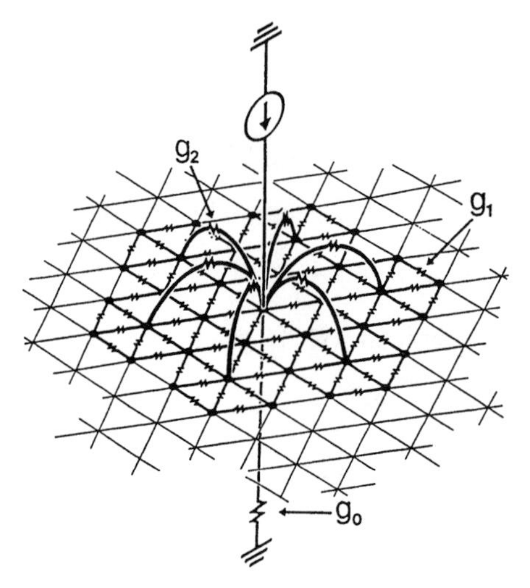

Figure 10 Architecture of a second-order regularization chip. Reprinted with permission from T. Matsumoto, H. Kobayashi, and Y. Togawa, *IEEE Trans. Neural Networks* 3:540–569, 1992 (©1992 IEEE).

the temporal dynamics converges to a unique stable equilibrium for any DC input.

(ii) At an equilibrium, the voltage distribution of the Pth layer, $1 \leq P \leq N$, simultaneously solves the Pth-order regularization with

$$\lambda_P = \frac{g_{s_1} \cdots g_{s_P}}{g_{m_1} \cdots g_{m_P}}, \tag{31}$$

$$\lambda_{P-1} = (g_{s_1} \cdots g_{s_{P-1}} g_{m_P} + g_{s_1} \cdots g_{s_{P-2}} g_{m_{P-1}} g_{s_P} + \cdots + g_{m_1} g_{s_2} \cdots g_{s_P})/(g_{m_1} \cdots g_{m_P}), \tag{32}$$

$$\lambda_{P-2} = (g_{s_1} \cdots g_{s_{P-2}} g_{m_{P-1}} g_{m_P} + g_{s_1} \cdots g_{s_{P-3}} g_{m_{P-2}} g_{m_{P-1}} g_{s_P} + \cdots + g_{m_1} g_{m_2} g_{s_3} \cdots g_{s_P})/(g_{m_1} \cdots g_{m_P}), \tag{33}$$

$$\lambda_1 = (g_{m_1} \cdots g_{m_{P-1}} g_{s_P} + g_{m_1} \cdots g_{m_{P-2}} g_{s_{P-1}} g_{m_P} + \cdots + g_{s_1} g_{m_2} \cdots g_{m_P})/(g_{m_1} \cdots g_{m_P}), \tag{34}$$

$$d_k = \frac{T_1 \cdots T_{P-1}}{g_{m_1} \cdots g_{m_P}} u_k. \tag{35}$$

(iii) The voltage distributions of all the layers are spatially stable in the sense of [40, 41, 45].

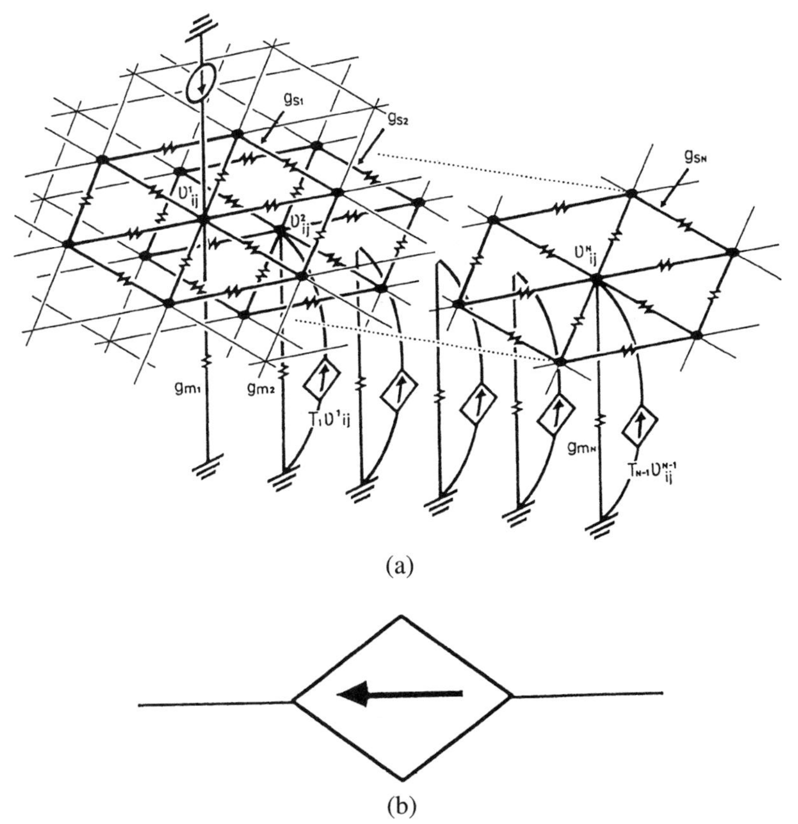

Figure 11 (a) The layered architecture. (b) Voltage-controlled current source. Reprinted from *Neural Networks* 6:327–350, H. Kobayashi, T. Matsumoto, T. Yagi, and T. Shimmi, "Image Processing Regularization Filters on Layered Architecture," Copyright 1993, with kind permission from Elsevier Science Ltd, The Boulevard, Langford Lane, Kidlington OX5 1GB, UK.

C. Two-Dimensional Problems

Although the basic idea of our layered architecture derived in the previous subsection is naturally carried over to two-dimensional problems, there are three issues which call for explanations. First, when there are two independent space variables, say x and y, there is more than one choice of the stabilizer Eq. (14). With $P = 2$, for instance, the stabilizer can be

$$\lambda \iint (v_{xx} + v_{yy})^2 \, dxy \tag{36}$$

or

$$\lambda \iint (v_{xx}^2 + 2v_{xy}^2 + v_{yy}^2)\, dxy \tag{37}$$

or other forms, where

$$v_{xx} = \frac{\partial^2 v}{\partial x^2}, \qquad v_{xy} = \frac{\partial^2 v}{\partial x \partial y}, \qquad v_{yy} = \frac{\partial^2 v}{\partial y^2}. \tag{38}$$

Second, natural boundary conditions get more involved. For instance, if $P = 2$, and $\lambda_1 = 0$, then the first variation of

$$G(v, d, \lambda) = \iint_D F(v(x, y), v_{xx}, v_{xy}, v_{yy}, x, y, d(x, y), \lambda)\, dx\, dy \tag{39}$$

on the boundary ∂D gives rise to

$$\int_{\partial D} \left[\psi_x \left(\frac{\partial F}{\partial v_{xx}} - \frac{1}{2} \frac{\partial F}{\partial v_{xy}} \right) - \psi \frac{\partial}{\partial x} \left(\frac{\partial F}{\partial v_{xx}} - \frac{1}{2} \frac{\partial F}{\partial v_{xy}} \right) \right] dy$$
$$- \int_{\partial D} \left[\psi_y \left(\frac{\partial F}{\partial v_{yy}} - \frac{1}{2} \frac{\partial F}{\partial v_{xy}} \right) - \psi \frac{\partial}{\partial y} \left(\frac{\partial F}{\partial v_{yy}} - \frac{1}{2} \frac{\partial F}{\partial v_{xy}} \right) \right] dx, \tag{40}$$

where $v(x, y)$ is perturbed to $v(x, y) + \psi(x, y)$. When one performs integration by parts on ∂D, one obtains, for instance, for Eq. (37),

$$-(v_{yy} + v_{xx}) + \left(v_{xx} x_\tau^2 + 2v_{xy} x_\tau y_\tau + v_{yy} y_\tau^2 \right) = 0, \tag{41}$$
$$\frac{\partial}{\partial n}(v_{yy} + v_{xx}) + \frac{\partial}{\partial \tau}\left(v_{xx} x_n x_\tau + v_{xy}(x_n y_\tau + x_\tau y_n) + v_{yy} y_n y_\tau \right) = 0, \tag{42}$$

on ∂D where x_n, y_n and x_τ, y_τ are the direction cosines of the outward normal and the tangent vectors, respectively. Approximation consistent with Eqs. (41) and (42) together with Euler equation

$$\frac{\partial F}{\partial v} + \frac{\partial^2}{\partial x^2} \frac{\partial F}{\partial v_{xx}} + \frac{\partial^2}{\partial x \partial y} \frac{\partial F}{\partial v_{xy}} + \frac{\partial^2}{\partial y^2} \frac{\partial F}{\partial v_{yy}} = 0 \tag{43}$$

will not be easy to justify rigorously.

Third, many of the vision chips implemented or proposed so far, including ours, are on a hexagonal grid because

(i) a network on a hexagonal grid has much better circular symmetry than on a square grid [3, 42, 43],
(ii) a hexagonal grid affords the greatest spatial sampling efficiency in the sense that the least number of nodes will attain a desired of the image [44].

We will handle the problem as a minimization problem on a finite-dimensional space as was in Eq. (23). It should be noted that in our arguments below, everything is rigorous insofar as the minimization is concerned.

On a hexagonal grid there are two labeling conventions: standard grid (Fig. 12a) and alternate grid (Fig. 12b). We will use the standard grid. Let

$$\mathbf{v} := (v_{11}, v_{12}, \ldots, v_{1n}, v_{21}, v_{22}, \ldots, v_{2n}, v_{n1}, v_{n2}, \ldots, v_{nn}) \in \mathcal{R}^{n\times n}, \tag{44}$$

and let $\mathbf{d}$ be similarly defined.

(i) $P = 1$. The most reasonable function to minimize is

$$G(\mathbf{v}, \mathbf{d}, \lambda_1) = \|\mathbf{v} - \mathbf{d}\|^2 + \lambda_1(\|\mathbf{D}_1\mathbf{v}\|^2 + \|\mathbf{D}_2\mathbf{v}\|^2 + \|\mathbf{D}_3\mathbf{v}\|^2), \tag{45}$$

where the (i, j)th components of $\mathbf{D}_1\mathbf{v}$, $\mathbf{D}_2\mathbf{v}$, and $\mathbf{D}_3\mathbf{v}$ are, respectively, given by

$$(\mathbf{D}_1\mathbf{v})_{ij} = v_{ij} - v_{i-1j}, \tag{46}$$

$$(\mathbf{D}_2\mathbf{v})_{ij} = v_{ij} - v_{ij-1}, \tag{47}$$

$$(\mathbf{D}_3\mathbf{v})_{ij} = v_{ij} - v_{i-1j+1}. \tag{48}$$

Appropriate modifications must be made on the boundary. Differentiation of Eq. (45) with respect to $\mathbf{v}$ gives

$$\mathbf{v} - \mathbf{d} - \lambda_1\mathbf{L}\mathbf{v} = \mathbf{0}, \tag{49}$$

where

$$\mathbf{L} := -(\mathbf{D}_1^T\mathbf{D}_1 + \mathbf{D}_2^T\mathbf{D}_2 + \mathbf{D}_3^T\mathbf{D}_3). \tag{50}$$

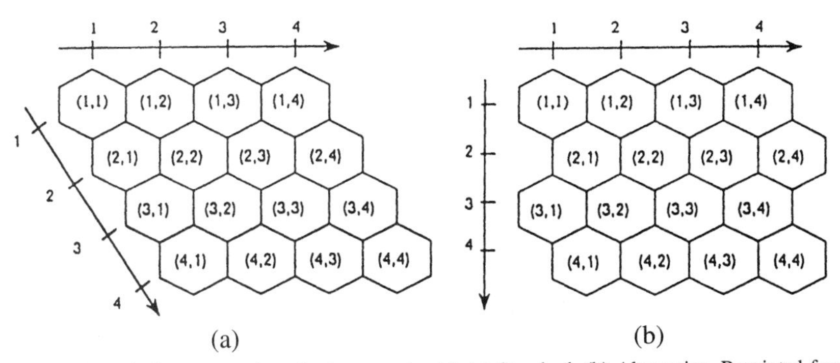

Figure 12 Labeling conventions for hexagonal grid. (a) Standard. (b) Alternative. Reprinted from *Neural Networks* 6:327–350, H. Kobayashi, T. Matsumoto, T. Yagi, and T. Shimmi, "Image Processing Regularization Filters on Layered Architecture," Copyright 1993, with kind permission from Elsevier Science Ltd, The Boulevard, Langford Lane, Kidlington OX5 1GB, UK.

The (i, j)th component of $\mathbf{Lv}$ in the interior reads

$$v_{i-1j} + v_{i+1j} + v_{ij-1} + v_{ij+1} + v_{i-1j+1} + v_{i+1j-1} - 6v_{ij}, \tag{51}$$

which is a reasonable approximation of the Laplacian on a hexagonal grid. One can easily show that Eq. (49) corresponds to the KCL of the network given in Fig. 12 with $P = 1$.

(ii) $P = 2$. As was remarked earlier, there is more than one reasonable choice of G.

(iia)

$$G(\mathbf{v}, \mathbf{d}, \lambda_1, \lambda_2) = \|\mathbf{v}-\mathbf{d}\|^2+\lambda_1(\|\mathbf{D}_1\mathbf{v}\|^2+\|\mathbf{D}_2\mathbf{v}\|^2+\|\mathbf{D}_3\mathbf{v}\|^2)+\lambda_2\|\mathbf{Lv}\|^2, \tag{52}$$

where $\mathbf{L}$ is defined by Eq. (50). The solution to this problem is given by

$$\mathbf{v} - \mathbf{d} - \lambda_1\mathbf{Lv} + \lambda_2\mathbf{L}^2\mathbf{v} = \mathbf{0}, \tag{53}$$

which, again, is of the form Eq. (23). The (i, j)th component of $\mathbf{L}^2\mathbf{v}$ in the interior reads

$$\begin{aligned} &v_{i-2j} + v_{i+2j} + v_{ij-2} + v_{ij+2} + v_{i-2j+2} + v_{i+2j-2} \\ &\quad + 2(v_{i-1j-1} + v_{i+1j+1} + v_{i-1j+2} + v_{i+1j-2} + v_{i-2j+1} + v_{i+2j-1}) \\ &\quad - 10(v_{i-1j} + v_{i+1j} + v_{ij-1} + v_{ij+1} + v_{i-1j+1} + v_{i+1j-1}) + 42v_{ij}, \end{aligned} \tag{54}$$

which is a reasonable approximation of the biharmonic operator on a hexagonal grid. Note that the third term $\lambda_2\|\mathbf{Lv}\|^2$ in Eq. (53) corresponds to a solution with Eq. (36) which is called the square Laplacian (Grimson 1981). The question as to what would be a good approximation of the quadratic variation Eq. (37) [47] on a hexagonal grid may not be easy to answer. We will not pursue this subject since it is not our purpose in the present paper. Grimson [47] observed a difference between solutions to a particular visual reconstruction problem (not regularization problem) with contraint Eq. (36) and constraint Eq. (37). We have, so far, observed no strange behavior to the solution to Eq. (52) on a hexagonal grid.

(iib) Another choice of G for $P = 2$ is

$$\begin{aligned} G(\mathbf{v}, \mathbf{d}, \lambda_1, \lambda_2) &= \|\mathbf{v} - \mathbf{d}\|^2 + \lambda_1(\|\mathbf{D}_1\mathbf{v}\|^2 + \|\mathbf{D}_2\mathbf{v}\|^2 + \|\mathbf{D}_3\mathbf{v}\|^2) \\ &\quad + \lambda_2(\|\mathbf{L}_1\mathbf{v}\|^2 + \|\mathbf{L}_2\mathbf{v}\|^2 + \|\mathbf{L}_3\mathbf{v}\|^2), \end{aligned} \tag{55}$$

where

$$\mathbf{L}_1 := -\mathbf{D}_1^T\mathbf{D}_1, \qquad \mathbf{L}_2 := -\mathbf{D}_2^T\mathbf{D}_2, \qquad \mathbf{L}_3 := -\mathbf{D}_3^T\mathbf{D}_3. \tag{56}$$

The solution is given by

$$\mathbf{v} - \mathbf{d} - \lambda_1\mathbf{Lv} + \lambda_2(\mathbf{L}_1^T\mathbf{L}_1 + \mathbf{L}_2^T\mathbf{L}_2 + \mathbf{L}_3^T\mathbf{L}_3)\mathbf{v} = \mathbf{0}. \tag{57}$$

Note that the last term $(\mathbf{L}_1^T\mathbf{L}_1 + \mathbf{L}_2^T\mathbf{L}_2 + \mathbf{L}_3^T\mathbf{L}_3)\mathbf{v}$ in Eq. (57) is not $\mathbf{Lv}$ and it reads [compare with Eq. (54)]

$$\begin{aligned} &v_{i-2j} + v_{i+2j} + v_{ij-2} + v_{ij+2} + v_{i-2j+2} + v_{i+2j-2} \\ &\quad - 4(v_{i-1j-1} + v_{i+1j} + v_{ij+1} + v_{ij+1} + v_{i-1j+1} + v_{i+1j-1}) + 18v_{ij}, \end{aligned} \tag{58}$$

which is a rather crude approximation of $\mathbf{L}^2\mathbf{v}$. The network given in Fig. 10 and hence $\mathbf{v}$ in Fig. 12 minimizes Eq. (55) with $\lambda_1 = 0$, $\lambda_2 > 0$.

(iii) $P = 3$. A possible choice of G will be

$$\begin{aligned} G(\mathbf{v}, \mathbf{d}, \lambda_1, \lambda_2, \lambda_3) = {} & \|\mathbf{v} - \mathbf{d}\|^2 + \lambda_1(\|\mathbf{D}_1\mathbf{v}\|^2 + \|\mathbf{D}_2\mathbf{v}\|^2 + \|\mathbf{D}_3\mathbf{v}\|^2) + \lambda_2\|\mathbf{Lv}\|^2 \\ & + \lambda_3(\|\mathbf{D}_1\mathbf{Lv}\|^2 + \|\mathbf{D}_2\mathbf{Lv}\|^2 + \|\mathbf{D}_3\mathbf{Lv}\|^2). \end{aligned} \tag{59}$$

Note that the third term corresponds to one of the penalty terms considered in [46] for the continuous two-dimensional problem. The solution is given by

$$\mathbf{v} - \mathbf{d} - \lambda_1\mathbf{Lv} + \lambda_2\mathbf{L}^2\mathbf{v} - \lambda_3\mathbf{L}^3\mathbf{v} = \mathbf{0}. \tag{60}$$

We will stop here and formalize the argument in the following.

Fact 2. Consider the minimization problem on a hexagonal array:

$$\begin{aligned} G(\mathbf{v}, \mathbf{d}, \lambda_1, \ldots \lambda_P) &= \|\mathbf{v} - \mathbf{d}\|^2 \\ &+ \sum_{r=1}^{P} \begin{cases} \lambda_r\|\mathbf{L}^{r/2}\mathbf{v}\|^2, & r\text{: even}, \\ \lambda_r(\|\mathbf{D}_1\mathbf{L}^{(r-1)/2}\mathbf{v}\|^2 + \|\mathbf{D}_2\mathbf{L}^{(r-1)/2}\mathbf{v}\|^2 + \|\mathbf{D}_3\mathbf{L}^{(r-1)/2}\mathbf{v}\|^2), & r\text{: odd}, \end{cases} \end{aligned} \tag{61}$$

where $\mathbf{L}$, $\mathbf{D}_1$, $\mathbf{D}_2$, and $\mathbf{D}_3$ are defined by Eqs. (50), (46), (47), and (48), respectively. Then the statements of Fact 1 are valid.

D. The SCE Filter

1. Theory

The following fact provides a theory for our smoothing contrast-enhancement (SCE) filter.

Fact 3. Consider the double-layer network given in Fig. 13. Let

$$x_k = \frac{T_3}{g_{m3}}v_k^1 + \frac{T_4}{g_{m3}}v_k^2,$$

i.e., x_k is a linear combination of v_k^1 and v_k^2.

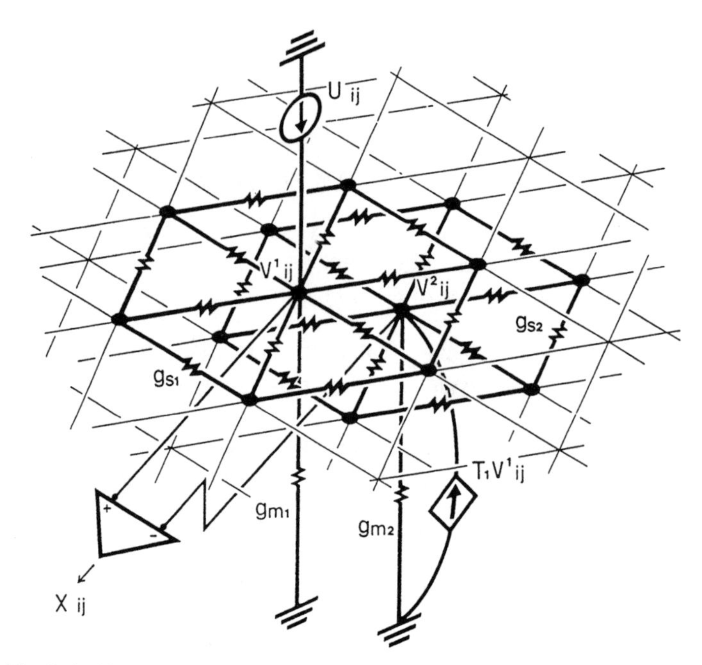

Figure 13 A double-layer network. Reprinted from *Neural Networks* 6:327–350, H. Kobayashi, T. Matsumoto, T. Yagi, and T. Shimmi, "Image Processing Regularization Filters on Layered Architecture," Copyright 1993, with kind permission from Elsevier Science Ltd, The Boulevard, Langford Lane, Kidlington OX5 1GB, UK.

(i) Then $\mathbf{x} := (x_1, \ldots, x_n)$ minimizes

$$G(\mathbf{x}, \mathbf{u}, \lambda_1, \lambda_2) := \sum_k \left(x_k - R_0(-u_{k-1} - u_{k+1} + 2u_k) - \nu R_0 u_k\right)^2 + \lambda_1 \sum_k (x_k - x_{k-1})^2 + \lambda_2 \sum_k (x_{k-1} + x_{k+1} - 2x_k)^2,$$

where

$$R_0 = \frac{g_{s2}}{g_{m1} g_{m2}} \frac{T_3}{g_{m3}}, \qquad \nu = \frac{1}{g_{s2}} \left(g_{m2} + \frac{T_4}{T_3} T_1 \right),$$

$$\lambda_1 = \frac{g_{m1} g_{s2} + g_{m2} g_{s1}}{g_{m1} g_{m2}}, \qquad \lambda_2 = \frac{g_{s1} g_{s2}}{g_{m1} g_{m2}}.$$

(ii) Consider the uniform input $u_k \equiv u$ for all k. If

$$g_{m_2} + \frac{T_4}{T_3} T_1 = 0, \tag{62}$$

then

$$x_k \equiv 0 \quad \text{for all } k. \tag{63}$$

Remarks. (i) This filter naturally has an impulse response similar to the one shown in Fig. 9a. Consider, that the input given by Fig. 14a, which is a rectangular

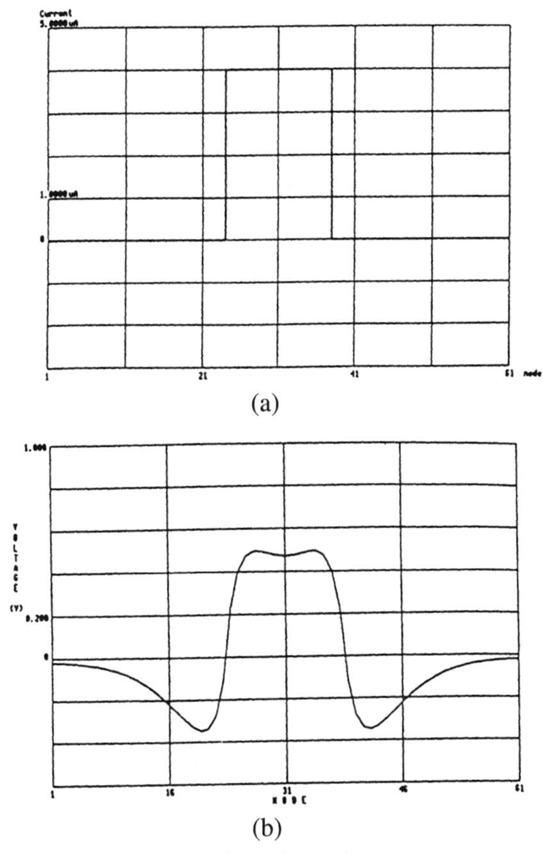

Figure 14 Responses to noisy input. (a) Noiseless input where

$$u'_k = \begin{cases} 4\ \mu\text{A}, & 24 \le k \le 38, \\ 0, & \text{elsewhere}. \end{cases}$$

(b) Responses to (a). (c) Input is corrupted by a white Gaussian noise with $3\sigma = 1\ \mu$A. (d) Response v_k^1 and v_k^2. (e) Response x_k. (f) Responses x_k when all the circuit parameters are perturbed by Gaussian around the nominal values with $3\sigma = 20\%$. Reprinted from *Neural Networks* 6:327–350, H. Kobayashi, T. Matsumoto, T. Yagi, and T. Shimmi, "Image Processing Regularization Filters on Layered Architecture," Copyright 1993, with kind permission from Elsevier Science Ltd, The Boulevard, Langford Lane, Kidlington OX5 1GB, UK.

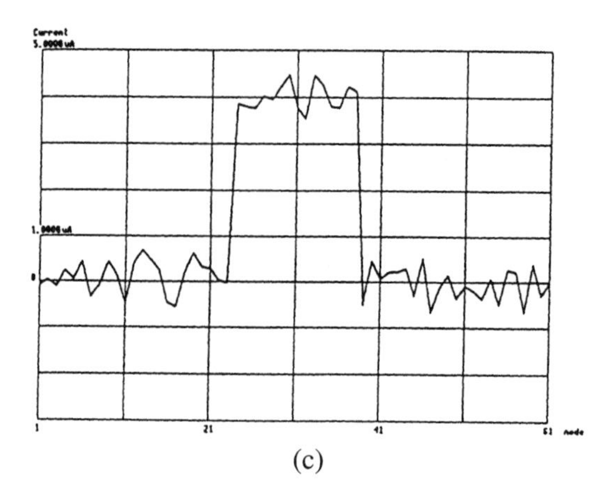

(c)

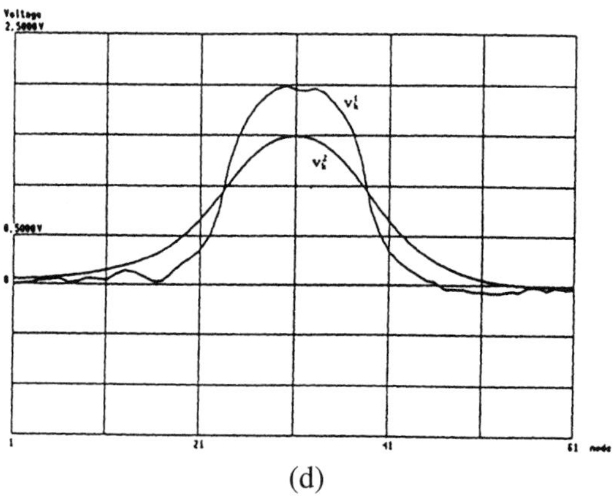

(d)

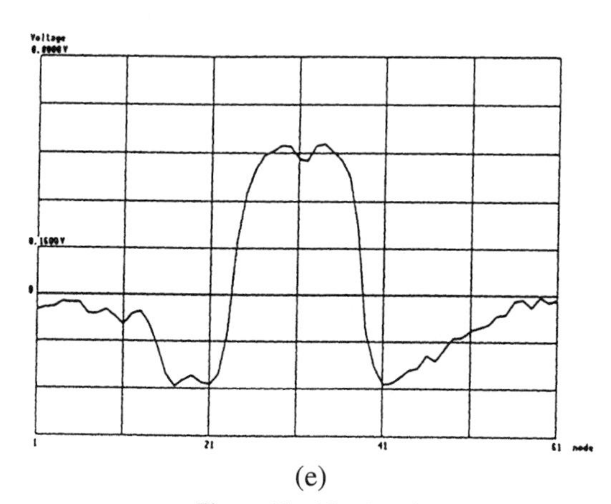

(e)

Figure 14 (*Continued*)

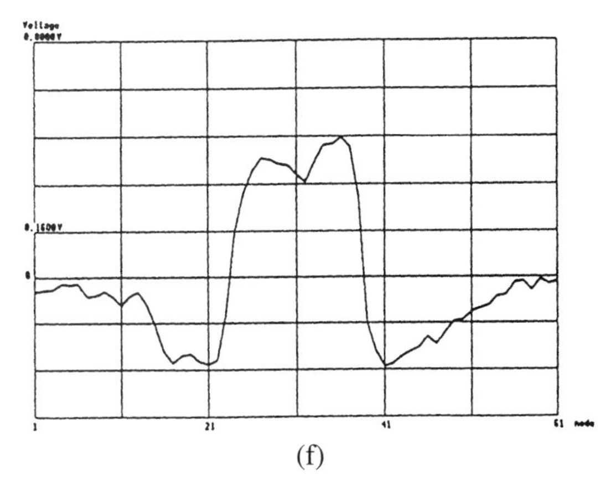

(f)

Figure 14 *(Continued)*

"image,"

$$u_k = \begin{cases} 4\ \mu\text{A}, & 24 \le k \le 38, \\ 0, & \text{elsewhere}, \end{cases} \tag{64}$$

is corrupted by a Gaussian noise n_k with $3\sigma = 1\ \mu\text{A}$, i.e.,

$$u'_k = u_k + n_k. \tag{65}$$

Figure 14b gives the filter response when $T_3/g_{m3} = 1$, $T_4/g_{m3} = -1$.

(ii) In engineering terms, this network can be regarded as a noncausal[1] IIR (infinite impulse response) implementation of a $\nabla^2 G$-like filter and it enhances contrast after smoothing. Speaking roughly, our filter output $\mathbf{x}$ is $(\mathbf{L}^{-1} - \mathbf{L}^{-2})\mathbf{u}$ where $\mathbf{L}$ is as defined by Eq. (5). We are avoiding the term "edge detection" simply because a zero-crossing of $\nabla^2 G$ is not necessarily an edge [49]. Note, however, that in the particular situation given in Fig. 14f, our SCE filter correctly identifies the two edges against noise and parameter variations, if one checks the zero-crossings.

(iii) Statement (i) in Fact 3 is straightforward. In order to prove statement (ii) on Fact 3, note that the input being uniform implies that no current flow through g_{s_1}, and hence $v_k^1 \equiv u/g_{m_1}$. Similarly, $v_k^2 \equiv (T_1/g_{m1}g_{m2})u$ which yields $x_k = (T_3/g_{m3})v_k^1 + (T_4/g_{m3})v_k^2 = (u/(g_{m1}g_{m3}))(T_3 + T_1T_4/g_{m2}) \equiv 0$. Thus Eq. (62) implies Eq. (63). This means that if Eq. (62) holds, then x_k does not respond to the "Dc component," namely, x_k responds only to intensity differences and is

[1] Noncausal is referred to the fact that the voltage at a particular node depends on the node voltages "to the right" as well as on those "to the left."

insensitive to absolute values. This is important from the information processing viewpoint.

(iv) That the voltage-controlled current source $T_1 v^1$ is a unilateral element is important. Namely, while the first-layer voltage v^1 does affect the second layer via $T_1 v^1$, the second-layer voltage v^2 has no effect on the first layer. Thus, if $T_1 v_k^1$ were replaced with a passive resistor (a bilateral element), then $v_k^1 > v_k^2$ always and hence Eq. (63) could never be satisfied. It is also clear that there would be no antagonistic surround.

2. Circuit Design

As this formation of the second-order regularization network requires only nearest neighbor connections, its principal virtue is the ease of implementation on an integrated circuit. Compared to an earlier implementation of a network with a Gaussian impulse response [42, 43], no resistor connections are required to second-nearest neighbors, nor are negative impedance converters necessary at every node. However, two independent resistor networks must now coexist on the same IC, so the compact design and layout of the unit cell at each node remains a most important consideration.

The quality of signal processing from all-analog parallel image processors has usually been inferior to that from digital implementations. The dynamic range is limited at the input transducer, and offsets, noise, and transistor mismatches often corrupt circuit action so profoundly that only a vague semblance remains between the experimentally obtained output and that predicted by theory or simulation. We used this filter as a means to access the potential of image processing with parallel analog circuits by designing individual circuits so that the well-known sources of imperfection are suppressed within reasonable bounds. Some key considerations were:

(i) To bias all FETs well above threshold, so that local random mismatches in threshold voltage or large-scale gradients across the chip do not introduce or distortion in the output reconstructed image. The bias values were constrained by the requirement of a 1-V signal swing, and operation with a single 5-V power supply.

(ii) To keep the chip power dissipation to a minimum, so that the chip surface is almost at constant temperature. Too large a temperature gradient across the chip will produce a nonuniform profile in dark currents in the photosensor, and warp the input image. This requirement is reconciled with (i) above by use of the smallest possible FET W/L ratio. Compactness in layout further requires that both W and L should be small, so almost all FETs were of the minimum channel length.

(iii) To place the photosensors on a hexagonal grid, so that no spatial distortion arises in sampling the input image. Although all unit cells and their associated wiring lie on a Manhattan geometry, the aspect ratio of the abutted rectangular cells was chosen so that their centers come to rest on a hexagonal grid.

a. Photoreceptor

The network was driven by the voltage output of the photoreceptor, in a Thevenin equivalent of the circuit of Fig. 15. An advantage over current drive is that when the network is uniformly illuminated, no current flows in the network resistors, so they dissipate zero power. A minimum differential pair with unity feedback buffers the photoreceptor from the network resistors.

b. Network Resistors

To keep power dissipation small, the network uses large-value resistors. Nominal values are $1/g_{m1} = 600\,\text{k}\Omega$, $1/g_{s1} = 400\,\text{k}\Omega$, $1/g_{s2} = 20\,\text{k}\Omega$–$200\,\text{k}\Omega$. These are most compactly implemented with FETs, rather than as diffused resistors. In this way, the variable resistor which must use FETs will track the fixed resistors over process and temperature.

The network uses a variant of a well-known circuit [3, 50] to cancel the quadratic nonlinearity between two FET resistors (Fig. 16a). FET sizes are $3 \times 10\ \mu\text{m}^2$ for $1/g_{m1}$, and $3 \times 7\ \mu\text{m}^2$ for $1/g_{s1}$. The circuit affords an acceptable

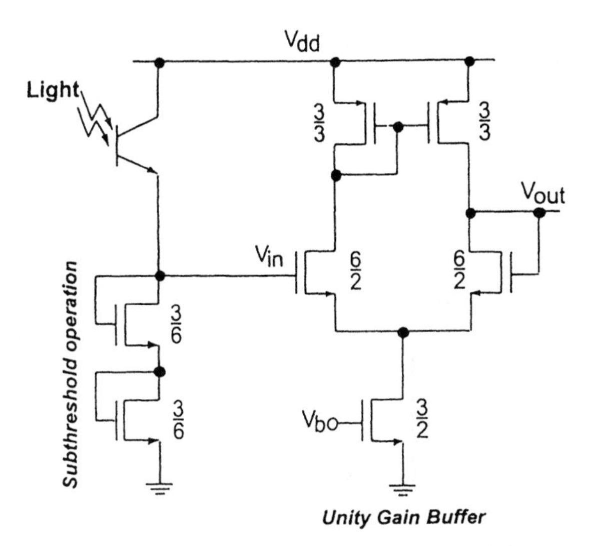

Figure 15 Photosensor circuit. Photocurrent is converted to voltage by diode-connected MOS FETs.

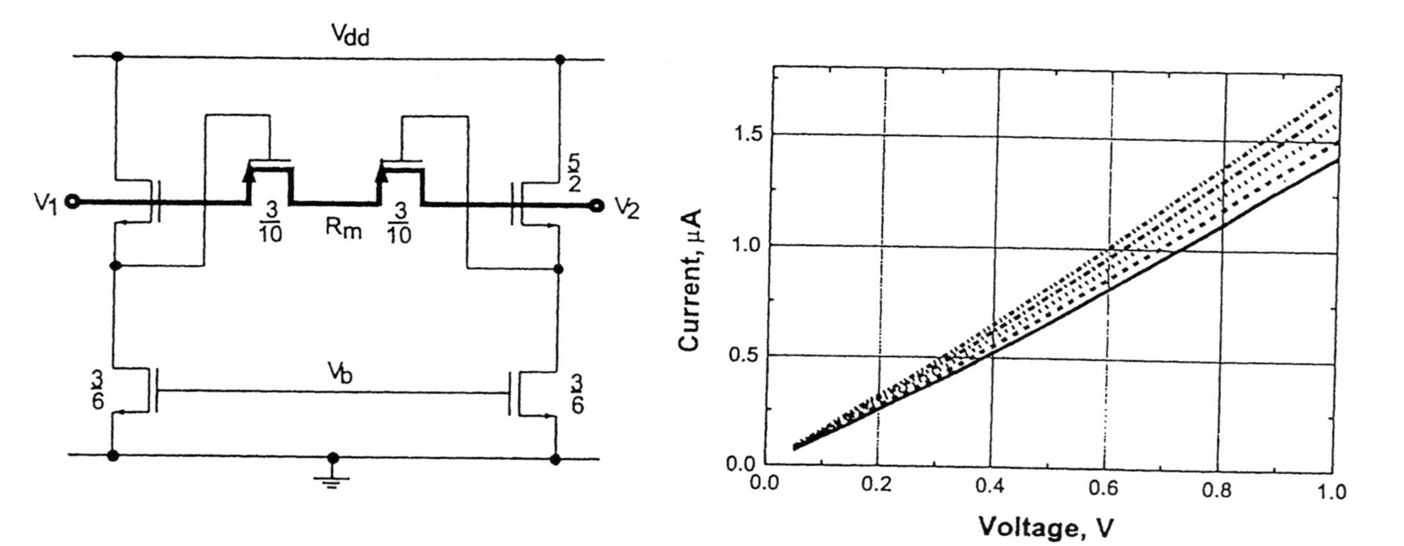

Figure 16 (a) Floating resistor circuit. Its resistance is variable with bias current. (b) Simulated I-V characteristics of resistor for common-mode voltage from 1.6 to 2 V (typical dynamic range of node voltages).

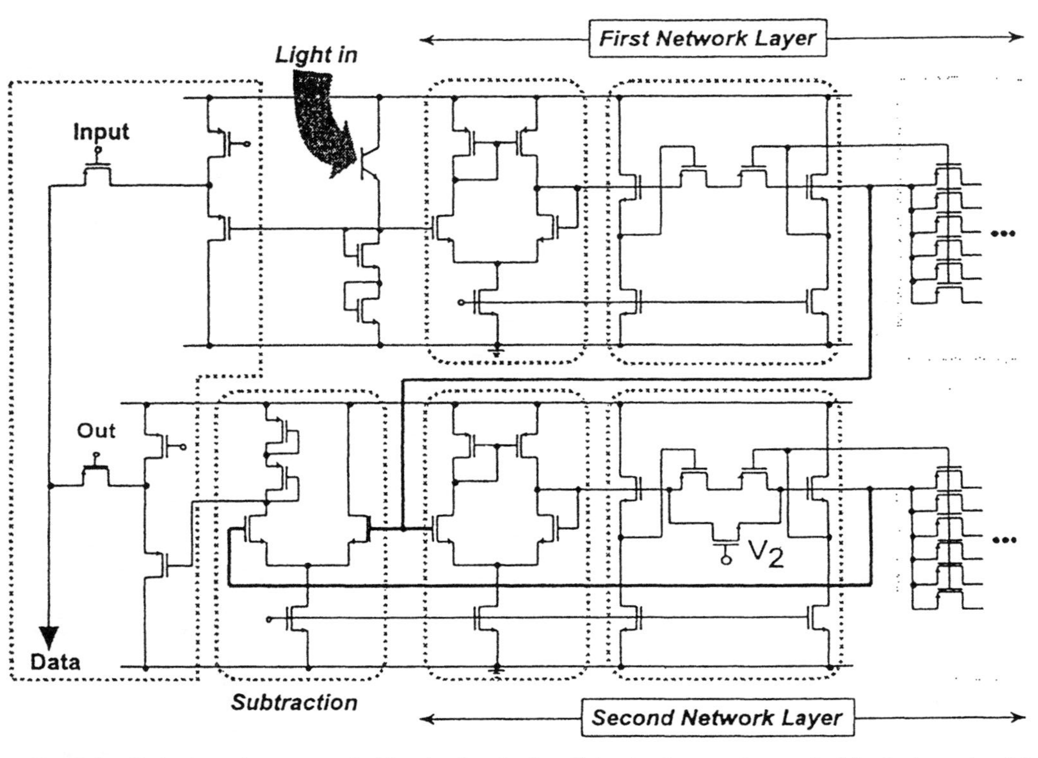

Figure 17 Unit cell circuit used at every node. The circuit embodies all the signal processing required for the two network layers in cascade.

linearity (Fig. 16b) over the maximum 1-V swing. The variable resistance g_{s2} is set by the gate voltage of a single FET in parallel with the two main resistor FETs (Fig. 16a).

c. Unit Cell

The network is assembled from these and other subsidiary components in each unit cell (Fig. 17). Using once again the Thevenin equivalent of the network prototype, the output voltage from the first mesh is buffered and applied as a voltage input to the second mesh. The output voltages from the two networks are subtracted in a differential pair. The pair NMOS FETs are biased at a $V_{gs} - V_t$ of 1 V and use a PMOS FET load to obtain an almost linear voltage input-output relation. Either the network input (the log compressed sampled light signal) or the output may be multiplexed on to a single line through addresable PMOS switches. Addressing is arranged to scan out one column at a time.

d. Layout

The unit cell size of $138 \times 160\ \mu\text{m}^2$ following 2-μm CMOS two-layer design rules is dominated by wiring (Fig. 18a). Centers of rectangles with this aspect ratio of $2 : \sqrt{3}$, when assembled in a checkerboard pattern, will coincide with the centers on a hexagonal grid (Fig. 18b). An array of 52×53 unit cells fits on a $7.9 \times 9.2\text{-mm}^2$ die (Fig. 19); this was thought to be the smallest sized array required to sense images of simple objects with a useful resolution.

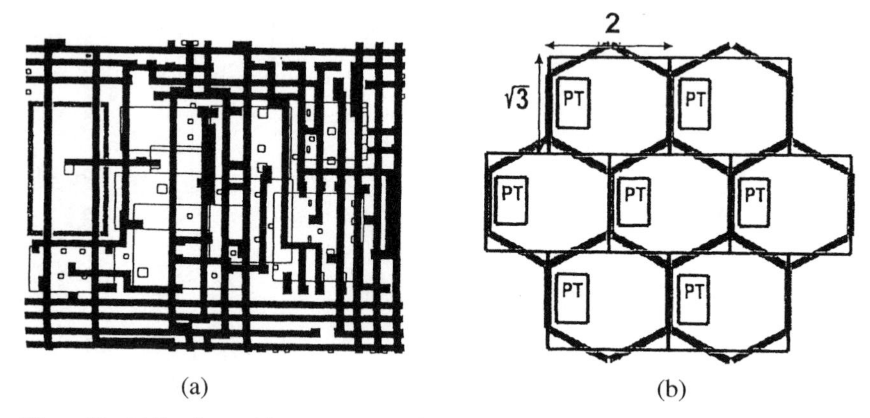

Figure 18 (a) Two-layer wiring pattern over unit cell layout. Cell size is dominated by wiring. (b) Arrangement of unit cell centers on a hexagonal grid by appropriate choice of cell aspect ratio.

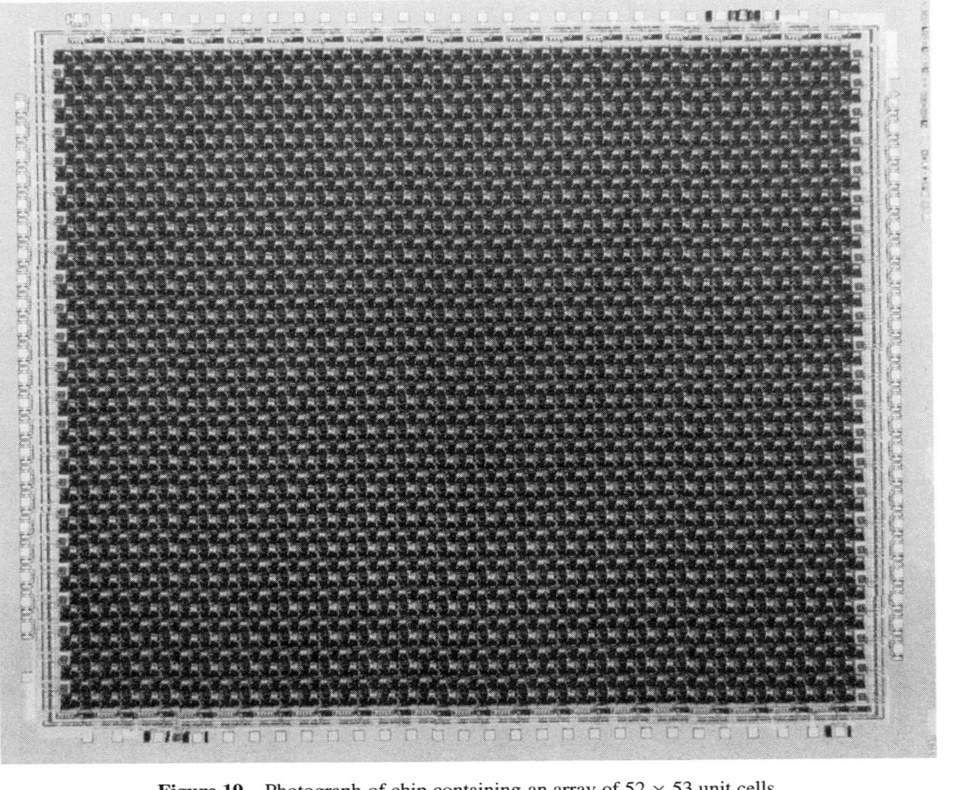

Figure 19 Photograph of chip containing an array of 52×53 unit cells.

3. Experimental Results

a. Measurement Method

As the network senses a 2D input, the incident image, and produces a 2D output, the smoothed image with enhanced contrast, it does not obtain any data reduction. A fairly elaborate interface is thus required to read and reconstruct its output (Fig. 20).

The voltage outputs from 52 columns are each digitized to 12 bits off-chip, buffered, and the entire processed image is reconstructed after a computer has addressed all the rows on the chip. The images shown in the next section were captured from the computer display, and were not subject to any subsequent numerical smoothing or enhancement.

b. Test Results

The fabricated chip was packaged in a pin-grid-array, and dissipated 300 mW from a 5-V supply. The network spatial impulse response was measured by illuminating the chip through a package lid with a pinhole in the middle. The measured output clearly shows the axis undershoot surrounding the peak and good circular symmetry. It closely matches a 2D simulated impulse response (Fig. 21). The

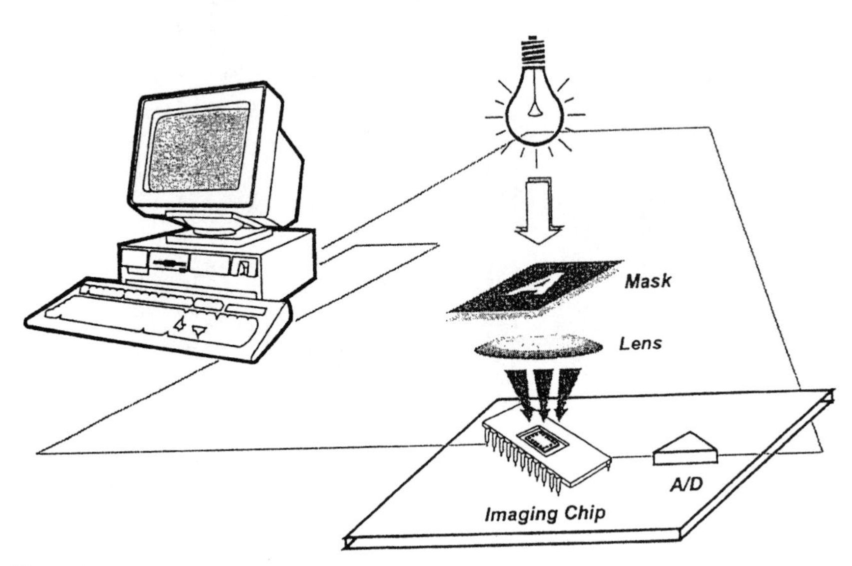

Figure 20 Optical input to chip is 2D; elaborate interface required to acquire and reconstruct 2D chip output.

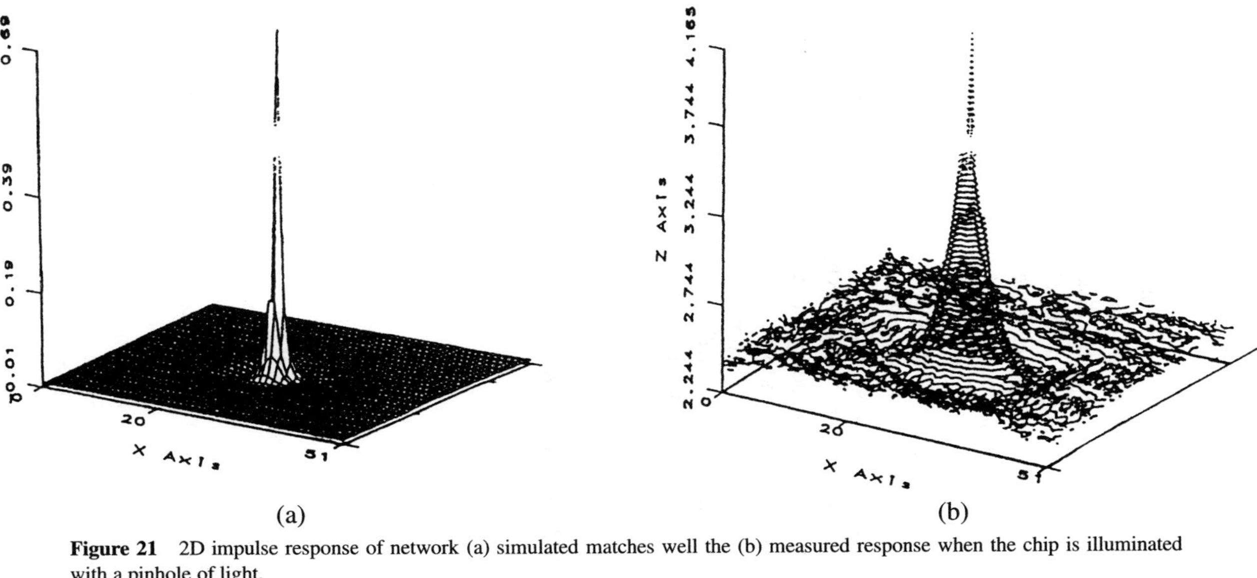

Figure 21 2D impulse response of network (a) simulated matches well the (b) measured response when the chip is illuminated with a pinhole of light.

small ripple on the baseline away from the peak relative to the height of the peak is a measure of the useful network dynamic range, in this case about 100:1.

Images of simple objects were also focused on the chip. The input image as sampled by the photoreceptor array is compared with the network output after image smoothing and contrast-enhancement. The image of a disk of light (Fig. 22a) appears at the output as a disk surrounded by a halo (Fig. 22b). This halo enhances the contrast at the edge of the disk. Most dramatic is the network action on a styrofoam coffee cup imaged on the chip (Fig. 23).

A halo surrounds the cup, enhancing the contrast of its outline, but more interestingly, streaks of light on the curved surface of the cup which were not noticeable on the incident images appear prominently after enhancement (Fig. 23). In all cases, the sensed and filtered images are remarkably clear, in fact the best obtained in our knowledge from a signal sensor and analog processor of this genre. Note that for edge detection, one locates the zero-crossings of the $\nabla^2 G$-filtered image, which is not necessarily "better" to human eyes.

The filter scale, as determined by the width at half maximum of the impulse response, is experimentally seen to be variable by almost 2:1. A new image will be smoothed by the network in the time interval required for every node to relax to its final equilibrium, set by the RC time constant of the network resistors and the associated capacitance of the FETs and interconnect wires.

More details are found in [51, 52].

E. Light-Adaptive Architecture

1. Theory

In all the vision chip architectures implemented/proposed so far that we know of, the hyperparameters λ_r are **fixed**. Our architecture proposed below makes λ_r variable so that adaptation can be incorporated. Most generally, λ_r can depend on $\mathbf{v}$, $\mathbf{d}$, and k. The dependency of λ_r on $\mathbf{v}$ makes Eq. (23) nonquadratic and the general analytical form corresponding to Eq. (24) can be nonlinear, which we do not pursue at least in the present paper. Although the dependency of λ_r on k does not alter the quadratic nature of the problem, the generalization in this direction does not, so far, find interesting enough applications. Therefore, we will consider the minimization of Eq. (23) where λ_r is now $\lambda_r(\mathbf{d})$. Although this requires only a straightforward modification in Eq. (24), i.e., λ_r should be replaced with $\lambda_r(\mathbf{d})$, it leads to rather interesting adaptation networks. Among many possible adaptive networks, the SCE (smoothing contrast-enhancement) filter network [1, 2, 5, 6] has probably one of the most interesting structures suited for this adaptation.

The following fact is a straightforward consequence of Fact 3 and the argument preceding it.

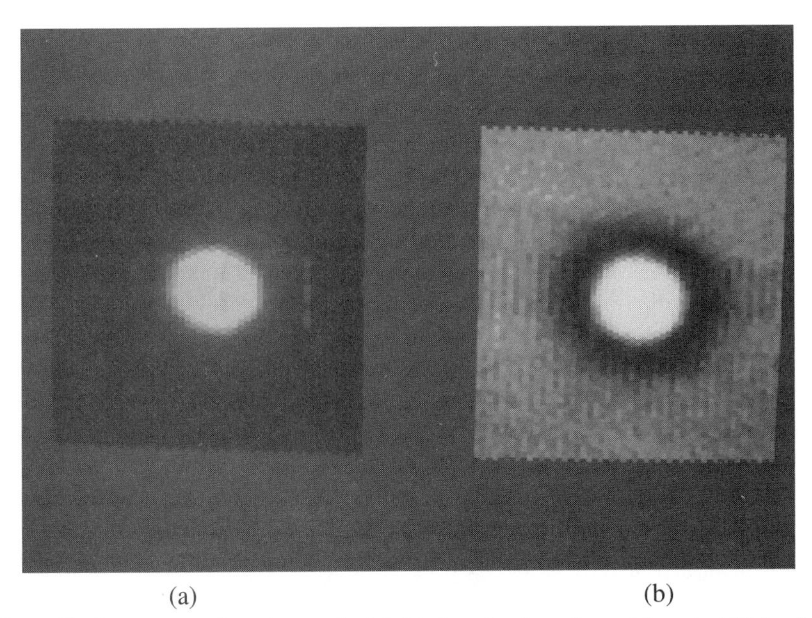

(a) (b)

Figure 22 In response to input image of a disk (a), the network produces at the output (b) the disk surrounded by a halo.

(a) (b)

Figure 23 Network accurately acquires (a) images of a styrofoam cup, and produces at its output (b) the filtered image, with major features enhanced.

Fact 4. Consider the double-layer network given in Fig. 13, where the second-layer horizontal conductance g_{s2} has an adaptation mechanism described by

$$g_{s2}(\mathbf{u}) := \frac{1}{G\sum_k u_k}, \qquad G > 0, \tag{66}$$

where G is a constant and u_k is a photocurrent induced at node k. Then

(i) the second-layer voltage distribution v_k^2 solves the second-order regularization problem with

$$\lambda_1(\mathbf{u}) = \frac{g_{s1}}{g_{m1}} + \frac{g_{s2}(\mathbf{u})}{g_{m2}}, \qquad \lambda_2(\mathbf{u}) = \frac{g_{s1}g_{s2}(\mathbf{u})}{g_{m1}g_{m2}},$$

so that the weight ratio is given by

$$\frac{\lambda_2(\mathbf{u})}{\lambda_1(\mathbf{u})} = \frac{1}{g_{m1}/g_{s1} + g_{m2}G(\sum_k u_k)}, \qquad d_k = \frac{T_1}{g_{m1}g_{m2}}u_k. \tag{67}$$

Statements (ii) and (ii) of Fact 3 are still valid.

Remarks. (i) When the total input current $\sum_k u_k$ gets larger, which amounts to the fact that the environment is **light**, the second-layer horizontal conductance g_{s2} dereases. Although the decrease of g_{s2} changes both $\lambda_1(\mathbf{u})$ and $\lambda_2(\mathbf{u})$, the ratio $\lambda_2(\mathbf{u})/\lambda_1(\mathbf{u})$ decreases [Eq. (67)]. This means that when $\sum_k u_k$ is large, the emphasis of the network on the second-order derivative decreases. This adaptation mechanism has rather interesting implications. Suppose that $u_k = u_k^0 + x_k$, where u_k^0 is the noiseless image while x_k stands for noise. Suppose also that the mean of the noise has been absorbed into u_k^0 so that x_k has zero mean. If $x_{\min} \leq x_k \leq x_{\max}$ where $x_{\min}$ and $x_{\max}$ are independent of u_k^0, then $\sum_k u_k$ large means that effect of noise is less significant than when $\sum_k u_k$ is smaller. Thus when $\sum_k u_k$ is smaller, noise is more significant and the network puts more emphasis on the second-order derivative penalty. This architecture is endowed with the capability shown in Fig. 9.

Figure 24 shows the effect of the adaptation mechanism. The input image is the sum of a (one-dimensional) restangular "image"

$$u_k^0 = \begin{cases} 1\ \mu\text{A}, & 61 \leq k \leq 141, \\ 0, & \text{otherwise,} \end{cases}$$

and the Gaussian white noise with mean 300 pA, $3\sigma = 600$ pA. Figure 24a shows the network response x_k, where

$$\begin{aligned} 1/g_{s2} &= 5\ \text{M}\Omega, \qquad 1/g_{s1} = 30\ \text{M}\Omega, \qquad 1/g_{m1} = 1/g_{m2} = 1\ \text{G}\Omega, \\ T_1 &= 10^{-9}\ \text{siemens}. \end{aligned} \tag{68}$$

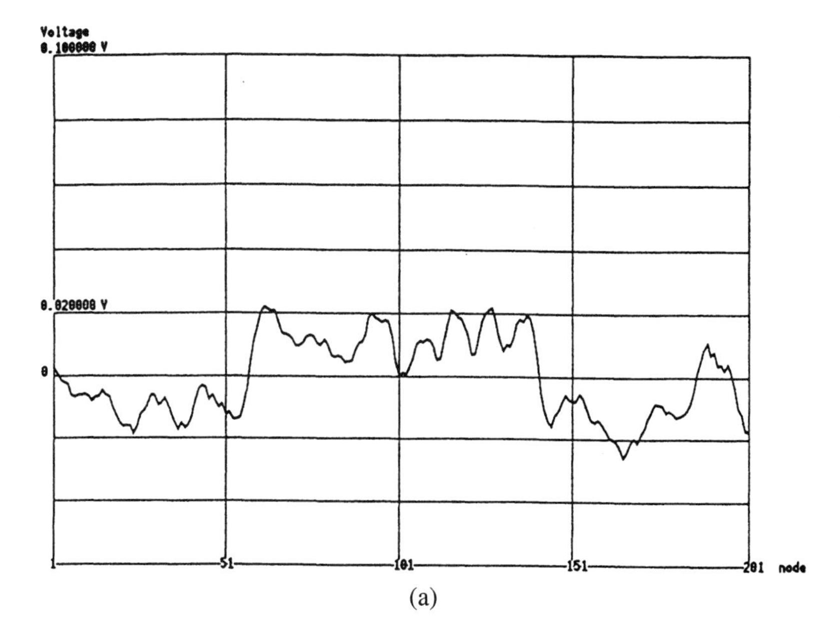

(a)

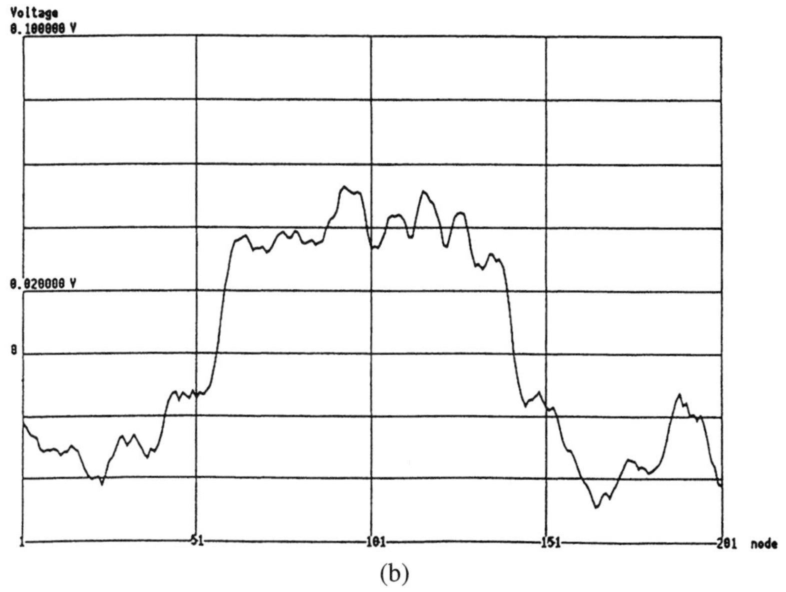

(b)

Figure 24 Responses of the network in Fig. 13. (a) Adaptation is *not* incorporated ($1/g_{s2} = 5\,\mathrm{M\Omega}$). (b) Adaptation of Eq. (66) is incorporated with $G = 1.0 \times 10^{13}$. Reprinted from *Neural Networks* 8:87–101, H. Kobayashi, T. Matsumoto, T. Yagi, and K. Tanaka, "Light-Adaptive Architectures for Regularization Vision Chips," Copyright 1995, with kind permission from Elsevier Science Ltd, The Boulevard, Langford Lane, Kidlington OX5 1GB, UK.

A dramatic effect is discernible when the g_{s2}-adaptation Eq. (66) is incorporated where

$$G = 1.0 \times 10^{13}.$$

It is known that the $\nabla^2 G$ filter identifies edges of an object by its zero-crossings even though not every zero-crossing corresponds to an edge [49]. Observe that while Fig. 24a gives no information about the edges of the original object, Fig. 24b, which is the network response with the g_{s2}-adaptation given by Eq. (66), correctly identifies the edge of the original image by its zero-crossings.

(ii) In [5, 6] the g_{s2} values are changed manually.

(iii) Since the photocurrent u_k is always positive, one does not have to square it or one does not have to take the absolute value. In fact, v_k^1 and v_k^2 are also positive. The output $x_k = v_k^1 - v_k^2$, however, can be negative.

2. CMOS Circuits for Light Adaptation

Figure 25 shows a possible configuration and note that the input circuit in Fig. 17 is the Thevenin equivalent of the current source in Fig. 11. Let us denote this equivalent voltage by

$$v_k^0 := g_{m1} u_k.$$

In Fig. 25, this voltage v_k^0 is first converted into current I_k by the V-I converter so that I_k is proportional to v_k^0. The summation of all these currents can be obtained *for free* by simply connecting the wires together because of the Kirchhoff current law, and the summed current I is given by

$$I := \sum_k I_k \propto \sum_k v_k^0.$$

The current I is fed into the bias voltage generator which produces a bias voltage v_c so that the g_{s2} value is inversely proportional to I. Figure 26 shows a circuit design example of the V-I converter, g_{s2}, and the bias generator. The V-I converter is designed with a differential pair and g_{s2} is implemented with two parallel MOS FETs [50] whose value becomes larger as v_c increases. In the bias generator, the summed current I is subtracted from a bias current I_b and the resultant current $I_b - I$ flows into a resistor R and a diode-connected NMOS which generate a bias voltage v_c. Thus as I becomes smaller, v_c (and then g_{s2}) increases. Figure 27 shows SPICE simulation results of g_{s2} characteristics at several different values of $\sum_k v_k^0$ and we see that as $\sum_k v_k^0$ becomes larger, g_{s2} decreases. It should be noted that perfect linearity is not necessary at all.

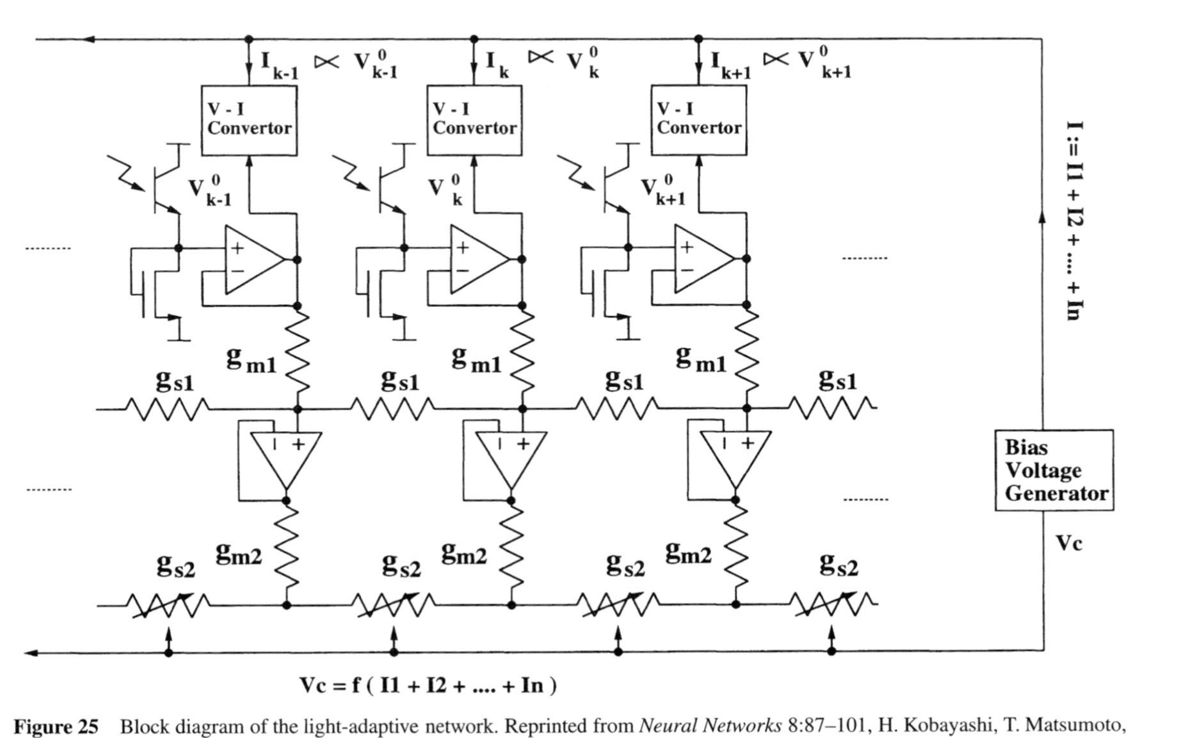

Figure 25 Block diagram of the light-adaptive network. Reprinted from *Neural Networks* 8:87–101, H. Kobayashi, T. Matsumoto, T. Yagi, and K. Tanaka, "Light-Adaptive Architectures for Regularization Vision Chips," Copyright 1995, with kind permission from Elsevier Science Ltd, The Boulevard, Langford Lane, Kidlington OX5 1GB, UK.

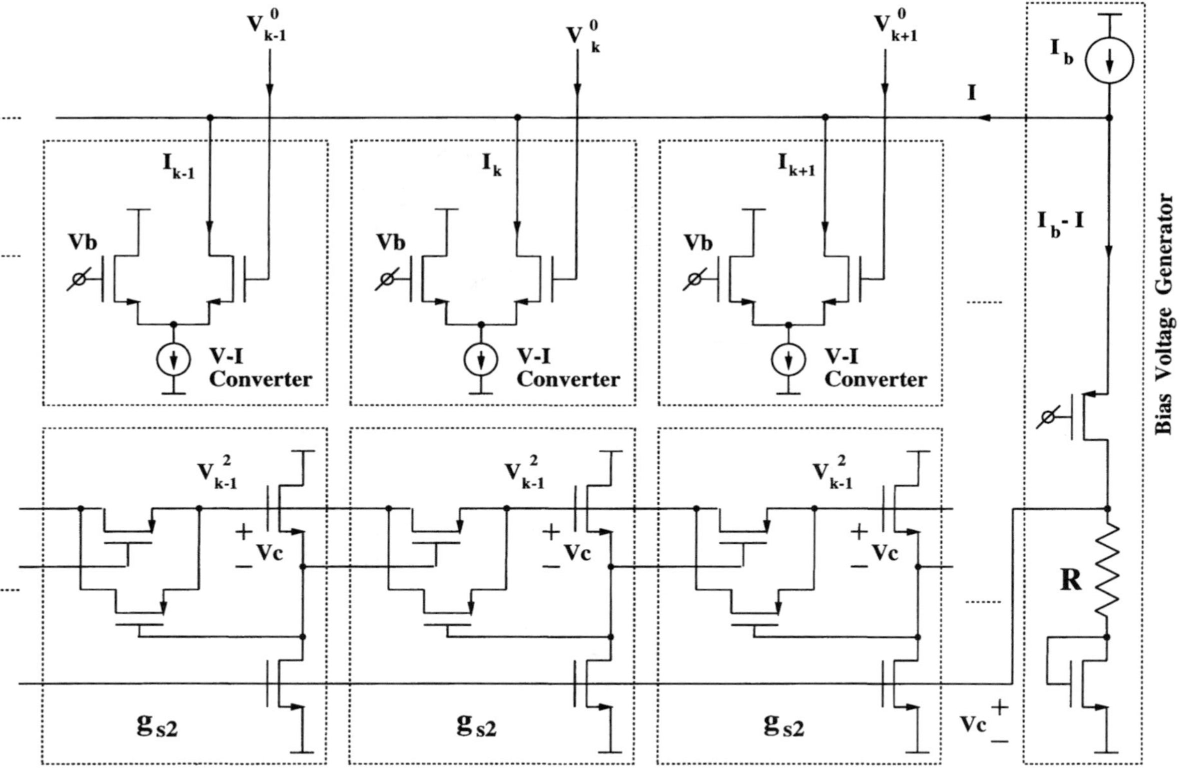

Figure 26 Circuit design of *V-I* converters, bias generators, and variable conductances g_{s2} in Fig. 25. Reprinted from *Neural Networks* 8:87–101, H. Kobayashi, T. Matsumoto, T. Yagi, and K. Tanaka, "Light-Adaptive Architectures for Regularization Vision Chips," Copyright 1995, with kind permission from Elsevier Science Ltd, The Boulevard, Langford Lane, Kidlington OX5 1GB, UK.

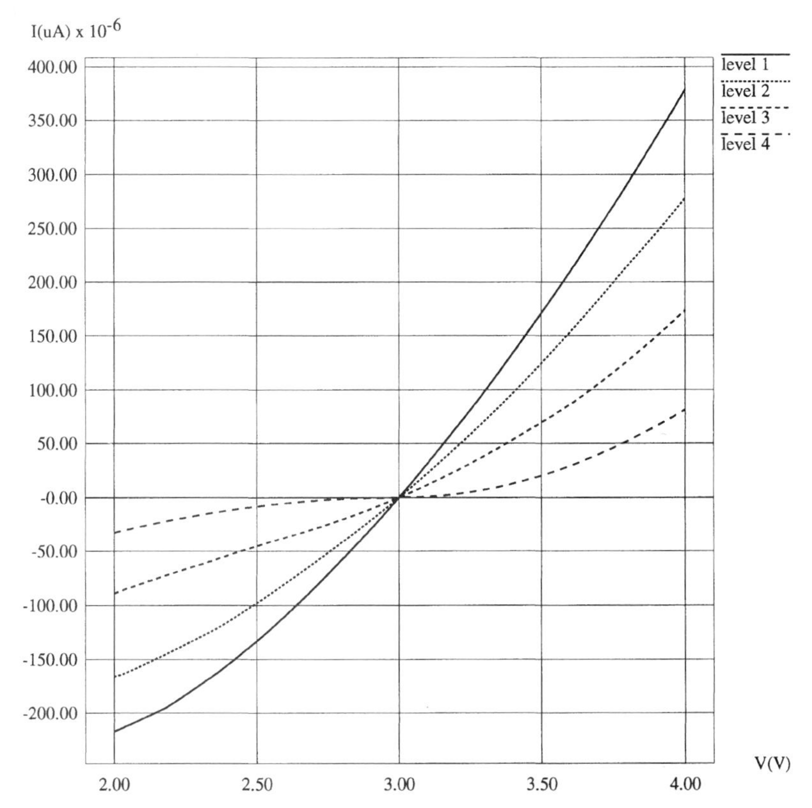

Figure 27 Simulation results of Figs. 25 and 26. V-I characteristics of g_{s2} are shown at several different values of $\sum_k v_k^0$. The "higher the level," the greater the value of $\sum_k v_k^0$. Reprinted from *Neural Networks* 8:87–101, H. Kobayashi, T. Matsumoto, T. Yagi, and K. Tanaka, "Light-Adaptive Architectures for Regularization Vision Chips," Copyright 1995, with kind permission from Elsevier Science Ltd, The Boulevard, Langford Lane, Kidlington OX5 1GB, UK.

3. Other Adaptations

a. Local Adaptation

The adaptation Eq. (66) is global in that the g_{s2} value changes according to the global information $\sum_k u_k$. If

$$g_{s2(k,k+1)} := \frac{1}{L(v_k^1 + v_{k+1}^1)}, \qquad L > 0, \tag{69}$$

where L is a constant, then the second-layer horizontal conductance value $g_{s2(k,k+1)}$ between node k and node $k+1$ is inversely proportional to the sum of

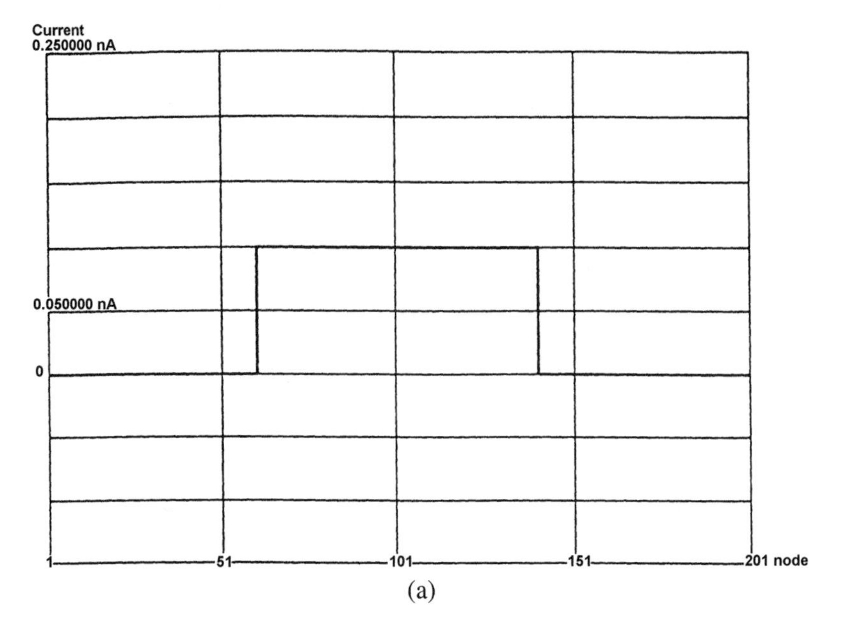

(a)

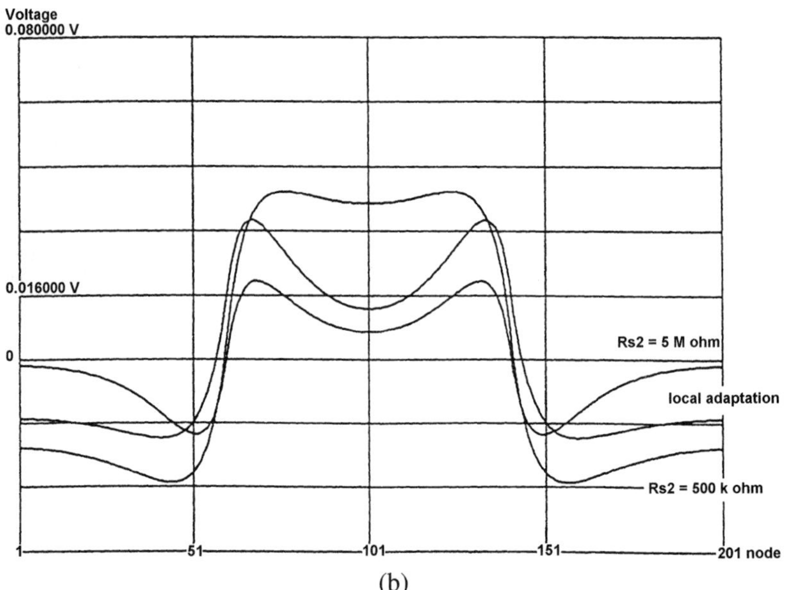

(b)

Figure 28 Response of the locally adaptive network. (a) A rectangular input image with 81 pixel width. (b) Responses of the networks with $1/g_{s2} = 5\ \mathrm{M\Omega}$ (no adaptation), $1/g_{s2} = 500\ \mathrm{k\Omega}$ (no adaptation), and $1/g_{s2(k,k+1)} = 2 \times 10^7 (v_k^1 + v_{k+1}^1)$ (local adaptation). Reprinted from *Neural Networks* 8:87–101, H. Kobayashi, T. Matsumoto, T. Yagi, and K. Tanaka, "Light-Adaptive Architectures for Regularization Vision Chips," Copyright 1995, with kind permission from Elsevier Science Ltd, The Boulevard, Langford Lane, Kidlington OX5 1GB, UK.

the first-layer voltages v_k^1 and v_{k+1}^1. Figure 28a is a simple rectangular input while Fig. 28b compares the response incorporating the local adaptation Eq. (69) where $L = 2 \times 10^7$ with those responses without adaptations where $1/g_{s2} = 5\ \mathrm{M}\Omega$ and $1/g_{s2} = 500\ \mathrm{k}\Omega$, respectively. Even though the effect of the local adaptation is not as dramatic as in Fig. 24, where the global adaptation is incorporated, one can see that where the input intensity is high, the response with Eq. (69) is closer to that with $1/g_{s2} = 5\ \mathrm{M}\Omega$. On the other hand, where the intensity is low, the adapted response behaves in a manner similar to the one with $1/g_{s2} = 500\ \mathrm{k}\Omega$. Therefore with Eq. (69) contrast is even more enhanced where interesting difference exists.

Figure 29 shows a possible circuit block diagram for the local adaptation and Fig. 30 shows a circuit design of locally adaptive conductances g_{s2} and bias generators in Fig. 29. The bias voltage generator at node k outputs v_k^c inversely proportional to the first-layer node voltage v_k^1, and $g_{s2(k,k+1)}$ is implemented with two parallel MOS FETs whose value is roughly proportional to $v_k^c + v_{k+1}^c$, and then this approximates Eq. (69). Figure 31 shows SPICE simulation results of $g_{s2(k,k+1)}$ characteristics at several different values of $v_k^1 + v_{k+1}^1$. One sees that as $v_k^1 + v_{k+1}^1$ becomes larger, $g_{s2(k,k+1)}$ decreases.

b. Maximum Value Adaptation

Consider

$$u_k^* := \frac{u_k}{M \max_i(u_i)}, \qquad M > 0, \tag{70}$$

which is implemented by the network in Fig. 32 where it senses the maximum input voltage and changes the gain of PGAs (programmable gain amplifiers) uniformly to as high a value as possible without overloading the network. Since there are all kinds of noises in a chip, one obtains a better signal-to-noise ratio if the input signal is amplified as much as possible without overloading the network. A similar method is widely used in A/D converters, where one can obtain a good signal-to-noise ratio if the converter is preceded by a PGA which amplifies small input signals so that the input signal stays within the full input range of the A/D converter.

Remarks. (i) When looked at as a regularization filter, the local adaptation mechanism Eq. (69) changes λ_1 and λ_2 according to v_k^1 and its local values so that they are described as $\lambda_1(\mathbf{v}^1, k)$ and $\lambda_2(\mathbf{v}^1, k)$ which are nonlinear.

(ii) Equation (70) corresponds to a different, though still linear, regularization problem. Namely, the function minimized is of the form

$$G(\mathbf{v}, \mathbf{d}^*(\mathbf{d})) = \|\mathbf{v} - \mathbf{d}^*(\mathbf{d})\|^2 + \lambda_1 \|\mathbf{D}\mathbf{v}\|^2 + \lambda_2 \|\mathbf{L}\mathbf{v}\|^2,$$

where $\mathbf{d}^*(\mathbf{d})$ indicates Eq. (70).

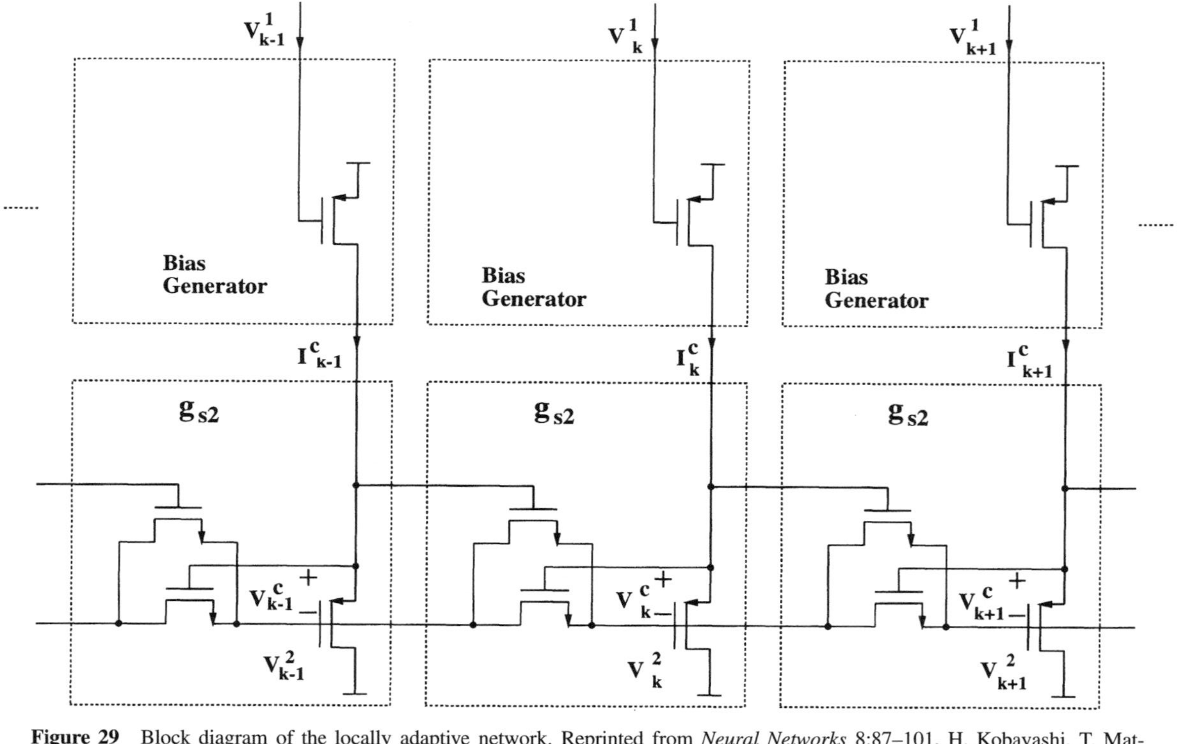

Figure 29 Block diagram of the locally adaptive network. Reprinted from *Neural Networks* 8:87–101, H. Kobayashi, T. Matsumoto, T. Yagi, and K. Tanaka, "Light-Adaptive Architectures for Regularization Vision Chips," Copyright 1995, with kind permission from Elsevier Science Ltd, The Boulevard, Langford Lane, Kidlington OX5 1GB, UK.

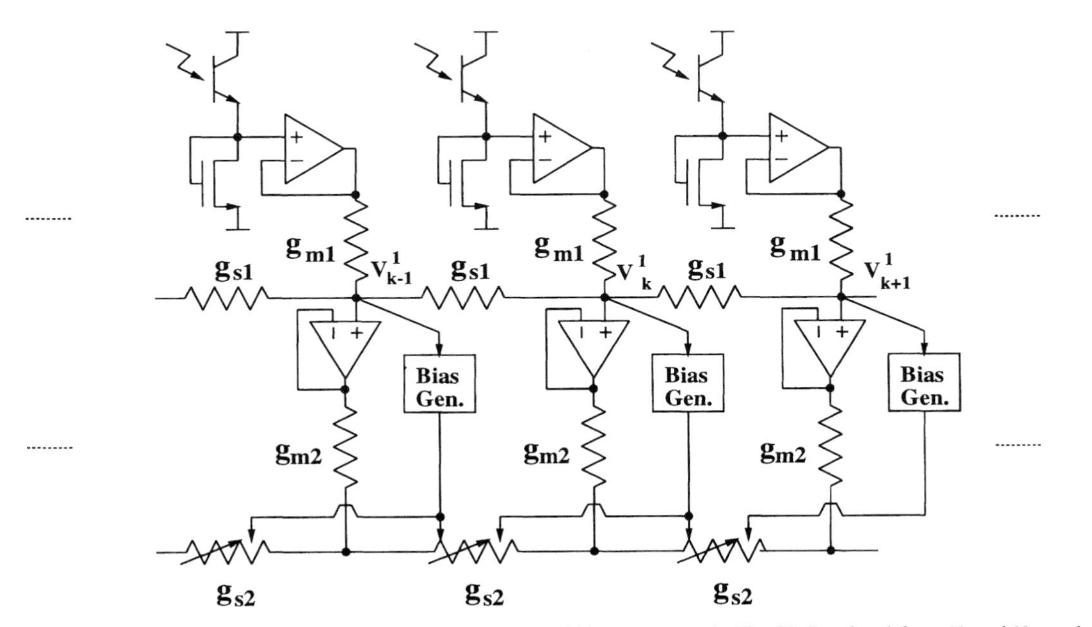

Figure 30 Circuit design of locally adaptive conductances g_{s2} and bias generators in Fig. 29. Reprinted from *Neural Networks* 8:87–101, H. Kobayashi, T. Matsumoto, T. Yagi, and K. Tanaka, "Light-Adaptive Architectures for Regularization Vision Chips," Copyright 1995, with kind permission from Elsevier Science Ltd, The Boulevard, Langford Lane, Kidlington OX5 1GB, UK.

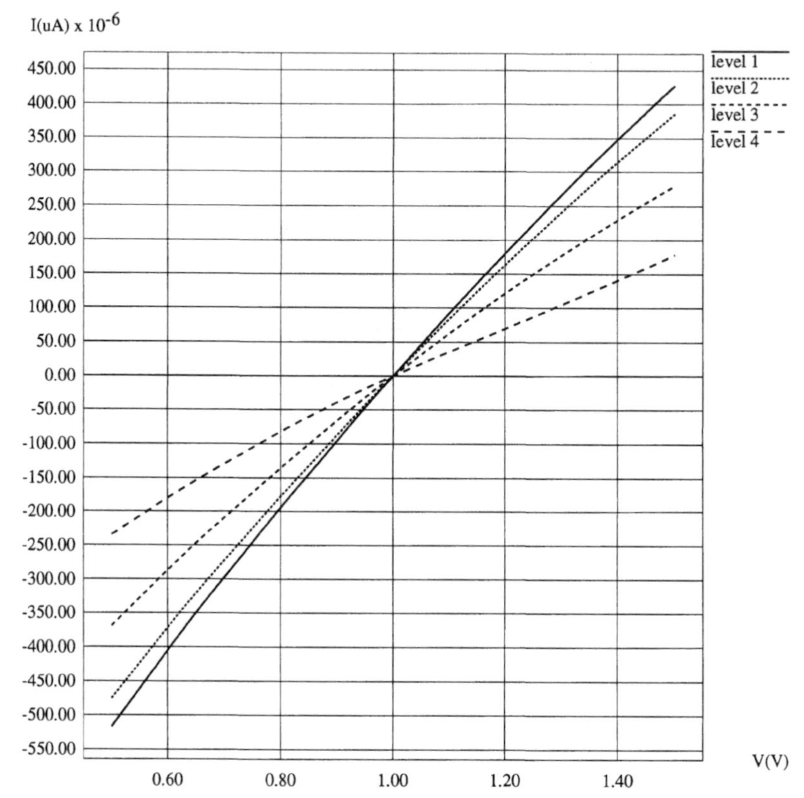

Figure 31 Simulation results of Figs. 29 and 30. V-I characteristics of $g_{s2(k,k+1)}$ are shown at several different values of $v_k^1 + v_{k+1}^1$. The "higher the level," the greater the value of $v_k^1 + v_{k+1}^1$. Reprinted from *Neural Networks* 8:87–101, H. Kobayashi, T. Matsumoto, T. Yagi, and K. Tanaka, "Light-Adaptive Architectures for Regularization Vision Chips," Copyright 1995, with kind permission from Elsevier Science Ltd, The Boulevard, Langford Lane, Kidlington OX5 1GB, UK.

F. Wiring Complexity

Wiring complexity is repeatedly emphasized in ([3, pp. 7, 116, 276–277]) as the **single most important** issue. It is indeed critical for implementing vision chips because, although each computing unit has relatively simple circuitry, there are thousands of computing units placed regularly so that the routing can be extremely difficult when the network architecture demands complicated interconnections among computing units.

Figure 10 shows a **unit cell** wiring for (an approximated) second-order regularization filter [42, 43], while Fig. 33 shows the actual implementation where every

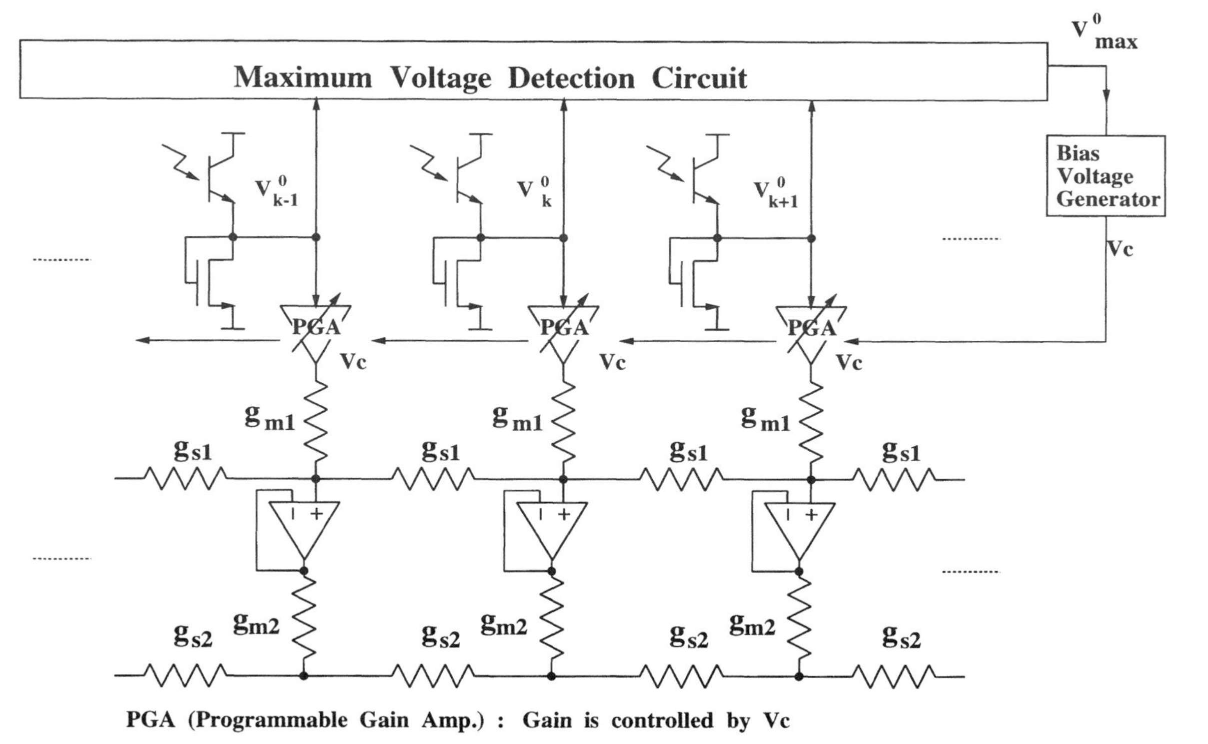

Figure 32 Maximum value adaptation network. Reprinted from *Neural Networks* 8:87–101, H. Kobayashi, T. Matsumoto, T. Yagi, and K. Tanaka, "Light-Adaptive Architectures for Regularization Vision Chips," Copyright 1995, with kind permission from Elsevier Science Ltd, The Boulevard, Langford Lane, Kidlington OX5 1GB, UK.

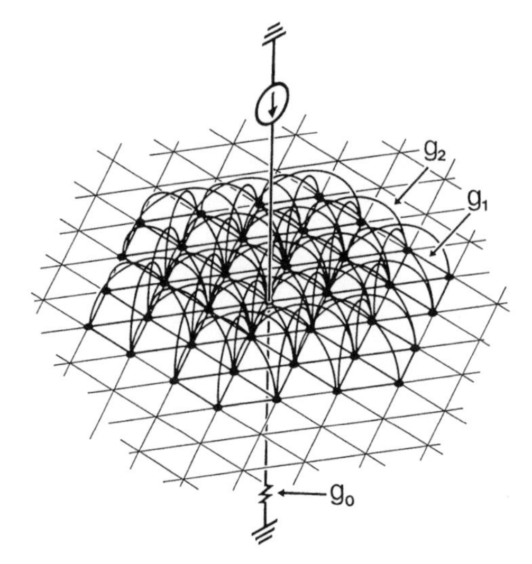

Figure 33 Actual implementation of the circuit in Fig. 10 demands connections with every second-nearest neighbor in addition to the immediate-neighbor connections. Reprinted from *Neural Networks* 6:327–350, H. Kobayashi, T. Matsumoto, T. Yagi, and T. Shimmi, "Image Processing Regularization Filters on Layered Architecture," Copyright 1993, with kind permission from Elsevier Science Ltd, The Boulevard, Langford Lane, Kidlington OX5 1GB, UK.

node must be connected with its second-nearest neighbors in addition to the nearest neighbors. Complexity of wiring was a serious problem in the layout phase of [42, 43] and yet this is a crude approximation to the second-order regularization filter.

If one wants to implement Eq. (53), the wiring gets even more serious. Let us look at, for instance, Fig. 34 which implments Eq. (53) (g_0 and input are not shown) provided that

$$g_1 : g_2 : \tilde{g}_2 = 10 + \frac{\lambda_1}{\lambda_2} : -2 : -1, \tag{71}$$

because the KCL reads

$$\begin{aligned} -(g_0 + 6g_1 + 6g_2 + 6\tilde{g}_2)v_{ij} &+ g_1(v_{i-1j} + v_{i+1j} + v_{ij-1} + v_{ij+1} + v_{i-1j+1} \\ &+ v_{i+1j-1}) + g_2(v_{i-2j} + v_{i+2j} + v_{ij-2} + v_{ij+2} + v_{i-2j+2} + v_{i+2j-2}) \\ &+ \tilde{g}_2(v_{i-1j-1} + v_{i+1j+1} + v_{i-1j+2} + v_{i+1j-2} + v_{i-2j+1} + v_{i+1j-1}) \\ &+ u_{ij} = 0, \end{aligned} \tag{72}$$

where u_{ij} is the input current source. Thus the network of Fig. 10 corresponds to $\tilde{g}_2 = 0$ in Fig. 35.

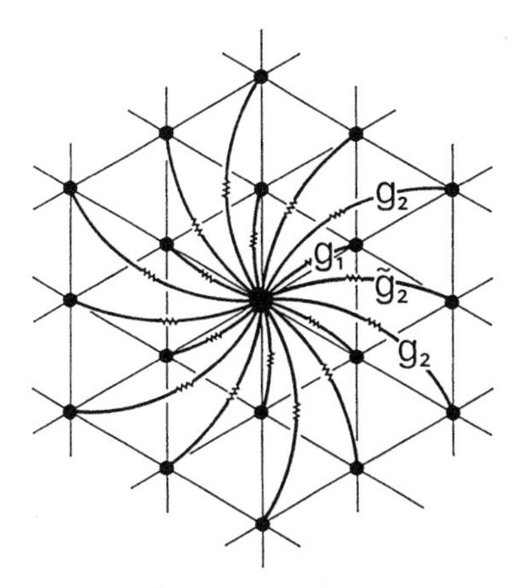

Figure 34 A network implementing $\mathbf{L}^2$. g_0 and input are not shown. Reprinted from *Neural Networks* 6:327–350, H. Kobayashi, T. Matsumoto, T. Yagi, and T. Shimmi, "Image Processing Regularization Filters on Layered Architecture," Copyright 1993, with kind permission from Elsevier Science Ltd, The Boulevard, Langford Lane, Kidlington OX5 1GB, UK.

Since Fact 1 claims that the layered network of Fig. 11 with only immediate neighbor connections, there must be a significant reduction of wiring complexity. This section tries to quantify the wiring complexity.

Let us first note that there are basically three categories in vision chip wiring:

Class 1: conductance interconnections between unit cells
Class 2: power supply lines and bias voltage lines
Class 3: data lines and address lines for data readout

Even though these are not completely independent of each other, we will pay particular attention to Class 1 because it is the dominant one and is critically dependent on the architecture of the signal processing part. Class 2 depends much more heavily on circuit design than the architecture. Class 3 essentially depends on the data readout mechanism.

Since a precise technical definition of wiring complexity is not given in [3], we will try to give a reasonable one here. Naturally we do not claim this is the best, nor only definition. In order to quantify wiring complexity, several simplifications are necessary. As far as wiring complexity is concerned, the following assumption will be made.

Assumption. The lateral conductances are regarded as pure wires, while the vertical conductances as well as the input circuit are regarded as a "unit cell."

Remark. Conductances g_1 and g_2 in Fig. 10 will be regarded as pure wires whereas g_0 and the input circuit are regarded as a unit cell. Similarly, g_{s1} and

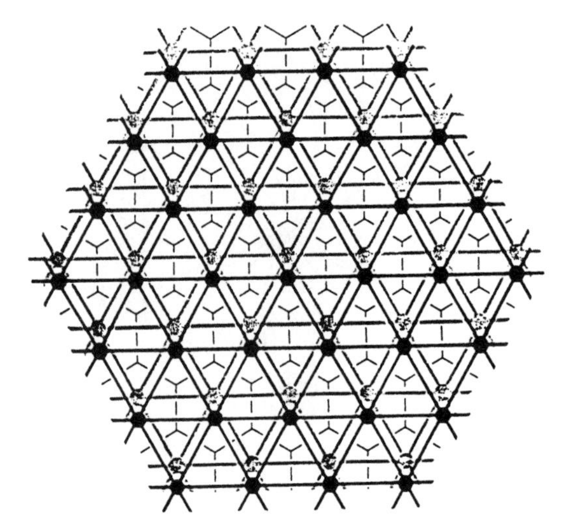

Figure 35 Wiring complexity of the layered network with $P = 2$ amounts to 6. A hexagon stands for a unit cell. Reprinted from *Neural Networks* 6:327–350, H. Kobayashi, T. Matsumoto, T. Yagi, and T. Shimmi, "Image Processing Regularization Filters on Layered Architecture," Copyright 1993, with kind permission from Elsevier Science Ltd, The Boulevard, Langford Lane, Kidlington OX5 1GB, UK.

g_{s2} in Fig. 13 are regarded as pure wires whereas g_{m1}, g_{m2}, and the input circuit consitute a unit cell.

A natural question arises. Does not the unit cell of a multilayered network need more chip area than that of a single-layered network? Not necessarily. Let us compare, for instance, Fig. 10 with Fig. 13. First note that in actual implementation, one-half of each lateral resistor $1/g_r$ or $1/g_{s_r}$ is realized in each unit cell area. Second, since g_2 in Fig. 10 is negative, it demands more transistors. In [42, 43], g_2 necessitates a transconductance amplifier and six transistors per node. In Fig. 13, the voltage-controlled current source is realized by a differential amplifier together with g_{m2} and hence six transistors are enough per node. Thus the unit cell area of a layered network would not be any larger. Hence the wiring complexity of a chip is the complexity of wiring among unit cells. We assume, therefore, that the unit cell area is normalized to 1×1.

DEFINITION. The wiring complexity of a vision chip is defined as the number of wires which **cross** a unit cell.

Remarks. (i) The unit cell defined above correponds to a pixel.

(ii) For the wiring complexity, one has to count not only the wires connecting a particular unit with another unit but also those which **pass through** a unit cell for the purpose of connecting **other** cells together.

(iii) If the unit cell size is normalized to 1×1, our definition of wiring complexity means the wire length. Observe that for a chip implementation, a wire which comes into a unit cell area contributes to the same complexity whether or not there is an electrical contact at the unit cell because one simply places a "via" (hole) if there is an electrical contact.

Fact 5. Consider the layered network of Fig. 11 on a hexagonal grid. If the number of layers is P, then

$$\text{wiring complexity} = 3P. \tag{73}$$

Proof. Since each layer has only immediate neighbor connections, three wires cross each unit cell represented by a hexagon. ■

Figure 35 shows the case with $P = 2$. As for a single-layer network with general P on a hexagonal grid, the wiring complexity formula itself gets complicated. We will give formulas up to $P = 3$ which is enough for the present purpose.

Fact 6. (i) For the single-layer network which implements Eq. (49) ($P = 1$),

$$\text{wiring complexity} = 3. \tag{74}$$

(ii) For the single-layer network of Fig. 34 which implements Eq. (53) ($P = 2$),

$$\text{wiring complexity} = 15. \tag{75}$$

(iii) For the single-layer network of Fig. 34 with $\tilde{g}_2 = 0$, which implements Eq. (57) ($P = 2$),

$$\text{wiring complexity} = 9. \tag{76}$$

(iv) For the single-layer network of Fig. 35 which implements Eq. (60) ($P = 3$),

$$\text{wiring complexity} = 33. \tag{77}$$

Proof. For $P = 1$, the single-layer network and the "multilayer network" coincide. Consider the network of Fig. 34 which implements Eq. (53). There are three classes of wires which cross a unit cell represented by a hexagon:

(a) The g_1 connections which give rise to three wires crossing a unit cell (Fig. 36). The g_2 connections demand six wires, not three, because, in addition to the three wires which connect each unit cell with its second neighbors, there is another set of three wires connecting between the neighboring nodes.

(b) In order to see the complexity of the g_2 connections, let us look at Fig. 37.

In order to avoid an obvious technical difficulty in drawing the figure, four different textures are used for wires. Where a circle is placed with a particular texture, there is an electrical contact by a wire with that particular texture.

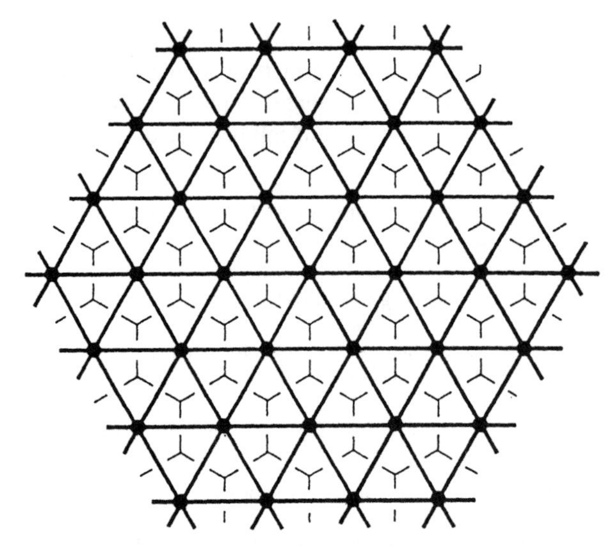

Figure 36 Wiring complexity of the g_1 connections contributes 3. Reprinted from *Neural Networks* 6:327–350, H. Kobayashi, T. Matsumoto, T. Yagi, and T. Shimmi, "Image Processing Regularization Filters on Layered Architecture," Copyright 1993, with kind permission from Elsevier Science Ltd, The Boulevard, Langford Lane, Kidlington OX5 1GB, UK.

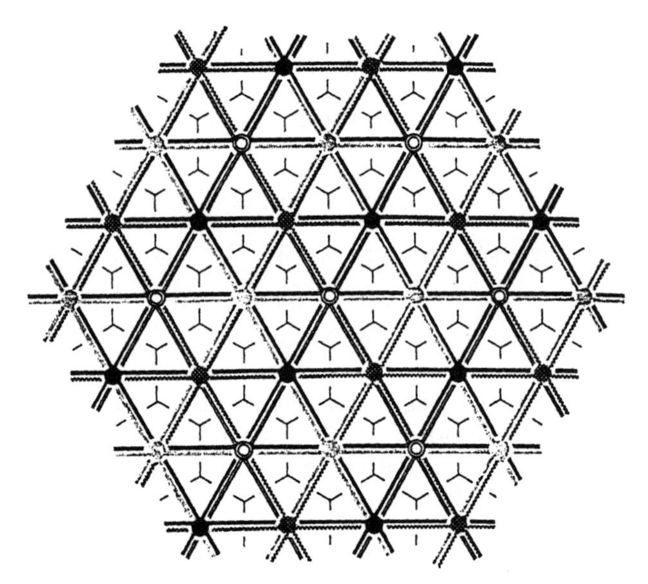

Figure 37 Wiring complexity of the g_2 connections is 6. Three wires connect a cell with its second-nearest neighbor while another three wires pass through each cell. Reprinted from *Neural Networks* 6:327–350, H. Kobayashi, T. Matsumoto, T. Yagi, and T. Shimmi, "Image Processing Regularization Filters on Layered Architecture," Copyright 1993, with kind permission from Elsevier Science Ltd, The Boulevard, Langford Lane, Kidlington OX5 1GB, UK.

(c) The $\tilde{g}_2$ connections which also demand six wires. In order to demonstrate this, let us look at Fig. 33. First, note that the wires drawn in this figure are not present in Fig. 37. For instance, there are no "horizontal" connections in Fig. 38, while "vertical" connections are present which are not present in Fig. 37. Thus, in addition to the three wires which cross a unit cell "in the middle," there are another six wires passing through the "boundary" of a unit cell represented by a hexagon. Since a wire must pass through somewhere, by an appropriate "splitting," one sees that the complexity contribution from these wires is 3.

Therefore, $3+6+6=15$ wires contribute to the complexity which is Eq. (75). If $\tilde{g}_2 = 0$, then one has nine wires, which is Eq. (76). Using a similar argument, one can show that the g_3 connections and the $\tilde{g}_3$ connections of Fig. 37 demand 18 wires which must be added to 15 and hence the complexity is 33. ∎

Reduction of the wiring complexity by the layered architecture is significant. Let us call the ratio between the wiring complexity of a layered network and the wiring complexity of a single-layer network, the **complexity ratio**.

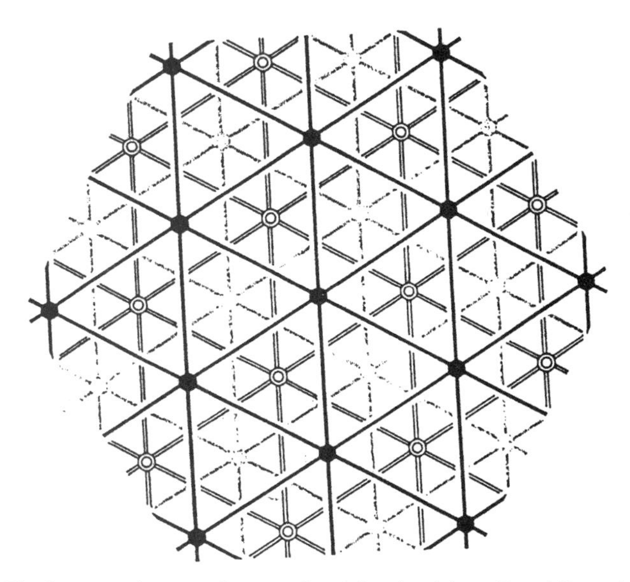

Figure 38 The $\tilde{g}_2$ connections contribute another 6. Reprinted from *Neural Networks* 6:327–350, H. Kobayashi, T. Matsumoto, T. Yagi, and T. Shimmi, "Image Processing Regularization Filters on Layered Architecture," Copyright 1993, with kind permission from Elsevier Science Ltd, The Boulevard, Langford Lane, Kidlington OX5 1GB, UK.

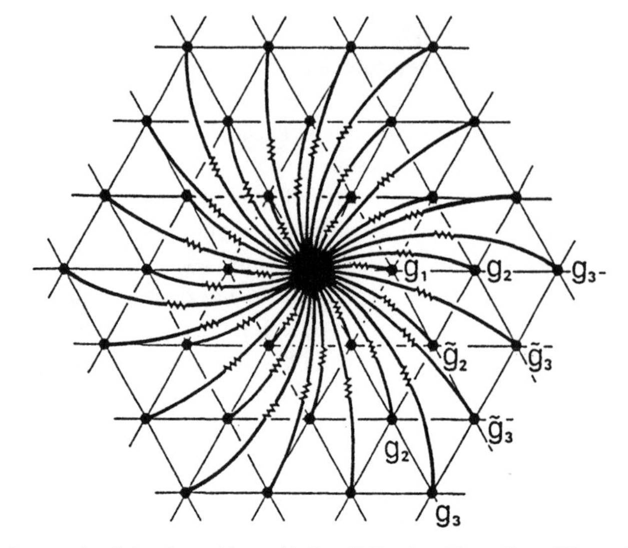

Figure 39 A network solving the problem with $P = 3$. Reprinted from *Neural Networks* 6:327–350, H. Kobayashi, T. Matsumoto, T. Yagi, and T. Shimmi, "Image Processing Regularization Filters on Layered Architecture," Copyright 1993, with kind permission from Elsevier Science Ltd, The Boulevard, Langford Lane, Kidlington OX5 1GB, UK.

Fact 7. (i) For the network of Fig. 34 ($P = 2$),

$$\text{complexity ratio} = \tfrac{2}{5}. \tag{78}$$

(ii) For the network of Fig. 39 ($P = 3$),

$$\text{complexity ratio} = \tfrac{3}{11}. \tag{79}$$

IV. SPATIO-TEMPORAL STABILITY OF VISION CHIPS

A. Introduction

Vision chip architecture sometimes demands *negative* conductance values. For instance, exact implementation of the second-order regularization

$$6v_k - 4(v_{k-1} + v_{k+1}) + (v_{k+2} + v_{k-2})$$

necessitates negative conductance values [2]. Whenever negative conductance is present, there are potential stability problems. This section has been motivated by the temporal versus spatial stability issues of an image smoothing vision chip [42, 43]. The function of the chip is to smooth a two-dimensional image in an extremely fast manner. It consists of the 45 × 40 hexagonal array of very simple "cell" circuits, described in Fig. 10. An image is projected onto the chip through a lens (Fig. 40), and the photosensor represented by the current source inputs the signal to the processing circuit. The output (smoothed) image is represented as the node voltage distribution of the array. With an appropriate choice of $g_0 > 0$, $g_1 > 0$, and $g_2 < 0$, the chip performs a regularization with second-order constraints and closely approximates the Gaussian convolver. Since the *negative conductance* $g_2 < 0$ is involved, two stability issues naturally arise:

(i) Because the chip is fabricated by a CMOS process, parasitic capacitors induce the dynamics with respect to time. This raises the *temporal stability* issue with respect to whether the network converges to a stable equilibrium point.

(ii) Because a processed (smoothed) image is given as the node voltage distribution of the array, the *spatial stability* issue also arises even if the temporal dynamics does converge to a stable equilibrium point. In other words, the node voltage distribution may behave wildly, e.g., oscillate.

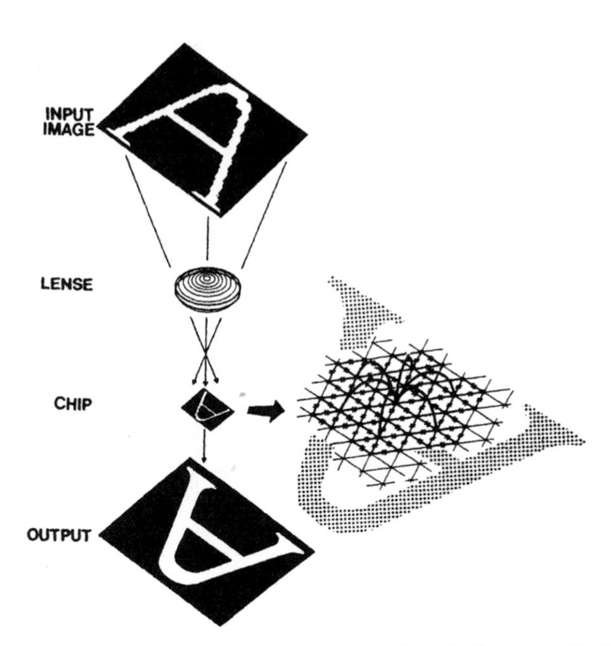

Figure 40 A chematic diagram. Reprinted with permission from T. Matsumoto, H. Kobayashi, and Y. Togawa, *IEEE Trans. Neural Networks* 3:540–569, 1992 (©1992 IEEE).

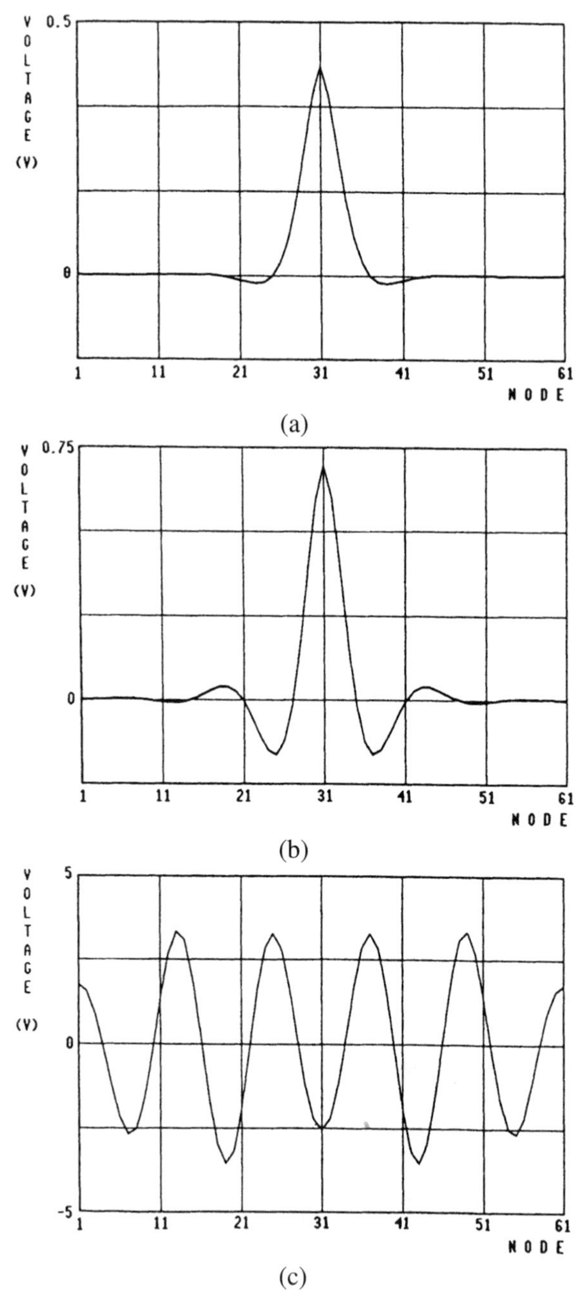

Figure 41 Spatial impulse responses with $N = 61$, $m = 2$, $1/g_0 = 200$ kΩ, $1/g_1 = 5$ kΩ, $u_{31} = 10$ μA, $u_k = 0$ for $k \neq 31$. (a) $1/g_2 = -20$ kΩ; stable. (b) $1/g_2 = -18$ kΩ; stable. (c) $1/g_2 = -17$ kΩ; unstable. Reprinted with permission from T. Matsumoto, H. Kobayashi, and Y. Togawa, *IEEE Trans. Neural Networks* 3:540–569, 1992 (©1992 IEEE).

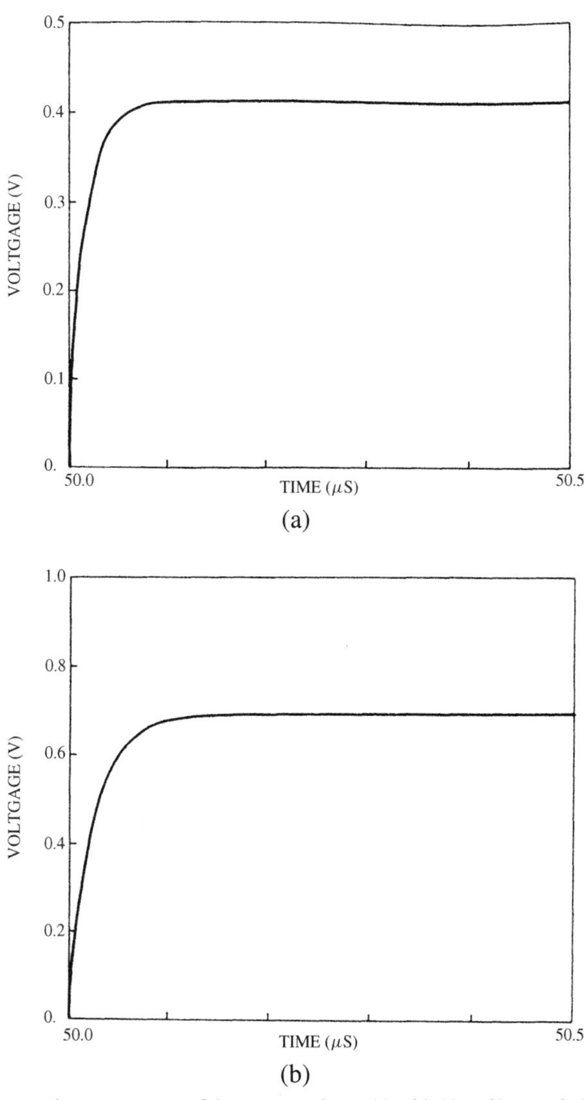

Figure 42 Temporal step responses of the center node $v_{31}(t)$ with $N = 61$, $m = 2$, $1/g_0 = 200$ kΩ, $1/g_1 = 5$ kΩ, $c_0 = 0.1$ pF, $u_k(t) \equiv 0$ for $k \neq 31$, and $u_{31}(t) = 0$ when $t < 50$ μs, 10 μA when $t \geq 50$ μs. (a) $1/g_2 = -20$ kΩ; stable. (b) $1/g_2 = -18$ kΩ; stable. (c) $1/g_2 = -17$ kΩ; unstable. Reprinted with permission from T. Matsumoto, H. Kobayashi, and Y. Togawa, *IEEE Trans. Neural Networks* 3:540–569, 1992 (©1992 IEEE).

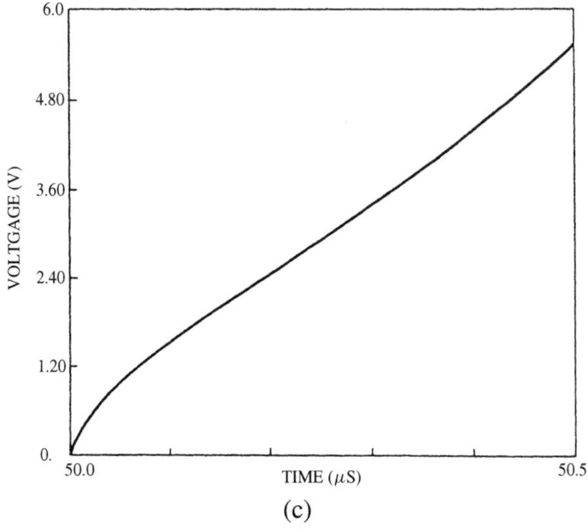

(c)

Figure 42 (*Continued*)

Our earlier numerical experiments investigating these issues were rather intriguing. The results suggested that the network is temporally stable "if and only if" it is spatially stable. Figure 41 shows spatial impulse responses at different sets of parameter values. The network has 61 nodes (linear array, for simplicity), and the impulse is injected at the center node. Figure 42 shows the corresponding temporal step responses of the center node. For simplicity, the only parasitic capacitors taken into account are those from each node to ground. The responses shown in Fig. 42a and b are temporally stable, while that in Fig. 42c is not. Figure 41c is spatially unstable because the response does not decay, which is highly undesirable for image processing. (A precise definition of spatial stability will be given later.) All of our earlier numerical experiments suggested the equivalence of the two stability conditions. However, there are no *a priori* reasons for them to be equivalent. As will be shown rigorously, the two stability conditions are *not* equivalent. The spatial stability condition is *stronger* than the temporal stability condition. Nevertheless, the set of parameter values (g_0, g_1, g_2) for which the two stability conditions disagree turns out to be a (Lebesgue) *measure zero* subset, which explains why our numerical experiments suggested equivalence between the two conditions. (A measure zero subset is difficult to "hit.") Explicit analytical conditions will be given for the temporal as well as the spatial stabilities in a general setting. Also given is an estimate of the speed of temporal responses of the networks. Since our results are proved in a general setting, they can be applied to other neural networks of a similar nature.

Remark 1. Due to the space limitation, many of the proofs and technical details cannot be included. The reader is referred to [40] for complete proofs and details. We note that the vision chip stability issues are descibed in [45, 51, 52] in a different problem setting and/or using different approaches.

B. Stability-Regularity

1. Formulation

Consider a neural network consisting of a linear array of N nodes, where each node is connected with its pth-nearest neighbor nodes, $p = 1, 2, \ldots, m < N$, via a (possibly negative) conductance g_p and a capacitance c_p. Figure 43 shows the case where $m = 3$. The network is described by

$$\sum_{p \in M} b_p \frac{dv_{i-p}}{dt} = \sum_{p \in M} a_p v_{i-p} + u_i, \qquad i = 1, 2, , \ldots, N, \tag{80}$$

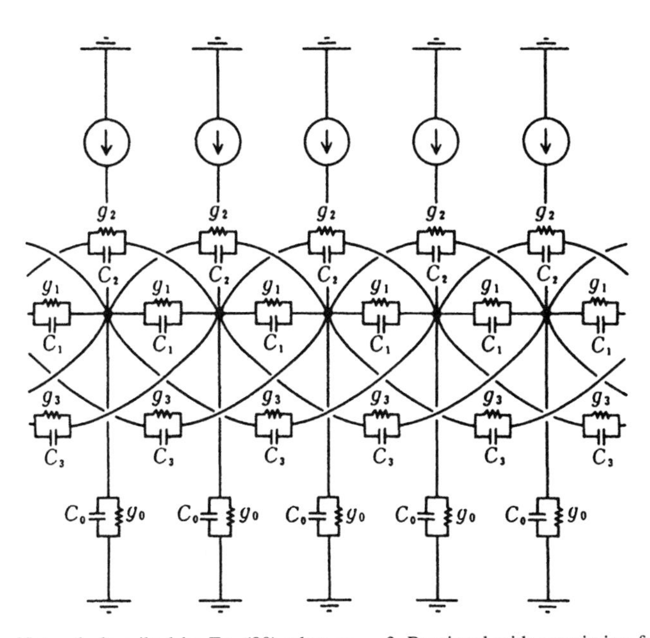

Figure 43 Network described by Eq. (80) when $m = 3$. Reprinted with permission from T. Matsumoto, H. Kobayashi, and Y. Togawa, *IEEE Trans. Neural Networks* 3:540–569, 1992 (©1992 IEEE).

where v_i and u_i are the voltage and the input current at the ith node, and

$$M = \{p\text{: integer } |p| \leq m\}, \tag{81}$$

$$a_0 = -\left(g_0 + 2\sum_{p=1}^{m} g_p\right), \qquad a_{\pm p} = g_p, \qquad 1 \leq p \leq m, \tag{82}$$

$$b_0 = c_0 + 2\sum_{p=1}^{m} c_p, \qquad b_{\pm p} = -c_p, \quad 1 \leq p \leq m. \tag{83}$$

Equation (80) is obtained simply by writing down the Kirchhoff's current law (KCL) at each node. Letting

$$\mathbf{v} := (v_0, v_1, \ldots, v_{N-1})^T \in \mathcal{R}^N \quad \text{and} \quad \mathbf{u} := (u_0, u_1, \ldots, u_{N-1})^T \in \mathcal{R}^N$$

(T denoting transpose), one can recast Eq. (80) as

$$\mathbf{B}\frac{d}{dt}\mathbf{v} = \mathbf{A}\mathbf{v} + \mathbf{u}, \tag{84}$$

where

$$\mathbf{A} := \{A(i,j)\} \in R^{N\times N}, \qquad i, j = 0, 1, \ldots, N-1,$$
$$A(i,j) := \begin{cases} a_k, & \text{when } i - j = \pm k,\ k = 0, \ldots, m, \\ 0, & \text{otherwise,} \end{cases} \tag{85}$$

$$\mathbf{B} := \{B(i,j)\} \in R^{N\times N}, \qquad i, j = 0, 1, \ldots, N-1,$$
$$B(i,j) := \begin{cases} b_k, & \text{when } i - j = \pm k,\ k = 0, \ldots, m, \\ 0, & \text{otherwise.} \end{cases} \tag{86}$$

Note $\mathbf{A}$ as well as $\mathbf{B}$ is symmetric and has a uniform band structure. If $\mathbf{B}$ is nonsingular, an equlibrium point of Eq. (84) satisfies

$$-\sum_{p\in M} a_p v_{i-p} = u_i, \tag{87}$$

which is a difference equation instead of a differential equation. Assuming that $a_m \neq 0$, one has

$$v_{i+m} = -\frac{1}{a_m}\left(\sum_{p\in M-\{m\}} a_p v_{i+p} + u_i\right). \tag{88}$$

Therefore, letting

$$\mathbf{F} := \begin{bmatrix} 0 & 1 & 0 & 0 & . & . & 0 \\ 0 & 0 & 1 & 0 & . & . & 0 \\ 0 & 0 & 0 & 1 & . & . & 0 \\ . & . & . & . & . & . & . \\ . & . & . & . & . & . & . \\ 0 & 0 & 0 & . & . & 0 & 1 \\ -1 & -a_{m-1}/a_m & -a_{m-2}/a_m & . & -a_0/a_m & . & -a_{m-1}/a_m \end{bmatrix} \in \mathcal{R}^{2m \times 2m} \tag{89}$$

with

$$\begin{aligned} \mathbf{x}_k &:= (v_{k-m}, v_{k-m+1}, \ldots, v_k, \ldots, v_{k+m-1})^T \in \mathcal{R}^{2m}, \\ \mathbf{y}_k &:= (0, \ldots, 0, -u_k/a_m)^T \in \mathcal{R}^{2m}, \end{aligned}$$

one can rewrite Eq. (88) as

$$\mathbf{x}_{k+1} = \mathbf{F}\mathbf{x}_k + \mathbf{y}_k. \tag{90}$$

Observe that subscript k in Eq. (90) is not time. Equation (90) represents the spatial dynamics induced by the temporal dynamics Eq. (84). Note also that $\dim \mathbf{v} = N$, the number of nodes, while $\dim \mathbf{x}_k = 2m$, the size of the neighborhood, which is independent of N.

In image processing, input is $\mathbf{u}$ while output is $\mathbf{v}(\infty)$, the stable equilibrium point of Eq. (84). Equation (90) describes how the coordinates of $\mathbf{v}(\infty)$ are distributed with respect to k. There are several issues that need care.

(i) The temporal dynamics given by Eq. (84) consitute an initial value problem while Eq. (87) or Eq. (90) is a boundary value problem. Namely, arbitrary $\mathbf{v}(0)$ and $\mathbf{u}(.)$ completely determine the solution to Eq. (84) while for Eq. (87) or Eq. (90), one *cannot* specify (for a given $\{\mathbf{y}_k\}$) an arbitrary $\{\mathbf{x}_0\}$ because a solution $\{\mathbf{x}_k\}$ must be consistent with the KCLs at the end points. Therefore the temporal dynamics (84) are *causal* while the spatial dynamics (90) are *noncausal.*

(ii) The stability of the spatial dynamics (90) must be carefully defined. That "Eq. (90) is stable iff all the eigenvalues of $\mathbf{F}$ lie inside the unit circle" does not work because $\mathbf{F}$ has a special structure [see Eq. (107)]:

$$\text{if } \lambda \text{ is an eigenvalue, so is } 1/\lambda.$$

Therefore, "$|\lambda| < 1$ for all λ" is never satisfied. Since $N = 2K + 1$ is *finite*, another standard definition of stability:

$$\sum_k \|\mathbf{y}_k\| < \infty \quad \text{implies} \quad \sum_k \|\mathbf{x}_k\|^2 < \infty \tag{91}$$

does not work either, because Eq. (91) is always satisfied. As was shown in Fig. 41c, $\mathbf{x}_k$ can behave in a wild manner even if $N = 2K + 1$ is finite, which is highly undesirable for image processing purposes.

2. Spatial Dynamics

Let λ_{s_i}, λ_{c_i}, and λ_{u_i} be the eigenvalues of $\mathbf{F}$ satisfying

$$|\lambda_{s_i}| < 1, \qquad |\lambda_{c_i}| = 1, \qquad |\lambda_{u_i}| > 1,$$

respectively, and let E^s, E^c, and E^u be the (generalized) eigenspaces corresponding to λ_{s_i}, λ_{c_i}, and λ_{u_i}, respectively. They are called stable, center, and unstable eigenspaces, respectively. Let $E = R^{2m}$. Then [53]

$$E = E^s \oplus E^c \oplus E^u, \tag{92}$$

where $\oplus$ denotes a direct sum decomposition, and

$$\mathbf{F}(E^\alpha) = E^\alpha, \qquad \alpha = s, c, u, \tag{93}$$

i.e., E^s, E^c, and E^u are *invariant* under $\mathbf{F}$.

Our task here is to give an appropriate definition of spatial stability while maintaining consistency with Eq. (91) when $N \to \infty$.

First, we remark that the boudnary conditions are crucial for the spatial stability as indicated by the following example.

EXAMPLE 1. Consider the simplest case, $m = 1$ in Eq. (90) with $g_0 = g$, $g_1 = 2g$, $g > 0$ (Fig. 44a). Then

$$\mathbf{F} := \begin{bmatrix} 0 & 1 \\ -1 & 5/2 \end{bmatrix},$$

and $\mathbf{F}$ is hyperbolic because eigenvalues are $\rho_1 = 1/2$ and $\rho_2 = 2$. Figure 45a shows the impulse response when $1/g = 50\ \mathrm{k}\Omega$, where the impulse is injected at the center node. Let us now replace the rightmost g_0 and the leftmost g_0 with $g_t = -g$ as in Fig. 44b. The impulse response is then given by Fig. 45b, which "explodes" in the negative direction.

There is another story about spatial responses. Our simulation results indicate that the spatial responses behave quite properly even if the g_t value is varied by a large amount. Namely, spatial impulse responses are very robust against variations of g_t from g_0. Thus, two fundamental questions concerning the spatial dynamics must be answered for the spatial stability definition:

(i) Why does a particular g_t value give rise to explosion of impulse responses even if the eigenvalues of $\mathbf{F}$ are off the unit circle?

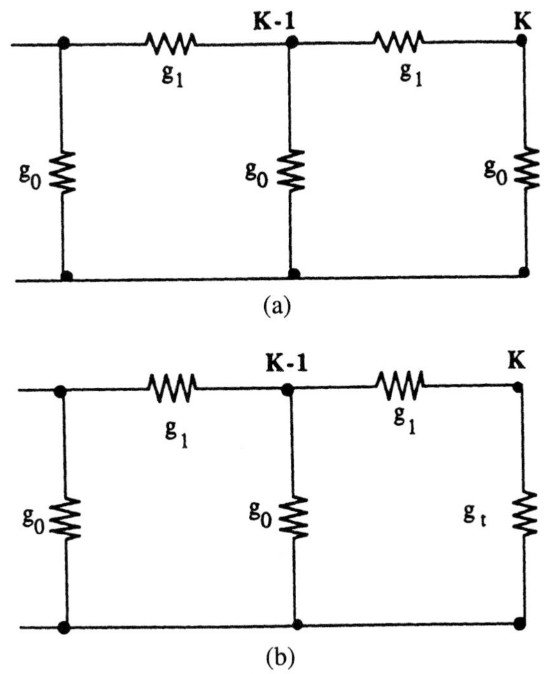

Figure 44 A network with $m = 1$. (a) Original network. (b) Modified boundary condition, where the rightmost g_0 is replaced by g_t. Reprinted with permission from T. Matsumoto, H. Kobayashi, and Y. Togawa, *IEEE Trans. Neural Networks* 3:540–569, 1992 (©1992 IEEE).

(ii) Why do impulse responses behave properly over a wide range of g_t values?

DEFINITION 1. Consider Eq. (90) and let $\{\mathbf{y}_k\}$ be nonzero only for $0 \leq k \leq d$. Then $\{\bar{\mathbf{x}}_k\}_{-\infty}^{+\infty}$ is said to be a *free-boundary solution* if

$$\bar{\mathbf{x}}_{k+1} = \mathbf{F}\bar{\mathbf{x}}_k, \qquad k \leq 0, \tag{94}$$

$$\bar{\mathbf{x}}_d = \mathbf{F}^d\bar{\mathbf{x}}_0 + \sum_{k=0}^{d-1} \mathbf{F}^{d-k}\mathbf{y}_k, \tag{95}$$

$$\bar{\mathbf{x}}_{k+1} = \mathbf{F}\bar{\mathbf{x}}_k, \qquad k \geq d. \tag{96}$$

Remark 2. If $d = 1$, then $\{\mathbf{y}_k\}$ is an impulse. If one redefines the summation term in Eq. (95) as a new $\mathbf{y}_0$, then Eqs. (94), (95), (96) can be replaced by

$$\bar{\mathbf{x}}_{k+1} = \mathbf{F}\bar{\mathbf{x}}_k, \qquad k \neq 0, \tag{97}$$

$$\bar{\mathbf{x}}_1 = \mathbf{F}\bar{\mathbf{x}}_0 + \mathbf{y}_0. \tag{98}$$

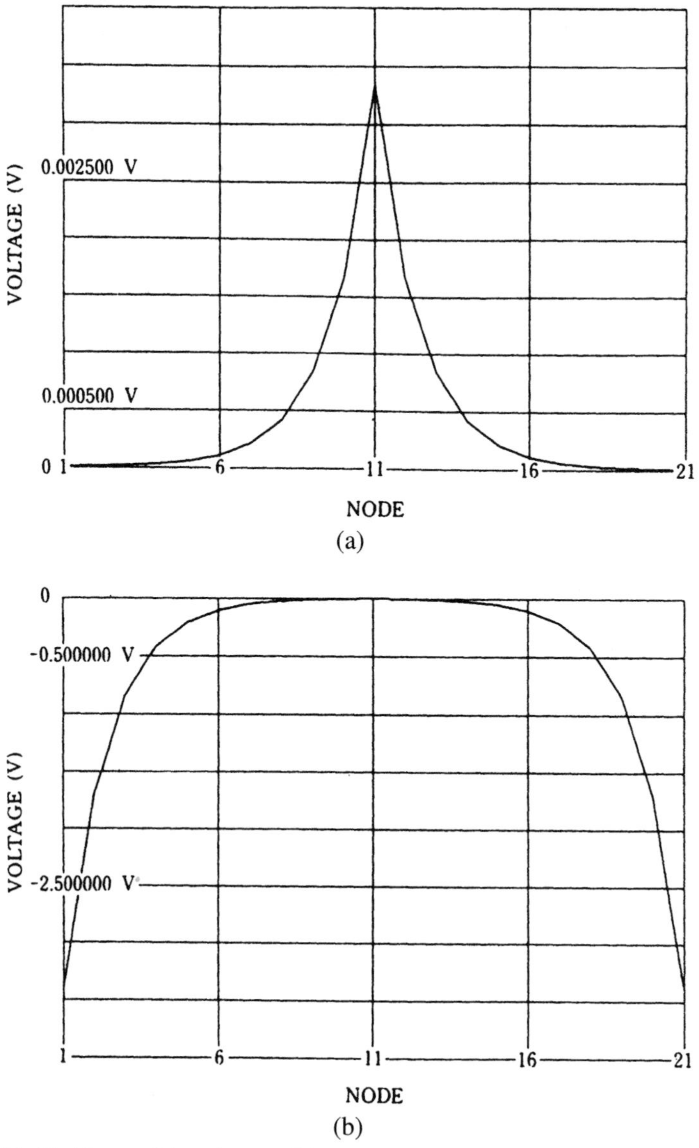

Figure 45 Significance of boundary conditions. (a) Impulse response for Fig. 44a with $g_0 = g$, $g_1 = 2g$, $1/g = 50\ \text{k}\Omega$, $u_{31} = 0.1\ \mu\text{A}$. (b) Impulse response for Fig. 44b with the same data except for $g_t = -g$, $u_{31} = 0.1\ \mu\text{A}$. Reprinted with permission from T. Matsumoto, H. Kobayashi, and Y. Togawa, *IEEE Trans. Neural Networks* 3:540–569, 1992 (©1992 IEEE).

Since no boundary conditions are imposed, $\{\bar{\mathbf{x}}_k\}_{-\infty}^{+\infty}$ is not unique.

DEFINITION 2 (Spatial stability). A neural network described by Eq. (90) is said to be spatially stable if and only if there is a unique free-boundary solution $\{\bar{\mathbf{x}}_k\}_{-\infty}^{+\infty}$ satisfying

$$\sum_{k=-\infty}^{+\infty} \|\bar{\mathbf{x}}_k\|^2 < \infty. \tag{99}$$

PROPOSITION 1. (i) *The network described in Eq. (90) is spatially stable if and only if the* $\mathbf{F}$ *matrix of the spatial dynamics is hyperbolic.*

(ii) *The unique free-boundary solution* $\{\bar{\mathbf{x}}_k\}_{-\infty}^{+\infty}$ *satisfying Eq. (90) is determined by*

$$\bar{\mathbf{x}}_1 \in E^s, \qquad \bar{\mathbf{x}}_0 \in E^u, \qquad \bar{\mathbf{x}}_1 = F^d\bar{\mathbf{x}}_0 + \mathbf{y}_0. \tag{100}$$

DEFINITION 3 (Stable free-boundary solution). The unique $\{\bar{\mathbf{x}}_k\}_{-\infty}^{+\infty}$ given in Proposition 1 is said to be the stable free-boundary solution.

Consider the spatial dynamics Eq. (87) and let T_+ (resp. T_-) be an m-dimensional linear subspace which describes the boundary conditions at the right (resp. left) end.

DEFINITION 4. Let $\{\mathbf{y}_k\}$ be nonzero for $0 \le k \le d$. Then $\{\mathbf{x}_k\}_{-K}^{+K}$ is said to be a solution for (T_+, T_-, K) if

$$\mathbf{x}_{k+1} = \mathbf{F}\mathbf{x}_k, \qquad -K \le k \le K, \quad k \ne 0, \tag{101}$$

$$\mathbf{x}_1 = \mathbf{F}^d\mathbf{x}_0, \tag{102}$$

$$\mathbf{x}_{-K} \in T_-, \qquad \mathbf{x}_K \in T_+. \tag{103}$$

The following result thoroughly answers the second question that arose in connection with spatial dynamics in a very general setting.

THEOREM 1. *Let a neural network described in Eq. (90) be spatially stable, i.e.,* $\mathbf{F}$ *be hyperbolic. If the boundary conditions* T_+ *and* T_- *satisfy*

$$T_+ + E^u = E, \qquad T_- + E^s = E, \tag{104}$$

then a solution $\{\mathbf{x}\}_{-K}^{+K}$ *for* (T_+, T_-, K) *converges to the stable free-boundary solution* $\{\bar{\mathbf{x}}\}_{-K}^{+K}$ *as* $K \to \infty$:

$$\lim_{K\to\infty} \sum_{k=-K}^{+K} \|\bar{\mathbf{x}}_k - \mathbf{x}_k\|^2 = 0. \tag{105}$$

3. Temporal Stability—Spatial Regularity

DEFINITION 5 (Spatial regularity). A neural network described by Eq. (90) is said to be spatially regular if there is a nonsingular $2m \times 2m$ matrix $\mathbf{T}$ such that

$$E = E^s \oplus E^c \oplus E^u,$$

$$\mathbf{TFT}^{-1} = \begin{bmatrix} \mathbf{F}_s & \mathbf{0} & \mathbf{0} & \mathbf{0} \\ \mathbf{0} & \mathbf{F}_c & \mathbf{G} & \mathbf{0} \\ \mathbf{0} & \mathbf{0} & \mathbf{F}_c & \mathbf{0} \\ \mathbf{0} & \mathbf{0} & \mathbf{0} & \mathbf{F}_s^{-1} \end{bmatrix}, \tag{106}$$

where a blank indicates a zero matrix, and elements of $\mathbf{G}$ consist of $+1$ or 0.

Remark 3. It can be easily shown that spatial stability *implies* spatial regularity, but not conversely.

We consider the temporal stability of the network for all N instead of a fixed N; if the temporal stability is defined for a fixed size of N, a designer has to recheck the stability when the network size is changed in response to certain design considerations. We also remark that for a fixed N, while it is easy to *say* that Eq. (84) is asymptotically stable iff $\mathbf{B}^{-1}\mathbf{A}$ is negative definite, it is very hard to *derive* analytical (*a priori*) iff conditions for negative definiteness even with $m = 2$. One can derive, however, an interesting analytical condition if one looks for negative definiteness of $\mathbf{B}^{-1}\mathbf{A}$ *for all* N, which will be shown in Section IV.D.

DEFINITION 6 (Temporal stability). A neural network described by Eq. (84) is said to be temporally stable if and only if it is asymptotically stable for all N.

PROPOSITION 2. *A neural network described by Eq. (84) is temporally stable if* $\mathbf{B}^{-1}\mathbf{A}$ *is negative definite for all* N.

The following standing assumptions are made throughout this chapter unless stated otherwise.

Standing Assumption 1. (i) $a_0 < 0,\ a_m \neq 0$;
(ii) $\mathbf{B}$ is positive definite for all N.

$\mathbf{A}$ must be negative definite (provided that $\mathbf{B}$ is positive definite), which is the inequality in (i). If $a_m = 0$, then the neighborhood M is of a smaller size. No restrictions will be imposed on the sign of a_p, $p \neq 0$. In image processing vision chips, c_p in Eq. (83) are parasitic capacitors of MOS processes, and positive definiteness of $\mathbf{B}$ is a mild condition. The following result establishes a fundamental relationship between the temporal and spatial dynamics.

THEOREM 2. *A neural network is temporally stable if and only if it is spatially regular.*

Proof. ($\Leftarrow$) Consider the characteristic polynomial of $\mathbf{F}$:

$$P_F(\lambda) := \det(\lambda\mathbf{1} - \mathbf{F}) = \lambda^m \left[\frac{a_0}{a_m} + \sum_{p=1}^{m} \frac{a_p}{a_m}(\lambda^p + \lambda^{-p}) \right]. \tag{107}$$

This implies that if λ_s (resp. λ_u) is a stable (resp. unstable) eigenvalue, i.e., $|\lambda_s| < 1$ (resp. $|\lambda_u| > 1$), then λ_s^{-1} (resp. λ_u^{-1}) is also an eigenvalue and unstable (resp. stable). $\mathbf{F}$ is nonsingular, therefore there are no zero eigenvalues. In order to discuss $\mathbf{F}|E^c$, let

$$\omega = \lambda + \lambda^{-1} \quad \text{or} \quad \lambda = \tfrac{1}{2}(\omega \pm \sqrt{\omega^2 - 4}). \tag{108}$$

By a repeated use of the binomial formula, one sees that

$$\frac{a_0}{a_m} + \sum_{p=1}^{m} \frac{a_p}{a_m}(\lambda^p + \lambda^{-p}) = \sum_{p=0}^{m} \alpha_p \omega^p := Q(\omega) \tag{109}$$

for real αs. Since $\mathbf{F}$ has no zero eigenvalues,

$$P_F(\lambda) = 0 \quad \text{iff} \quad Q(\omega) = 0, \tag{110}$$

where λ and ω are related via Eq. (108). Hence if λ_c is *real* and $|\lambda_c| = 1$, then Eq. (108) forces λ_c to be a *double* eigenvalue $\{\lambda_c, \lambda_c\}$ or its *multiple*. It is easy to show

$$\dim\ker(\lambda\mathbf{1} - \mathbf{F}) = 1 \tag{111}$$

for any eigenvalue λ, where "ker" denotes the kernel of a matrix. Thus, for each eigenvalue λ of $\mathbf{F}$, there is *only one elementary Jordan block* [53]. Therefore, the *real canonical form* of $\mathbf{F}|E^{\lambda_c}$, restriction of $\mathbf{F}$ to the eigenspace corresponding to λ_c, is given by

$$\begin{bmatrix} \lambda_c & 1 & 0 & \cdot & 0 \\ 0 & \lambda_c & 1 & \cdot & \cdot \\ \cdot & 0 & \cdot & \cdot & 0 \\ \cdot & \cdot & \cdot & \lambda_c & 1 \\ \cdot & \cdot & \cdot & 0 & \lambda_c \end{bmatrix} \in C^{2q \times 2q}, \tag{112}$$

where $2q$ is the multiplicity. This is clearly of the form Eq. (106).

So far, no use has been made of the negative definiteness of $\mathbf{B}^{-1}\mathbf{A}$ and yet we are already close to Eq. (106), the regularity. The situation, however, is slightly subtle when it comes to a *nonreal* λ_c with $|\lambda_c| = 1$, because λ_c^*, the complex conjugate, is also an eigenvalue [see Eq. (107)]. This last is of no use since $\mathbf{F}$ is a real matrix and λ_c^* also being an eigenvalue is automatic. We now assume that $\mathbf{B}^{-1}\mathbf{A}$ is negative definite for all N, $\mathbf{A}$ is negative definite for all N. It is known

[54], then, that there are $z_p \in R$, $p = 0, \ldots, m$, such that the elements of **A** satisfy

$$-a_p = \sum_{i=0}^{m-p} z_i z_{i+p}, \qquad p = 0, \ldots, m, \tag{113}$$

i.e., a_ps can be decomposed as in Eq. (113). Substitution of Eq. (113) into Eq. (107) yields

$$\begin{aligned} P_F(\lambda) &= -\frac{\lambda^m}{z_0 z_m}\left[\sum_{i=0}^{m} z_i^2 + \sum_{p=1}^{m}\sum_{i=0}^{m-p} z_i z_{i+p}(\lambda^i + \lambda^{-i})\right] \\ &= -\frac{\lambda^m}{z_0 z_m}\left(\sum_{i=0}^{m} z_i \lambda^{-i}\right)\left(\sum_{i=0}^{m} z_i \lambda^{-i}\right). \end{aligned} \tag{114}$$

Since $0 \neq a_m = -z_0 z_m$ and since **F** has no zero eigenvalues, one sees that

$$P_F(\lambda) = 0 \quad \text{iff} \quad R(\lambda)R(1/\lambda) = 0, \tag{115}$$

where

$$R(\lambda) = \sum_{i=0}^{m} z_i \lambda^i. \tag{116}$$

Therefore, if λ is a nonreal eignevalue with $|\lambda_c| = 1$, Eq. (115) forces the eigenvalue configuration to be of the form $\{\lambda_c, \lambda_c^*, \lambda_c, \lambda_c^*\}$ or its *multiple*. It follows from Eq. (111) that the real canonical form of **F** on this eigenspaces is given by

$$\begin{bmatrix} \alpha & -\beta & 1 & 0 & 0 & . & . \\ \beta & \alpha & 0 & 1 & 0 & . & . \\ 0 & 0 & \alpha & -\beta & 1 & 0 & . \\ 0 & 0 & \beta & \alpha & 0 & 1 & . \\ . & . & . & . & . & . & . \\ . & . & . & . & . & . & . \\ . & . & . & . & . & . & . \end{bmatrix} \in C^{2q' \times 2q'}, \tag{117}$$

where

$$\alpha^2 + \beta^2 = 1 \tag{118}$$

and $2q'$ is the multiplicity. This, again, is of the form Eq. (106).

($\Rightarrow$) If a neural network is spatially regular, the real canonical form, of the spatial dynamics **F** is equivalent to Eq. (106). The characteristic polynomial **F**, then, admits a decomposition of the form given by Eq. (114). Comparing Eq. (114) with Eq. (109), one sees that Eq. (113) holds. The condition is known [54] to be not only a necessary but also a sufficient condition for $\mathbf{A}$ to be negative definite for all N. Since $\mathbf{B}$ is positive definite and symmetric for all N, it follows from [55] that

$$\text{max. eigenvalue of } \mathbf{B}^{-1}\mathbf{A} = \max_{\mathbf{v}\neq\mathbf{0}} \frac{\mathbf{v}^T\mathbf{A}\mathbf{v}}{\mathbf{v}^T\mathbf{B}\mathbf{v}} < 0 \tag{119}$$

for any N which impies temporal stability. ■

Remark 4. (i) Consider Eq. (80) and let

$$W := \sum_{i=1}^{N} v_i u_i,$$

which is the *power* injected into the network. It follows from Eq. (80) that

$$\begin{aligned} W &= -\sum_i \sum_p v_i a_p v_{i-p} + \sum_i \sum_p v_i b_p \frac{dv_{i-p}}{dt} \\ &= -\mathbf{v}^T\mathbf{A}\mathbf{v} + \mathbf{v}^T\mathbf{B}\frac{d\mathbf{v}}{dt} := W_R + W_C. \end{aligned}$$

Thus the first term

$$W_R := -\mathbf{v}^T\mathbf{A}\mathbf{v} = \text{power dissipated by the resistive part of the network.}$$

Therefore, a neural network is temporally stable iff its resistive part is *strictly passive*, i.e.,

$$W_R > 0, \qquad \mathbf{v} \neq \mathbf{0} \quad \text{for all } N.$$

It follows from the previous remark that spatial stability demands more than strict passivity of the resistive part.

(ii) Observe that $\mathbf{v}^T\mathbf{B}\mathbf{v}/2$ = energy stored in the capacitors. Therefore Eq. (119) says that

$$\begin{aligned} \text{max. eigenvalues of } \mathbf{B}^{-1}\mathbf{A} &= \max\left(\frac{-\text{power dissipated by resistors}}{2 \times \text{energy stored in capacitors}}\right) \\ &= -\min\left(\frac{\text{power dissipated by resistors}}{2 \times \text{energy stored in capacitors}}\right). \end{aligned}$$

Since the temporal stability condition is equivalent to spatial stability, we will say, hereafter, that the *stability-regularity* condition is satisfied if a network is temporally stable or spatially regular. Recall $Q(\omega)$ defined by Eq. (109).

PROPOSITION 3. *The following are equivalent:*

(i) *Stability-Regularity.*
(ii) *Every nonreal eigenvalue ρ_c of* $\mathbf{F}$ *with* $|\rho_c| = 1$ *has an even multiplicity.*
(iii) *Every real zero ω_R of Q with $|\omega_R| < 2$ has an even multiplicity.*

For the sake of completeness, we will state the following.

PROPOSITION 4. *The following are equivalent:*

(i) *Spatial stability.*
(ii) *Eigenvalues of* $\mathbf{F}$ *are off the unit circle.*
(iii) *Q has no real zero on* $[-2, 2]$.

C. EXPLICIT STABILITY CRITERIA

Recall Q defined by Eq. (109). The following functions will be called the stability indicator functions:

$$\sigma_+(a_0, a_1, \ldots, a_m) := \max_{\omega \in [-2,2]} a_m Q(\omega),$$
$$\sigma_-(a_0, a_1, \ldots, a_m) := \min_{\omega \in [-2,2]} a_m Q(\omega). \tag{120}$$

PROPOSITION 5. *The network described in Eqs. (84) and (90) is stability-regular if and only if*

$$\sigma_+(a_0, a_1, \ldots, a_m) \le 0. \tag{121}$$

PROPOSITION 6. *The network described in Eq. (90) is spatially stable if and only if*

$$\sigma_+(a_0, a_1, \ldots, a_m) < 0. \tag{122}$$

The following fact gives upper and lower bounds for eigenvalues of the temporal dynamics $\mathbf{A}$.

PROPOSITION 7. (i) *Any eigenvalue μ of the temporal dynamics* $\mathbf{A}$ *for any N satisfies the following bounds:*

$$\sigma_-(a_0, a_1, \ldots, a_m) < \mu < \sigma_+(a_0, a_1, \ldots, a_m). \tag{123}$$

(ii) *The bounds (123) are optimal in the sense that if σ_+^* (respectively σ_-^*) is any number which satisfies*

$$\sigma_+^* < \sigma_-(a_0, a_1, \ldots, a_m) \quad [\text{respectively}\ \sigma_-(a_0, a_1, \ldots, a_m) < \sigma_-^*],$$

then there is an eigenvalue μ *of* $\mathbf{A}$ *for some* N *such that*

$$\sigma_+^* < \mu \quad (\textit{respectively } \mu < \sigma_-^*).$$

We would like to emphasize the *if and only if* nature of Propositions 5 and 6 and the *optimality* of Proposition 7 which indicate that σ_+ and σ_- are crucial to the stability issues of our interest.

PROPOSITION 8. *When* $m = 2$, *the stability indicator functions are given by*

$$\sigma_+(g_0, g_1, g_2) = \begin{cases} -g_0 - 2g_1 + 2|g_1|, & \textit{when } g_2 > 0 \textit{ or } \{g_2 < 0 \\ & \quad \textit{and } |g_1/g_2| \geq 4\}, \\ -g_0 - 2g_1 - 4g_2 - g_1^2/(4g_2), & \textit{when } g_2 < 0 \\ & \quad \textit{and } |g_1/g_2| \leq 4, \end{cases}$$
$$\sigma_-(g_0, g_1, g_2) = \begin{cases} -g_0 - 2g_1 - 2|g_1|, & \textit{when } g_2 < 0 \{\textit{or } g_2 > 0 \\ & \quad \textit{and } |g_1/g_2| \geq 4\}, \\ -g_0 - 2g_1 - 4g_2 - g_1^2/(4g_2), & \textit{when } g_2 > 0 \\ & \quad \textit{and } |g_1/g_2| \leq 4. \end{cases} \tag{124}$$

EXAMPLE 2. For a Gaussian-like convolver [42, 43],

$$g_1 > 0, \qquad g_2 < 0, \qquad g_1 = 4|g_2|. \tag{125}$$

Propositions 5 and 8 tell us that the stability-regularity is equivalent to

$$\sigma_+(g_0, g_1, g_2) = -g_0 \leq 0, \tag{126}$$

i.e., passivity of g_0. Furthermore, Proposition 6 says that the network is spatially stable iff

$$\sigma_+(g_0, g_1, g_2) = -g_0 < 0,$$

i.e., iff g_0 is strictly passive. Thus g_0 can be safely varied over any range as long as it is positive.

Remark 5. (i) Even when g_1 as well as g_2 is negative, a network can satisfy the stability-regularity or/and the spatial stability condition provided that g_0 is "sufficiently" passive because

$$\sigma_+(g_0, g_1, g_2) = \begin{cases} -g_0 + 4|g_1|, & \text{when } |g_1/g_2| > 4, \\ -g_0 - 2g_1 - 4g_2 - g_1^2/(4g_2), & \text{when } |g_1/g_2| \leq 4. \end{cases}$$

(ii) If $g_2 > 0$, then

$$\sigma_+(g_0, g_1, g_2) = \begin{cases} -g_0, & \text{when } g_1 > 0, \\ -g_0 + 4|g_1|, & \text{when } g_1 \leq 0. \end{cases}$$

(iii) Since Q is quadratic, conditions (ii) and (iii) of Proposition 3 are sharpened, respectively, to the following:

(ii)′ **F** has no simple nonreal eigenvalue on the unit circle.
(iii)′ Q has no real zero on $(-2, 2)$.

It follows Proposition 5 (resp. Proposition 7) that the set of parameter values (g_0, g_1, g_2) for which stability-regularity and the spatial stability hold are given, respectively, by

$$SR := \{(g_0, g_1, g_2)|\sigma_+(g_0, g_1, g_2) \leq 0, g_0 + 2g_1 + 2g_2 > 0\}, \quad (127)$$
$$SS := \{(g_0, g_1, g_2)|\sigma_+(g_0, g_1, g_2) < 0, g_0 + 2g_1 + 2g_2 > 0\}. \quad (128)$$

We will now give a fact which, as its by-product, explains why our numerical experiments suggested $SR = SS$, which is untrue. Let

$$G := \{(g_0, g_1, g_2)|g_2 < 0\},$$

on which our numerical experiments were performed.

PROPOSITION 9. (i) meas$[SS \cap G] > 0$.
(ii) meas$[(SR - SS) \cap G] = 0$, *where* meas[.] *denotes the Lebesgue measure on* R^3.

This proposition explains why our experiments suggested $SR = SS$ for a Lebesgue measure zero subset is "hard to hit."

Conjecture 1. Proposition 9 will be true for a general m.

Neural networks with $m = 1$ are used in an extensive manner [3]. Although those networks contain only positive conductances ($g_0, g_1 > 0$), it would be worth clarifying the temporal as well as the spatial stability issues when $g_1 < 0$ or $g_0 < 0$.

PROPOSITION 10. *When* $m = 1$, *the stability indicator functions are given by*

$$\sigma_+(g_0, g_1) = -g_0 - 2g_1 + 2|g_1|,$$
$$\sigma_-(g_0, g_1) = -g_0 - 2g_1 - 2|g_1|.$$

EXAMPLE 3. When $g_0 > 0$ but $g_1 < 0$, the network is temporally (resp. spatially) stable iff

$$-g_0 + 4|g_1| \leq 0 \quad (\text{resp. } - g_0 + |g_1| < 0).$$

Remark 6. The reader is referred to [40] for the proofs and explicit formula for $m = 3$.

D. Transients

This section gives an estimate of the "processing speed" of vision chips.

Corollary 1. *Consider the temporal dynamics Eq. (84) with* $\mathbf{v}(0) = \mathbf{0}$. *If Eq. (121) is satisfied and* $\mathbf{B}$ *is positive definite, then the solution* $\mathbf{v}(t)$ *of Eq. (84) satisfies the following bounds:*

$$\frac{\eta_-}{\sigma_-}\left[\exp\left(\frac{\eta_-}{\sigma_-}t - 1\right)\right]\left\|\mathbf{B}^{-1}\mathbf{u}\right\| \leq \left\|\mathbf{v}(t)\right\| \leq \frac{\eta_+}{\sigma_+}\left[\exp\left(\frac{\eta_+}{\sigma_+}t - 1\right)\right]\left\|\mathbf{B}^{-1}\mathbf{u}\right\|. \quad (129)$$

Remark 7. (i) The above corollary is obtained by the analysis of the capacitance matrix $\mathbf{B}$ in Eq. (83) using the method used for analyzing $\mathbf{A}$.

(ii) The result tells us how fast/slow a step response of Eq. (84) grows. Although there is no precise concept of the time constant RC for Eq. (84) ($\dim \mathbf{v} \gg 1$), Eq. (129) can be interpreted as

$$-\frac{\eta_-}{\sigma_-} \leq \text{"time constant"} \leq -\frac{\eta_+}{\sigma_+}. \quad (130)$$

(iii) Let us compute the upper bound in Eq. (130) for $m = 2$. It is not difficult to show that

$$\eta_+(c_0, c_1, c_2) = \begin{cases} c_0 + 2c_1 + 2|c_1|, & \text{when } c_2 < 0 \text{ or } c_2 > 0 \\ & \quad \text{and } |c_1/c_2| \geq 4, \\ c_0 + 2c_1 + 4c_2 + c_1^2/4c_2, & \text{when } c_2 > 0 \text{ and } |c_1/c_2| \leq 4. \end{cases}$$

If $g_0, g_1, c_0, c_1, c_2 > 0$, then it follows from Eq. (124) and

$$-\eta_+/\sigma_+ = -\eta_+/g_0 = \begin{cases} (c_0 + 4c_1)/g_0, & \text{when } |c_1/c_2| \geq 4, \\ (c_0 + 2c_1 + 4c_2 + c_1^2/4c_2)/g_0, & \text{when } |c_1/c_2| \leq 4. \end{cases}$$

Since it is difficult to estimate parasitic capacitances accurately, this is as much as one can tell from the corollary.

REFERENCES

[1] K. A. Boahen and A. G. Andreou. *Adv. Neural Inform. Process. Syst.* 4:764–772, 1992.

[2] H. Kobayashi, T. Matsumoto, T. Yagi, and T. Shimmi. *Neural Networks* 6:327–350, 1993.

[3] C. Mead and M. Mahowald. *Neural Networks* 1:91–97, 1988.

[4] C. Mead. *Analog VLSI and Neural Systems.* Addison-Wesley, Reading, MA, 1989.

[5] T. Shimmi, H. Kobayashi, T. Yagi, T. Sawaji, T. Matsumoto, and A. A. Abidi. In *Proceedings of European Solid-State Circuits Conference*, pp. 163–166, 1992.

[6] T. Matsumoto, T. Shimmi, H. Kobayashi, A. A. Abidi, T. Yagi, and T. Sawaji. In *Proceedings of IJCNN 92*, Beijing, Vol. 3, pp. 188–197, 1992.

[7] C. Koch and H. Li (Eds.) *Vision Chips: Implementing Vision Algorithms with Analog VLSI Circuits,* IEEE Computer Soc. Press, Los Alamitos, CA, 1995.

[8] C. D. Nilson, R. B. Darling, and R. B. Pinter. *IEEE J. Solid-State Circuits* 29:1291–1296, 1994.
[9] J. E. Dowling. *The Retina: An Approachable Part of the Brain*, Belknap Press, Cambridge, MA, 1987.
[10] T. D. Lamb and E. J. Simon. *J. Physiol.* 263:256–286, 1976.
[11] V. Torre and W. G. Owen. *Biophys. J.* 41:305–324, 1983.
[12] T. D. Lamb. *J. Physiol.* 263:239–255, 1976.
[13] P. B. Detwiler and A. L. Hodgkin. *J. Physiol.* 291:75–100, 1979.
[14] T. Yagi, F. Ariki, and Y. Funahashi. In *Proceedings of International Joint Conference on Neural Networks*, Vol. 1, pp. 787–789, 1989.
[15] S. Ohshima, T. Yagi, and Y. Funahashi. *Vision Res.* 35:149–160, 1995.
[16] J. E. Dowling and B. Ehinger. *Proc. R. Soc. London* B 201:7–26, 1978.
[17] T. Teranishi, K. Negishi, and S. Kato. *Nature* 301:234–246, 1983.
[18] O. P. Hamill, A. Marty, E. Neher, B. Sakmann, and F. J. Sigworth. *Pflügers Arch.* 391:85–100, 1981.
[19] M. Tachibana. *J. Physiol.* 345:329–351, 1983.
[20] D. A. Baylor and M. G. F. Fuortes. *J. Physiol.* 207:77–92, 1970.
[21] G. Fain and J. E. Dowling. *Science* 180:1178–1181, 1973.
[22] D. A. Baylor, A. L. Hodgkin, and T. D. Lamb. *J. Physiol.* 242:685–727, 1974.
[23] D. A. Baylor, M. G. F. Fuortes, and P. M. O'Bryan. *J. Physiol.* 214:256–294, 1971.
[24] E. A. Schwartz. *J. Physiol.* 257:379–406, 1976.
[25] M. Tessier-Lavigne and D. Attwell. *Proc. R. Soc. London* B 234:171–197, 1988.
[26] K. I. Naka and W. A. H. Rushton. *J. Physiol.* 192:437–461, 1967.
[27] T. Yagi. *J. Physiol.* 375:121–135, 1986.
[28] T. Kujiraoka and T. Saito. *Proc. Nat. Acad. Sci. USA* 83:4063–4066, 1986.
[29] A. Kaneko. *J. Physiol.* 207:623–633, 1970.
[30] D. Marr and E. Hildreth. *Proc. Roy. Soc. London* B 207:187–217, 1980.
[31] T. Shigematsu and M. Yamada. *Neuro. Res. Suppl.* 8:s69–s80, 1988.
[32] T. Yagi, S. Ohshima, and Y. Funahashi. *Biol. Cybern.*, to appear.
[33] T. Poggio and C. Koch. *Proc. Royal Soc. London* B 226:303–323, 1985.
[34] A. N. Tikhonov. *Sov. Math. Dokl.* 4:1035–1038, 1963.
[35] A. N. Tikhonov. *Sov. Math. Dokl.* 4:1624–1627, 1963.
[36] A. N. Tikhonov. *Sov. Math. Dokl.* 6:559–562, 1965.
[37] G. Whaba. In *Inverse and Ill-Posed Problems* (H. W. Engl and C. W. Groetsch, Eds.). Academic, New York, 1987.
[38] D. J. C. MacKay. Bayesian methods for adaptive models. Ph.D. Thesis, California Institute of Technology, 1991.
[39] Takeuchi, D. J. C. MacKay, and T. Matsumoto. In *Proceedings of 1994 International Symposium on Artificial Neural Networks*, Taiwan, pp. 419–428, 1994.
[40] T. Matsumoto, H. Kobayashi, and Y. Togawa. *IEEE Trans. Neural Networks* 3:540–569, 1992.
[41] T. Matsumoto, H. Kobayashi, and Y. Togawa. In *Proceedings IJCNN 91*, Seattle, Vol. 2, pp. 283–295, 1991.
[42] H. Kobayashi, J. L. White, and A. A. Abidi. *IEEE J. Solid-State Circuits* 26:738–748, 1991.
[43] H. Kobayashi, J. L. White, and A. A. Abidi. *ISSCC Dig. Tech. Pap.* pp. 216–217, 1990.
[44] J. Harris. In *IEEE Conference on Neural Information Processing Systems—Natural and Synthetic*, 1988.
[45] H. Kobayashi, T. Matsumoto, and J. Sanekata. *IEEE Trans. Neural Networks* 6:1148–1164, 1995.
[46] D. Dudgeon and R. Mersereau. *Multidimensional Signal Processing*. Prentice-Hall, Englewood Cliffs, NJ, 1984.
[47] W. E. L. Grimson. *From Images to Surfaces*. MIT Press, Cambridge, MA, 1986.

[48] T. Poggio, H. Voorhees, and A. Yuille. Artificial Intelligence Laboratory, Memo 833, Massachusetts Institute of Technology, 1985.
[49] J. Clark. *IEEE Trans. Pattern Anal. Machine Intell.* 11:43–57, 1989.
[50] M. Banu and Y. Tsividis. *Electron. Lett.* 18:678–679, 1982.
[51] D. L. Standley and J. L. Wyatt Jr. *IEEE Trans. Circuits Syst.* 36:675–681, 1989.
[52] J. L. White and A. N. Wilson Jr. *IEEE Trans. Circuits Syst.* 39: 734–743, 1992.
[53] M. W. Hirsh and S. Smale. *Differential Equations, Dynamical Systems and Linear Algebra.* Academic, New York, 1974.
[54] E. L. Allgower. *Numer. Math.* 16:157–162, 1970.
[55] F. R. Gantmacher. *The Theory of Matrices*. Chelsea, New York, 1960.

Algorithmic Techniques and Their Applications

Rudy Setiono
Department of Information Systems and Computer Science
National University of Singapore
Kent Ridge, Singapore 119260, Republic of Singapore

I. INTRODUCTION

Pattern recognition is an area where neural networks have been widely applied with much success. The network of choice for pattern recognition is a multilayered feedforward network trained by a variant of the gradient descent method known as the back-propagation learning algorithm. As more applications of these networks are found, the shortcomings of the back-propagation network become apparent. Two drawbacks often mentioned are the need to determine the architecture of a network before training can begin and the inefficiency of the back-propagation learning algorithm. Without proper guidelines on how to select an appropriate network for a particular problem, the architecture of the network is usually determined by trial-and-error adjustments of the number of hidden layers and/or hidden units. The back-propagation algorithm involves two parameters: the learning rate and the momentum rate. The values of these parameters have significant effect on the efficiency of the learning process. However, there have been no clear guidelines for selecting their optimal values. Regardless of the values of the parameters, the back-propagation method is generally slow to converge and prone to get trapped at a local minimum of the error function.

When designing a neural network system, the choice of a learning algorithm for training the network is very crucial. As problems become more complex, larger networks are needed and the speed of training becomes critical. Instead of the gradient descent method, more sophisticated methods with faster convergence rate can be used to speed up network training. In Section II of this chapter,

Image Processing and Pattern Recognition
Copyright © 1998 by Academic Press. All rights of reproduction in any form reserved.

we describe a variant of the quasi-Newton method that we have used to reduce the network training time significantly.

Another important aspect of the feedforward neural network learning is the selection of a suitable network architecture for solving the problem in hand. There is no doubt that the performance of a neural network system can be greatly affected by the network architecture. When building a neural network system, there are several components of the network that need to be determined:

1. the number of input and output units,
2. the number of hidden layers,
3. the number of hidden units in each layer, and
4. the connectivity patterns among the units in the network.

Most of the remaining sections of this chapter are devoted to the issues of finding the optimal number of units in each layer of a feedforward network and of finding the relevant connectivity patterns among these units. In order to achieve optimal performance, network systems designed for different problem domains require different network architectures. We describe some algorithms that have been developed to automatically construct a suitable network architecture. These algorithms have been shown to be very successful in finding appropriate network architectures for a wide variety of problems.

We shall consider only a particular network architecture, namely, layered feedforward networks. Layered feedforward networks are among the most commonly used network architectures at present. We also restrict the number of hidden layers to one and hence we consider feedforward networks with only three layers of units. Theoretically, it has been proved that a network with a single hidden layer is capable of forming arbitrary decision boundaries if there are a sufficient number of units in the hidden layer [1, 2]. Experimental studies have also shown that there is no advantage to using four-layered networks over three-layered networks [3]. Section III discusses the selection of the right number of output units in a network. Neural network construction algorithms which dynamically add units in the hidden layer are described in Section IV. By making use of the cross-entropy error measure, we show how the addition of a hidden unit to the network is guaranteed to decrease the error function. We also present the results of applying a neural network construction algorithm on the well-known spiral problem [4]. Section V presents an algorithm that we have developed to determine the required number of input units by pruning. Section VI presents an algorithm that removes redundant or irrelevant connections from a fully connected network. Section VII discusses the potential applications of the techniques for constructing a neural network system discussed in this chapter to data mining. Data mining is a multidisciplinary field which in recent years has been attracting a great deal of attention from researchers in data base, machine learning, and statistics. It is concerned with discovering interesting patterns that are hidden in data bases. In this section, we discuss how a neural network system can be used as a tool to extract rules that distinguish between benign and malignant samples in a breast cancer data set. Finally, a summary is given in Section VIII.

We briefly describe now our notation. For a vector x in the n-dimensional real space $\mathbb{R}^n$, the norm $\|x\|$ denotes the Euclidean distance of x from the origin, that is, $\|x\| = (\sum_{i=1}^n x_i^2)^{1/2}$. For a matrix $A \in \mathbb{R}^{m\times n}$, A^T will denote the transpose of A. The superscript T is also used to denote the scalar product of two vectors in $\mathbb{R}^n$, that is, $x^T y = \sum_{i=1}^n x_i y_i$. For a twice-differentiable function $f(x)$, the gradient of $f(x)$ is denoted by $\nabla f(x)$, while its Hessian matrix is denoted by $\nabla^2 f(x)$.

II. QUASI-NEWTON METHODS FOR NEURAL NETWORK TRAINING

The problem of training a feedforward neural network can be cast as an unconstrained optimization problem. Consider the three-layered network with one output unit depicted in Fig. 1. The optimization problem that is usually solved when training this network is the minimization of the squared-error function [5]:

$$f(w,\zeta,v,\tau) := \sum_{i=1}^{k}\left(\sigma\left(\sum_{j=1}^{h}\sigma((x^i)^T w^j - \zeta^j)v^j - \tau\right) - t^i\right)^2, \tag{1}$$

where

$$\begin{aligned}
h &= \text{integer number of hidden units,}\\
k &= \text{fixed integer number of given samples } x^i \in \mathbb{R}^n,\\
t^i &= 0 \text{ or } 1 \text{ target value for } x^i,\ i = 1,2,\ldots,k,\\
\tau &= \text{real number threshold of output unit,}\\
v^j &= \text{real number weights of outgoing arcs from hidden units,}\\
&\quad\ j = 1,2,\ldots,h,\\
\zeta^j &= \text{real number thresholds of hidden units, } j = 1,2,\ldots,h,\\
w^j &= n\text{-vector weights of incoming arcs to hidden units, } j = 1,2\ldots,h,\\
x^i &= \text{given } n\text{-dimensional vectors samples, } i = 1,2,\ldots,k,\\
\sigma(\xi) &= 1/(1+e^{-\xi}) \text{ is the sigmoid activation function.}
\end{aligned}$$

If we let $z = (w,\zeta,v,\tau)$, then given an initial approximation z^k, each epoch of the back-propagation method can be viewed as an attempt to minimize an approximation of the function $f(w)$ by the linear function

$$f_k(z) = f(z^k) + \nabla f(z^k)^T(z - z^k),$$

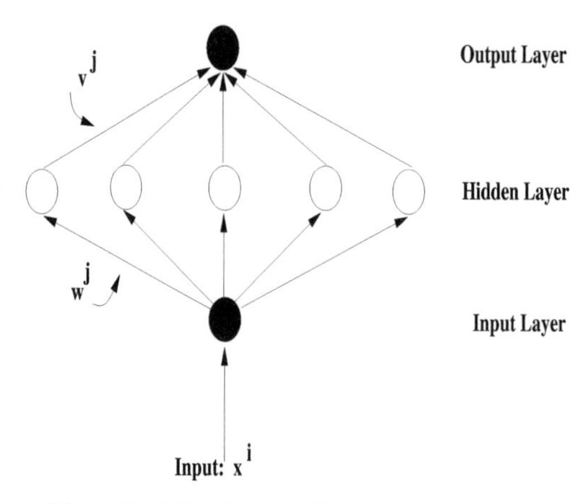

Figure 1 A three layer feedforward neural network.

subject to the constraint $\|z - z^k\| \leq 1$. The solution of this auxiliary problem is

$$z - z^k = -\nabla f(z^k) / \|\nabla f(z^k)\|.$$

The steepest descent algorithm proceeds by performing a line search along the descent direction $-\nabla f(z^k)/\|\nabla f(z^k)\|$, or equivalently along the direction of the negative of the gradient at z^k, $d^k = -\nabla f(z^k)$. The algorithm thus generates the sequence

$$z^{k+1} = z^k + \lambda^k d^k,$$

where λ^k is a solution of the line search problem

$$\min_{\lambda \geq 0} f(z^k + \lambda d^k). \tag{2}$$

The step length λ^k is commonly referred to as the learning rate and the simplest variant of the steepest descent method holds the value of this step length constant, i.e., $\lambda^k \equiv \lambda$, $\forall k$ for some small positive value of λ.

A momentum term can be added when updating z to include contribution from the previous iteration. With a momentum parameter $\alpha \in (0, 1)$, the new weight z^{k+1} is computed as

$$z^{k+1} = z^k + \lambda^k d^k + \alpha(z^k - z^{k-1}).$$

Newton's method is obtained when a quadratic approximation instead of a linear approximation of the function $f(z)$ is used. For Newton's method, the next approximate solution is obtained as a point that minimizes the quadratic function

$$f_k(z) = f(z^k) + \nabla f(z^k)^T (z - z^k) + \tfrac{1}{2}(z - z^k)^T \nabla^2 f(z^k)(z - z^k).$$

Hence, we obtain the sequence

$$z^{k+1} = z^k - [\nabla^2 f(z^k)]^{-1} \nabla f(z^k).$$

The step length λ^k can also be incorporated in the method to generate the damped Newton sequence

$$z^{k+1} = z^k - \lambda^k [\nabla^2 f(z^k)]^{-1} \nabla f(z^k),$$

where λ^k is a solution of the line search problem (2) with the search direction $d^k = -[\nabla^2 f(z^k)]^{-1} \nabla f(z^k)$. The main advantage of the Newton's method is that it has a quadratic convergence rate, while the steepest descent method has a much slower, linear convergence rate. However, each step of the Newton's method requires a large amount of computation. Assuming that the dimensionality of the problem is n, then an $O(n^3)$ floating point operation is needed to compute the search direction d^k.

A method that uses an approximate Hessian matrix in computing the search direction is the quasi-Newton method. Let B^k be an $n \times n$ symmetric matrix that approximates the Hessian matrix $\nabla^2 f(z^k)$; then the search direction for the quasi-Newton method is obtained by minimizing the quadratic function

$$f_k(z) = f(z^k) + \nabla f(z^k)^T (z - z^k) + \tfrac{1}{2}(z - z^k)^T B^k (z - z^k).$$

If B^k is invertible, then a descent direction can be obtained from the solution of the above quadratic program:

$$d^k := z - z^k = -(B^k)^{-1} \nabla f(z^k). \tag{3}$$

Since we would like to have the matrix B^k to approximate the Hessian of the function $f(z)$ at z^k, it needs to be updated from iteration to iteration by incorporating the most recent gradient information. One of the most widely used quasi-Newton methods is the BFGS method, where the matrix B^k is updated according to the following equation:

$$B^{k+1} = B^k - \frac{B^k \delta^k (\delta^k)^T B^k}{(\delta^k)^T B^k \delta^k} + \frac{\gamma^k (\gamma^k)^T}{(\delta^k)^T \gamma^k}, \tag{4}$$

where

$$\begin{aligned} \delta^k &= z^{k+1} - z^k, \\ \gamma^k &= \nabla f(z^{k+1}) - \nabla f(z^k). \end{aligned}$$

This updating formula was independently proposed by Broyden, Fletcher, Goldfarb, and Shanno [6–9].

The BFGS method is an example of a rank-2 method, since the matrix B^{k+1} differs from the matrix B^k by a symmetric matrix of rank at most 2. A quasi-Newton method that updates the matrix B^k by adding only a rank-1 matrix is the SR1 (symmetric rank-1) method. It updates the matrix B^k as follows:

$$B^{k+1} = B^k - \frac{(\gamma^k - B^k\delta^k)(\gamma^k - B^k\delta^k)^T}{(\gamma^k - B^k\delta^k)^T\delta^k}. \tag{5}$$

It can be shown that the matrix B^{k+1} defined by either the BFGS update (4) or the SR1 update (5) satisfies the quasi-Newton condition

$$B^{k+1}\delta^k = \gamma^k.$$

A minor modification to the BFGS update (4) was proposed by Biggs [10]. A scalar variable t^k is introduced into the update formula as follows:

$$B^{k+1} = B^k - \frac{B^k\delta^k(\delta^k)^T B^k}{(\delta^k)^T B^k\delta^k} + t^k\frac{\gamma^k(\gamma^k)^T}{(\delta^k)^T\gamma^k}, \tag{6}$$

where t^k is defined by

$$t^k = \frac{2}{(\delta^k)^T\gamma^k}\left[3f(z^k) - 3f(z^{k+1}) + (\delta^k)^T\left(\nabla f(z^k) + 2\nabla f(z^{k+1})\right)\right]. \tag{7}$$

It was shown that for some functions, this update resulted in faster convergence than the original BFGS update where $t^k = 1$.

The search direction d^k given by Eq. (3) can be obtained by either

1. computing the inverse of the matrix B^k and then multiplying the inverse by the negative of the gradient, or
2. finding the Cholesky factorization of B^k, that is, computing the lower triangular matrix L such that $LL^T = B^k$ and then computing the direction d^k via backward and forward substitutions.

For a function with n variables, either one of these two approaches requires an $O(n^3)$ floating point operation.

This cost of computing d^k can be reduced to $O(n^2)$ if, instead of B^k, the inverse of the matrix is kept. Suppose, at iteration k, we have a matrix H^k that is equal to $(B^k)^{-1}$; then the search direction d^k is equal to

$$d^k = -H^k\nabla f(z^k). \tag{8}$$

For the SR1 method, it can be easily verified that if we define

$$H^{k+1} = H^k + \frac{(\delta^k - H^k\gamma^k)(\delta^k - H^k\gamma^k)^T}{(\delta^k - H^k\gamma^k)^T\gamma^k}, \tag{9}$$

then $H^{k+1} = (B^{k+1})^{-1}$, where B^{k+1} is the update defined by Eq. (5). Similarly, we can show that the inverse of the matrix B^{k+1} for the BFGS update given by Eq. (6) is

$$H^{k+1} = \left(I - \frac{\delta^k(\gamma^k)^T}{(\delta^k)^T\gamma^k}\right)H^k\left(I - \frac{\delta^k(\gamma^k)^T}{(\delta^k)^T\gamma^k}\right)^T + \frac{1}{t^k}\frac{\delta^k(\delta^k)^T}{(\delta^k)^T\gamma^k}. \tag{10}$$

Given a search direction d^k, an iterative one-dimensional optimization method can be applied to find a step length λ that solves the line search problem (2). However, this procedure may require an excessive number of function and/or gradient evaluations. In fact, it is well known [11, 12] that often inexact line searches are preferable to exact line search (2). A step length $\lambda^k > 0$ is considered acceptable if it satisfies the following two conditions:

$$f(z^k + \lambda^k d^k) \leq f(z^k) + c_1\lambda^k (d^k)^T \nabla f(z^k), \tag{11}$$

$$(d^k)^T \nabla f(z^k + \lambda^k d^k) \geq c_2 (d^k)^T \nabla f(z^k), \tag{12}$$

where c_1 and c_2 are two constants such that $0 < c_1 \leq c_2 < 1$ and $c_1 < 0.5$. The condition (11) is to ensure that the step length λ^k produces a sufficient decrease in the value of the function $f(z)$ at the new point z^{k+1}, while the second condition is to ensure that the step length is not too small. The values of c_1 and c_2 that have been suggested are 0.0001 and 0.9, respectively [13]. An iterative algorithm for finding a step length λ^k that satisfies both conditions (11) and (12) is given in [13].

A quasi-Newton method that allows a choice between a rank-1 update and a rank-2 update at each iteration is the SR1/BFGS algorithm [14]. This method is shown to be faster than the standard BFGS method for a wide range of nonlinear optimization problems. The SR1/BFGS quasi-Newton method with inexact line search can be summarized as follows.

SR1/BFGS algorithm for minimizing $f(z)$

Step 0. Initialization.

Choose any z^1 as a starting point. Let $H^1 = I$, set $k = 1$. Let $\epsilon > 0$ be a small terminating scalar.

Step 1. Iterative Step.

- Check for convergence:
  If $\|\nabla f(z^k)\| \leq \epsilon \max\{1, \|z^k\|\}$ then Stop.

Otherwise

1. Compute the search direction
$$d^k = -H^k \nabla f(z^k).$$
2. Calculate a step length λ^k such that both conditions (11) and (12) are satisfied and let
$$z^{k+1} = z^k + \lambda^k d^k.$$
3. Compute the value of t^k by Eq. (7).
$$\text{if } t^k < 0.5 \text{ then set } t^k = 0.5,$$
$$\text{else if } t^k > 100 \text{ then set } t^k = 100.$$
4. If $(\delta^k - H^k\gamma^k)^T\gamma^k > 0$, then compute H^{k+1} using Eq. (9),

   else compute H^{k+1} using Eq. (10).
5. Set $k = k + 1$ and repeat Step 1.

If the matrix H^k is positive definite and $(\delta^k - H^k\gamma^k)^T\gamma^k > 0$, then the matrix H^{k+1} computed using the SR1 update (9) will also be positive definite. If the matrix H^k is updated using the BFGS update (10) and if the condition
$$t^k(\delta^k)^T\gamma^k > 0$$
holds, then H^{k+1} will also be positive definite. The line search condition (12) and t^k in [0.5, 100] guarantee that $t^k(\delta^k)^T\gamma^k > 0$ holds at every iteration. It is important to have positive definite matrix H^k to ensure that direction d^k is a descent direction.

An iterative line search procedure may require more than one function and gradient evaluation before a step length λ^k that satisfies conditions (11) and (12) can be found. Hence, in general the total number of function and gradient evaluations required by the SR1/BFGS algorithm to find a minimum of the error function is more than the total number of iterations. While the total number of iterations reflects the total number of times that the weights of the network are updated, the total number of function/gradient evaluations is a more accurate indication of the cost of training the network. Since the gradient of the function is always computed when the function value is computed and vice versa, the number of function evaluations is equal to the number of gradient evaluations.

We note that for the steepest descent method with a fixed step length, only two n-dimensional vectors need to be stored: the current estimate of the minimum z^k and the gradient of the function at this point, $\nabla f(z^k)$. When a line search procedure is implemented in conjunction with this method, two more n-dimensional vectors are needed by the procedure to store the new estimate of the minimum: $z^k - \lambda \nabla f(z^k)$ for some $\lambda > 0$ and the gradient at this new estimate.

In addition to these four n-dimensional vectors, the quasi-Newton method requires extra storage for holding the vector $H^k y^k$ and the matrix H^k. Since this matrix is symmetric, an additional $n(n + 1)/2$ real words of storage will be sufficient. Hence, the total storage requirement for the quasi-Newton method is $n(n + 11)/2$ plus several scalar variables for storing the various constants and scalar products. Although this $O(n^2)$ storage requirement and the $O(n^2)$ floating point operations needed at each iteration to update the matrix H^k may seem to be major drawbacks of the quasi-Newton method, our experience with this method indicates that the number of iterations and the number of function and gradient evaluations required by this method are much fewer than those of the steepest descent method. The conjugate gradient method, which has also been used for neural network training [15] requires an $O(n)$ storage space. If the storage space is limited, this approach is suitable for a network with many units. However, in general, quasi-Newton methods converge faster than the conjugate gradient method [12]. The fast convergence of the quasi-Newton method should make it the method of choice for training a neural network when the storage space is not a restricting factor.

III. SELECTING THE NUMBER OF OUTPUT UNITS

The necessary number of output units in a network is usually the easiest to determine. For a pattern classification problem to distinguish between patterns from two classes, a single output unit would suffice. Each pattern that belongs to one class can be assigned a target of 1, while a pattern that belongs to the other class can be assigned a target of 0. If the classification involves patterns from $N > 2$ classes, a commonly used approach is to have N output units. Each pattern is labeled by an N-dimensional binary vector, where $N - 1$ bits are set to zero and exactly one bit is set to 1. The position of the 1-bit indicates the class to which the pattern belongs. When N is large, instead of having an N-dimensional target output for each pattern, we could use a binary encoding to represent class membership. Using binary encoding, only $\lceil \log N \rceil$ output units would be needed. With the smaller number of output units, however, more hidden units may be needed to represent the mapping of the input patterns and their binary encoded class labels.

The number of output units is generally much fewer than the number of input or hidden units. For applications other than pattern classification, however, a large number of output units may be needed. One such applications is image compression. Image compression using neural networks with one hidden layer can be considered as a learning problem where the target to be learned is actually the same as the input. Typically, an image is divided up into small patches of 4×4 or 8×8 pixels [16–18]. A patch of 8×8 pixels would require 64 input units and the same number of outputs. The connections between the input units and the hidden units act as an encoder which compresses the image, while the connections between the hidden units and the output units act as a decoder which will be needed to recover the original image. The activation values of the hidden units thus represent the coded image. These activation values, which are real numbers in the interval [0, 1] or [−1, 1] depending on the activation function used, are discretized into a small number of bits. If the number of bits is n, then the number of distinct activation values in a hidden unit can be up to 2^n. A small number of hidden units and a small number of bits used to represent the discretized hidden unit activation values result in a high compression ratio. For example, if four hidden units are present in the network and four bits are used to represent the activation values of an 8×8 input patch at each hidden unit, a compression ratio of $(8 \times 8)/(4 \times 4) = 4$ is achieved. Hence, it is desirable to have a network with a small number of hidden units and a small number of distinct discretized hidden unit activation values to achieve a high degree of compression. The goal of achieving a high compression ratio, however, must be balanced against the quality of the decoded image.

IV. DETERMINING THE NUMBER OF HIDDEN UNITS

While it is known that a network having a single hidden layer is capable of approximating any decision boundary, in general, it is not known how many units in the hidden layer are needed. The problem of selecting an appropriate number of hidden units in a network is a very challenging one. If the network has too many hidden units, it may overfit the data and result in poor generalization. On the other hand, a network with too few hidden units may not be able to achieve the required accuracy rate.

Two different approaches have been described in the literature to address the difficulty of finding the right number of hidden units of a network. The first approach begins with an oversized network and then prunes redundant units [19–22]. The second approach begins with a small network with one or two hidden units and adds more units only when they are needed to improve the learning capability of the network.

Algorithms which automatically build neural networks have been proposed by many researchers. These methods include the cascade correlation algorithm [23], the tiling algorithm [24], the self-organizing neural network [25], and the upstart algorithm [26]. For a given problem, these algorithms will generally build networks with many layers. The dynamic node creation method proposed by

Ash [27] is an algorithm which constructs neural networks with a single hidden layer. The method creates feedforward neural networks by sequentially adding hidden units to the hidden layer.

The neural network construction algorithm FNCAA [28] is similar to Ash's dynamic node creation algorithm. It starts with a single hidden layer network consisting of a single hidden unit and finds a set of optimal weights for this network. If the network with these weights does not achieve the required accuracy rate, then one hidden unit is added to the network and the network is retrained. The process is repeated until a network that correctly classifies all the input patterns or meets some other prespecified stopping criteria has been constructed. The outline of the algorithm is as follows:

Feedforward neural network construction algorithm (FNNCA)

1. Let h be the initial number of hidden units in the network. Set all the initial weights in the network randomly.
2. Find a point that minimizes the error function (1).
3. If this solution results in a network that meets the stopping condition, then stop.
4. Add one unit to the hidden layer and select initial weights for the arcs connecting this new node with the input units and the output unit. Set $h = h + 1$ and go to Step 2.

The difference between the dynamic node creation algorithm and FNNCA lies in the training of the growing network. In the dynamic node creation algorithm, the network is trained using the standard back-propagation method, while in FNNCA the growing network is trained by the SR1/BFGS method described in the previous section.

Interesting results were obtained when FNNCA was applied to solve the N-bit parity problem. This problem is a well-known difficult problem that has often been used for testing the performance of a neural network training algorithm. The input set consists of 2^n patterns in n-dimensional space and each pattern is an n-bit binary vector. The target value t^i is equal to 1 if the number of one's in the pattern is odd and it is 0 otherwise. To solve this problem by a feedforward neural network, the number of hidden units is usually set to N, the same as the number of input units. The initial number of hidden units in FNNCA was set to two. For the 4-bit parity problem, FNNCA terminated after 105 iterations and 132 function/gradient evaluations. The final number of hidden units was three. The algorithm required 168 iterations with 222 function/gradient evaluations to construct a network having four hidden units that correctly classified all 32 inputs of the 5-bit parity problem. A network with five hidden units was also found by the algorithm for the 7-bit parity problem after 943 iterations and 1532 function/gradient evaluations. Using the dynamic node creation algorithm, the 4-bit parity problem required more than 4000 iterations and the final network constructed had four hidden units.

Instead of the sum of the squared-error function (1), any function that attains its minimum or maximum when the output value from the network for each input pattern is equal to its target value can be used to compute these weights. The maximum likelihood neural network construction algorithm (MLNNCA) [29] is similar to FNNCA except that it trains the growing network by minimizing the cross-entropy error function

$$\min_{w,v} F^h(y,z) := -\sum_{i \in \mathcal{I}} \log S^i - \sum_{i \notin \mathcal{I}} \log\left(1 - S^i\right), \tag{13}$$

where

$$\begin{aligned} S^i &= \text{the predicted output for input } x^i,\ \sigma\left(\textstyle\sum_{j=1}^{h} \psi((x^i)^T w^j) v^j\right),\\ \psi(\eta) &= \text{the hyperbolic activation function, } (e^{\eta} - e^{-\eta})/(e^{\eta} + e^{-\eta}),\\ \mathcal{I} &= \{i \,|\, t^i = 1\}. \end{aligned}$$

The superscript h on the function F has been added to emphasize that the function corresponds to a network with h hidden units. The components of the gradient of the function $F^h(w,v)$ are as follows:

$$\begin{aligned} \frac{\partial F^h(w,v)}{\partial w_\ell^m} &= -\sum_{i \in \mathcal{I}} \left[(1 - S^i) \times v^m \times \left(1 - \psi\left(x^i w^m\right)^2\right) \times x_\ell^i\right] \\ &\quad + \sum_{i \notin \mathcal{I}} \left[S^i \times v^m \times \left(1 - \psi\left(x^i w^m\right)^2\right) \times x_\ell^i\right] \\ &= \sum_{i=1}^{k} \left[e^i \times v^m \times \left(1 - \psi(x^i w^m)^2\right) \times x_\ell^i\right], \\ \frac{\partial F^h(w,v)}{\partial v^m} &= -\sum_{i \in \mathcal{I}} \left[(1 - S^i) \times \psi\left(x^i w^m\right)\right] + \sum_{i \notin \mathcal{I}} \left[S^i \times \psi\left(x^i w^m\right)\right] \\ &= \sum_{i=1}^{k} \left[e^i \times \psi\left(x^i w^m\right)\right], \end{aligned}$$

for all $m = 1, 2, \ldots, h$ and $\ell = 1, 2, \ldots, n$, with the error $e^i = S^i - t^i$.

Let $(\overline{w}, \overline{v}) \in \mathbb{R}^{(n+1)\times h}$ be a point such that $\nabla F^h(\overline{w}, \overline{v}) = 0$ and suppose that the network with h hidden units corresponding to this set of weights does not meet the stopping condition of the network construction algorithm. Let $w^{h+1} \in \mathbb{R}^n$ be a randomly generated vector; it is clear that $F^{h+1}(\overline{w}, w^{h+1}, \overline{v}, 0) = F^h(\overline{w}, \overline{v})$. We wish to find $\nu \in \mathbb{R}$ such that the value of the cross-entropy error function for the network with an additional hidden unit is less than that of the original network;

that is,

$$F^{h+1}\left(\overline{w}, w^{h+1}, \overline{v}, \nu\right) < F^{h}(\overline{w}, \overline{v}).$$

The variable ν represents the weight of the connection from the new hidden unit to the output unit. For simplicity of derivations, we hold w^{h+1} constant and define a new function of a single variable

$$\begin{aligned} \mathcal{F}(v) &= F^{h+1}\left(\overline{w}, w^{h+1}, \overline{v}, v\right) \\ &= -\sum_{i\in\mathcal{I}} \log\left[\sigma\left(\Delta^{i} + \delta^{i} v\right)\right] - \sum_{i\notin\mathcal{I}} \log\left[1 - \sigma\left(\Delta^{i} + \delta^{i} v\right)\right], \end{aligned}$$

where

$$\begin{aligned} \Delta^{i} &= \sum_{j=1}^{h} \psi\left(x^{i}\overline{w}^{j}\right)\overline{v}^{j}, \\ \delta^{i} &= \psi\left(x^{i} w^{h+1}\right). \end{aligned}$$

It follows that the first and second derivatives of this function are

$$\begin{aligned} \mathcal{F}^{'}(v) &= -\sum_{i\in\mathcal{I}} \left[1 - \sigma\left(\Delta^{i} + \delta^{i} v\right)\right]\delta^{i} + \sum_{i\notin\mathcal{I}} \left[\sigma\left(\Delta^{i} + \delta^{i} v\right)\right]\delta^{i}, \\ \mathcal{F}^{''}(v) &= \sum_{i=1}^{k} \left(\delta^{i}\right)^{2}\left[\sigma\left(\Delta^{i} + \delta^{i} v\right)\left(1 - \sigma\left(\Delta^{i} + \delta^{i} v\right)\right)\right]. \end{aligned}$$

Hence we have that the derivative of this function at zero is

$$\begin{aligned} \mathcal{F}^{'}(0) &= -\sum_{i\in\mathcal{I}} \left[1 - \sigma\left(\Delta^{i}\right)\right]\delta^{i} + \sum_{i\notin\mathcal{I}} \sigma\left(\Delta^{i}\right)\delta^{i} \\ &= \sum_{i=1}^{k} e^{i}\psi\left(x^{i} w^{h+1}\right), \end{aligned}$$

and that the second derivative is bounded above as follows:

$$\left|\mathcal{F}^{''}(v)\right| \leq k/4, \qquad \forall v \in \mathbb{R}.$$

By definition of the function $\mathcal{F}$, we have that

$$F^{h+1}\left(\overline{w}, w^{h+1}, \overline{v}, \lambda v\right) = \mathcal{F}(\lambda v).$$

From the second-order Taylor expansion of this function, we have

$$\mathcal{F}(\lambda v) = \mathcal{F}(0) + \lambda\mathcal{F}^{'}(0)v + \tfrac{1}{2}(\lambda v)^{2}\mathcal{F}^{''}(\rho\lambda v), \qquad 0 < \rho < 1.$$

By letting $v = -\mathcal{F}'(0)$, we obtain

$$\begin{aligned}\mathcal{F}(\lambda v) &= \mathcal{F}(0) - \lambda\big(\mathcal{F}'(0)\big)^2 + \frac{1}{2}\lambda^2\big(\mathcal{F}'(0)\big)^2\mathcal{F}''(\rho\lambda v)\\ &\le \mathcal{F}(0) - \lambda\big(\mathcal{F}'(0)\big)^2 + \frac{k}{8}\lambda^2\big(\mathcal{F}'(0)\big)^2.\end{aligned}$$

The inequality above is obtained from the fact that the second derivative $\mathcal{F}''(v)$ is bounded by $k/4$. Now let us set $\lambda = 4/k$; we have

$$\begin{aligned}\mathcal{F}(\lambda v) &\le \mathcal{F}(0) - \frac{2}{k}\big(\mathcal{F}'(0)\big)^2\\ &= \mathcal{F}(0) - \frac{2}{k}\left[\sum_{i=1}^{k} e^i\psi\big(x^i w^{h+1}\big)\right]^2\\ &= F^h(\overline{w},\overline{v}) - \frac{2}{k}\left[\sum_{i=1}^{k} e^i\psi\big(x^i w^{h+1}\big)\right]^2.\end{aligned}$$

For a randomly generated w^{h+1}, it is very unlikely that the sum $\sum_{i=1}^{k} e^i \times \psi(x^i w^{h+1})$ will be zero. Thus, even before the new expanded network is retrained, if we pick w^{h+1} randomly and set $v^{h+1} = -4(\sum_{i=1}^{k} e^i\psi(x^i w^{h+1}))/k$, there is already a decrease in the function value.

If the sum of the squared-error function (1) is used for training the network instead of the cross-entropy error function (13), then the function $\mathcal{F}(v)$ will become

$$\mathcal{F}(v) = \sum_{i=1}^{k}\big(\sigma\big(\Delta^i + \delta^i v\big) - t^i\big)^2. \tag{14}$$

The derivative of this function at zero is

$$\mathcal{F}'(0) = 2\sum_{i=1}^{k}\big[e^i \times \sigma\big(\Delta^i\big) \times \big(1 - \sigma\big(\Delta^i\big)\big) \times \delta^i\big]. \tag{15}$$

Due to rounding error, the product $e^i\sigma(\Delta^i)(1-\sigma(\Delta^i))$ is often zero. This happens when each of the network outputs S^i is either very close to zero or very close to 1. When the training of a network with h hidden units converges to a point $(\overline{w},\overline{v})$ such that e^i is equal to 0, or 1, or -1 for all $i = 1, 2, \ldots, k$, then the derivative (15) will be zero and the point $(\overline{w}, w^{h+1}, \overline{v}, 0)$ with any $w^{h+1} \in \mathbb{R}^n$ is in fact a local minimum of the function $f(w, v)$ for a new expanded network with $h + 1$ hidden units. Since there is no decrease in the function value, the addition of a new hidden unit will be futile and the recognition rate of the network will not improve. It has also been observed that neural network training with a fixed number of hidden units requires less iterations if one substitutes the cross-entropy error function (13) for the sum of the squared-error function [4, 30].

MLNNCA was run 50 times using 50 different random starting points to solve the N-bit parity problems for N ranging from four to eight. The minimum number of hidden units in the constructed networks was $N/2+1$ for even N, and $(N+1)/2$ for odd N. Not all runs ended with the minimal network. However, regardless of the starting random weights, the algorithm was always successful in constructing a network that correctly classified all the input patterns.

The algorithm was also tested on the spiral problem [4]. The problem of distinguishing two intertwined spirals is a nontrivial one. The two spirals shown in Fig. 2 consist of a total of 970 patterns. Solutions to the spiral problem have been obtained by feedforward networks with several hidden layers having connections connecting every layer to all succeeding layers [4], by networks where there are connections among hidden units such as those generated by the cascade correlation algorithm [23], or by networks with connections among hidden units and

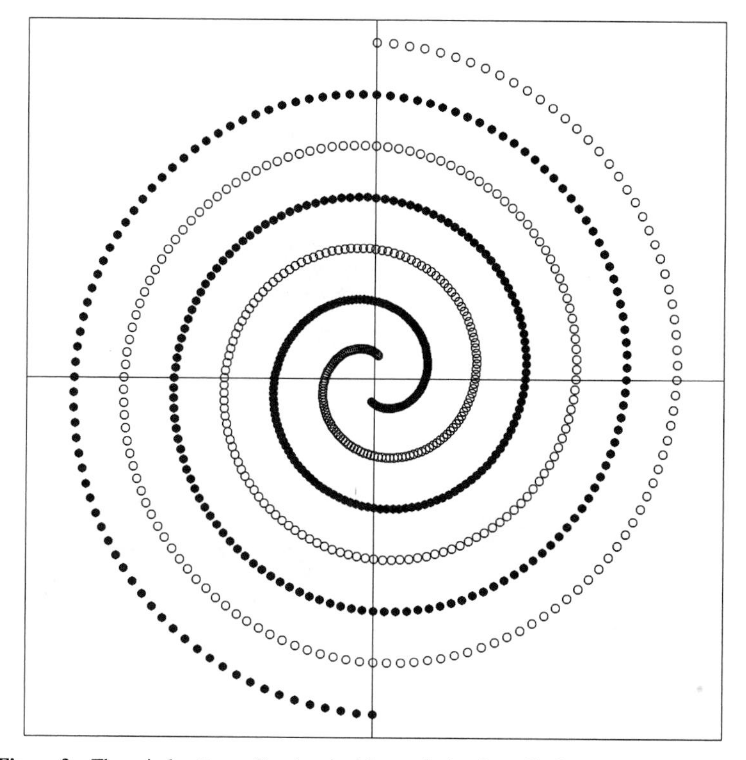

Figure 2 The spiral patterns. Reprinted with permission from Carfax Publishing Limited.

shortcut connections from the input units to the output units [31]. Solutions from the standard single hidden layer networks have been reported only for a substantially reduced problem [32].

FNCAA was run 10 times to solve the spiral problems. It constructed networks with a final number of hidden units ranging from 28 to 38. A two-dimensional classification graph of one of the networks is shown in Fig. 3. The graph shows the classification of the network at different growing stages. In each square, black

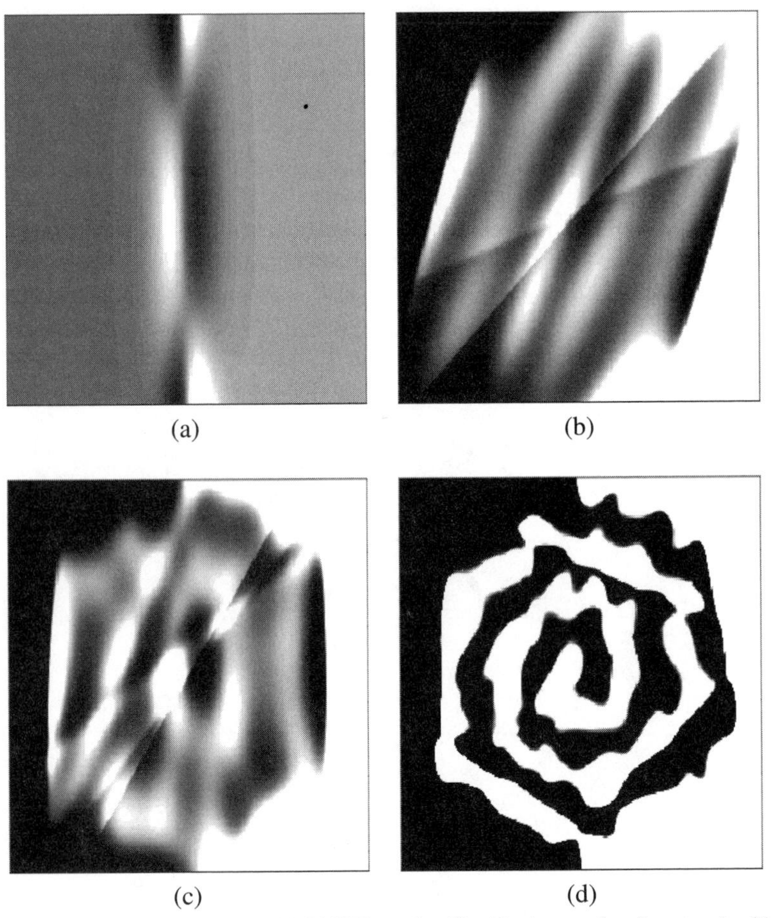

Figure 3 Evolution of a network with 34 hidden units. Classification graphs of a network with (a) 4 hidden units (b) 12 hidden units (c) 22 hidden units (d) 34 hidden units. Reprinted with permission from Carfax Publishing Limited.

represents an activation value of 0, white represents 1, and grey represents intermediate values between 0 and 1. The final number of hidden units for this particular run is 34. The classification graph of the network with 34 hidden units shows that most of the points in the square enclosing the training data are classified as 0 or 1.

V. SELECTING THE NUMBER OF INPUT UNITS

Finding the optimal number of input units is equivalent to selecting the set of attributes of the patterns that are useful for classification. While a great dea of research has been focused on algorithms that optimize the number of hidden units, there has not been much work that addresses the issue of optimal number of input units of a neural network classifier.

It is quite common that data sets collected contain many attributes that are redundant and/or irrelevant. By excluding these attributes from the classification process, a classifier with higher generalization capability, i.e., better predictive accuracy on new/unseen patterns, can often be found. The dimensionality of patterns with attributes that are highly correlated may be reduced with little or no loss of information. Hence, by collecting only values of the relevant attributes, the cost of future data collection may also be reduced.

Feature selection aims at selecting a subset of the attributes that are relevant for classification. Similar to selecting an optimal number of hidden units, there are two approaches that have been applied in feature selection. One can begin with no feature and start adding the relevant features one at a time, or one can begin with the entire feature set and remove those irrelevant features one by one.

Setiono and Liu [33] propose an algorithm for determining the relevant subset of attributes for classification using neural networks. A network is trained with the complete set of attributes as input. For each attribute $\mathcal{A}_i$ in the network, the accuracy of the network with all the weights of the connections associated with this attribute set to zero is computed. The attribute that gives the smallest decrease in the network accuracy is removed. The network is then retrained and the process is repeated. To facilitate the process of identifying the irrelevant attributes, the network is trained to minimize an augmented error function. The augmented error function consists of two components. The first component is a measure of network accuracy and the second component is a measure of the network complexity. The accuracy of the network is measured using the cross-entropy error function, while the complexity of the network is measured by a penalty term. A network weight with a small magnitude incurs almost no penalty, while a weight that falls in a certain allowable range incurs an almost constant penalty. The penalty of a large weight that falls outside this interval increases as a quadratic function of its magnitude.

Relevant and irrelevant inputs are distinguished by the strength of their connections from the input layer to the hidden layer in the network. The network is trained such that the connections from the irrelevant inputs to the hidden layer have small magnitude. These connections can be removed from the network without affecting the network accuracy. Since we are interested in finding the smallest subset of the attributes that still preserves the characteristics of the patterns, it is important that the network be trained such that only those connections from the necessary inputs have large magnitude. To achieve this goal, a penalty term $P(w)$ is added for each connection from the input layer to the hidden layer of the network. It is defined as follows:

$$P(w) = \epsilon_1 \left(\sum_{j=1}^{h} \sum_{\ell=1}^{n} \frac{\beta (w_\ell^j)^2}{1 + \beta (w_\ell^j)^2} \right) + \epsilon_2 \left(\sum_{j=1}^{h} \sum_{\ell=1}^{n} (w_\ell^j)^2 \right). \tag{16}$$

$\epsilon_1 > \epsilon_2 > 0$ are penalty parameters and β is a positive constant.

There are two components of the penalty function $P(w)$; the first component is to discourage the use of unnecessary connections and the second component is to prevent the weights of these connections from taking very large values. These two components have been used individually in conjunction with many pruning algorithms proposed in the past few years [34].

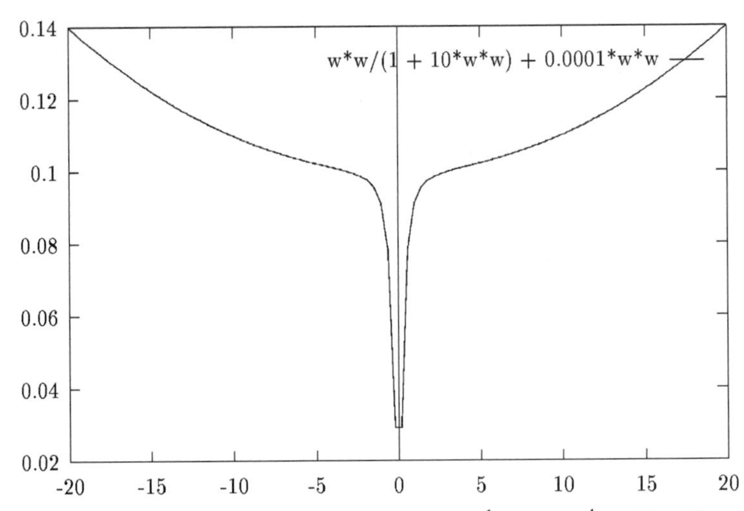

Figure 4 Plot of the function $f(w)$ with $\epsilon_1 = 10^{-1}$, $\epsilon_2 = 10^{-4}$, and $\beta = 10$.

A weight with small magnitude is encouraged to converge to zero as reflected by the steep drop in the function value near zero (Fig. 4). On the other hand, the weights of the network are prevented from taking values that are too large as reflected by the quadratic component of the penalty function which becomes dominant for large values of w.

Combining the cross-entropy error function and the penalty function, we minimize the following function during network training:

$$\theta(w, v) = -\left(\sum_{i=1}^{k} \sum_{p=1}^{C} t_p^i \log S_p^i + \left(1 - t_p^i\right) \log\left(1 - S_p^i\right) \right) + P(w), \qquad (17)$$

where C is the number of output units in the network, and S_p^i and t_p^i are the network output and the target output for pattern x^i at output unit p, respectively.

Features are selected for removal based on their saliency. Several saliency measures are reported by Belue and Bauer [35]. These measures of saliency of an attribute involve the derivative of the network error function, or the weights of the network, or both. In order to obtain a confidence interval for the mean value of the saliency of the attributes, the network needs to be retrained repeatedly starting from different random weights. It is suggested that the network be trained at least 30 times in order to find a reliable mean and standard deviation of the saliency measure. As network training can be very slow, the requirement that the network be trained many times makes their proposed scheme computationally unappealing. Instead of using a saliency measure that is a function of the network weights, we use a very simple criterion to determine which attribute is to be excluded from the network. This criterion is the network accuracy on the training data set. Given a trained network with the set of attributes $\mathcal{A} = \{\mathcal{A}_1, \mathcal{A}_2, \ldots, \mathcal{A}_N\}$ as its input, we compute the accuracy of the networks having one less attribute, i.e., the set $\mathcal{A} - \{\mathcal{A}_k\}$, for each $k = 1, 2, \ldots, N$, is an input attribute set. The accuracy rates are computed by simply setting the connection weights from input attribute $\mathcal{A}_k$ of the trained network to zero. The accuracy rates of these networks are then ranked. Starting with the network having the highest accuracy, the set of attributes to be retained is searched. The steps of the algorithm are outlined below.

Neural network feature selection algorithm

1. Let $\mathcal{A} = \{\mathcal{A}_1, \mathcal{A}_2, \ldots, \mathcal{A}_N\}$ be the set of all input attributes. Separate the patterns into two sets: the training set $\mathcal{S}_1$ and the cross-validation set $\mathcal{S}_2$. Let $\Delta\mathcal{R}$ be the allowable maximum decrease in accuracy rate on the set $\mathcal{S}_2$ and let $\epsilon_1(k)$ and $\epsilon_2(k)$ be the penalty parameters [cf. Eq. (16)] for the connections from input $\mathcal{A}_k$ to the hidden layer, for all $k = 1, 2, \ldots, N$.

2. Train network $\mathcal{N}$ to minimize the augmented error function $\theta(w, v)$ with the set $\mathcal{A}$ as input such that it achieves a minimum required accuracy rate on the set $\mathcal{S}_1$. Let $\mathcal{R}^2$ be the accuracy of the network on the set $\mathcal{S}_2$.
3. For all $k = 1, 2, \ldots, N$, let $\mathcal{N}_k$ be the network whose weights are set as follows:

   (a) From all inputs except for $\mathcal{A}_k$, set the weights of $\mathcal{N}_k$ equal to the weights of $\mathcal{N}$.
   (b) Set the weights from input $\mathcal{A}_k$ to zero.

   Compute $\mathcal{R}_k^1$ and $\mathcal{R}_k^2$, the accuracy rates of network $\mathcal{N}_k$ on the sets $\mathcal{S}_1$ and $\mathcal{S}_2$, respectively.
4. Rank the networks $\mathcal{N}_k$ according to their accuracy rates: $\mathcal{R}_{r(1)}^1 \geq \mathcal{R}_{r(2)}^1 \geq \cdots \geq \mathcal{R}_{r(N)}^1$. Let $\mathcal{R}_{ave}^1$ be the average of these rates.

   (a) Set $k = 1$.
   (b) Retrain the network $\mathcal{N}_{r(k)}$.
   (c) Let $\delta = (\mathcal{R}^2 - \mathcal{R}_{r(k)}^2)/\mathcal{R}^2$.
   (d) If $\delta \leq \Delta\mathcal{R}$, then

      - Update the penalty parameters for all attributes $j \neq r(k)$:
        - For each input attribute $\mathcal{A}_j$ with network accuracy rate $\mathcal{R}_j^1 \geq \mathcal{R}_{ave}^1$, set $\epsilon_1(j) := 1.1\epsilon_1(j)$ and $\epsilon_2(j) := 1.1\epsilon_2(j)$.
        - For each input attribute $\mathcal{A}_j$ with network accuracy rate $\mathcal{R}_j^1 < \mathcal{R}_{ave}^1$, set $\epsilon_1(j) := \epsilon_1(j)/1.1$ and $\epsilon_2(j) := \epsilon_2(j)/1.1$.
      - Reset the input attribute set to $\mathcal{A} - \{\mathcal{A}_{r(k)}\}$, and set $N := N - 1$.
      - Set $\mathcal{R}^2 := \max\{\mathcal{R}^2, \mathcal{R}_{r(k)}^2\}$.
      - Go to step 3.

   (e) If $k < N$, set $k := k + 1$ and go to Step 4(b).
   Else stop.

The available patterns for training are divided into two sets, $\mathcal{S}_1$ and $\mathcal{S}_2$. The set $\mathcal{S}_1$ consists of patterns that are actually used to obtain the weights of the neural networks. The set $\mathcal{S}_2$ consists of patterns that are used for cross-validation. By checking the accuracy of the networks on the set $\mathcal{S}_2$, the algorithm decides whether to continue or to stop removing more attributes. The best accuracy rate $\mathcal{R}^2$ of the networks on this set is kept by the algorithm. If there is still an attribute that can be removed such that the relative accuracy rate on $\mathcal{S}_2$ does not drop by more than $\Delta\mathcal{R}$, then this attribute will be removed. If no such attribute can be found among the inputs, the algorithm terminates.

At the start of the algorithm, the values of the penalty parameters $\epsilon_1(k)$ and $\epsilon_2(k)$ are set equal for all attributes $\mathcal{A}_k$, since it is not yet known which are the relevant attributes and which are not. In our experiments, we have set the initial values for $\epsilon_1(k)$ and $\epsilon_2(k)$ to 0.1 and 10^{-4}, respectively. After the network is trained, the relative importance of each attribute can be inferred from the accuracy rates of all the networks $\mathcal{N}_k$ having one less attribute. A high accuracy rate of $\mathcal{N}_k$ suggests that the attribute $\mathcal{A}_k$ can be removed from the attribute set. Step 4(d) of the algorithm updates the values of the penalty parameters for all the remaining attributes based on the accuracy of the networks. If the accuracy rate of network $\mathcal{N}_k$ is higher than the average, then the penalty parameters for the network connections from input attribute $\mathcal{A}_k$ are multiplied by a factor 1.1. It is expected that with larger penalty parameters, the connections from this input attribute will have smaller magnitudes after the network is retrained, and therefore the attribute can be removed in the next round of the algorithm. On the other hand, a below-average accuracy rate of the network $\mathcal{N}_k$ indicates that the attribute $\mathcal{A}_k$ is important for classification. For all such attributes, the penalty parameters are divided by a factor of 1.1.

Among the problems on which the neural network feature selection algorithm was tested are the monks problems [36]. There are three monks problems in which robots are described by six different attributes (Table I). The learning tasks of the three monks problems are of binary classification; each of them is given by the following logical description of a class:

- Problem Monks 1: (head_shape = body_shape) or (jacket_color = red). From 432 possible samples, 112 were randomly selected for the training set, 12 for cross-validation, and all 432 for testing.
- Problem Monks 2: Exactly two of the six attributes have their first value. From 432 samples, 152 were selected randomly for the training set, 17 for cross-validation, and all 432 for testing.
- Problem Monks 3: (Jacket_color is green and holding a sword) or (jacket_color is not blue and body_shape is not octagon). From 432 samples, 122 were selected randomly for training and among them there was 5% misclassification, i.e., noise in the training set. Twenty samples were selected for cross-validation, and all 432 samples formed the testing set.

Table I

Attributes of the Three Monks Problems

A_1:	head_shape	$\in$	round, square, octagon;
A_2:	body_shape	$\in$	round, square, octagon;
A_3:	is_smiling	$\in$	yes, no;
A_4:	holding	$\in$	sword, balloon, flag;
A_5:	jacket_color	$\in$	red, yellow, green, blue;
A_6:	has_tie	$\in$	yes, no.

In order to demonstrate the effectiveness of the feature selection algorithm, each possible value of the six attributes is treated as a single new attribute. For example, the attribute head_shape, which can be either round, square, or octagon, is represented by three new attributes. The three attributes are head_shape = round, head_shape = square and head_shape = octagon. Exactly two of the three attributes have values 0, while the third attribute has value 1. This representation of the original six attributes enables us not only to select the relevant attributes, but also to discover which particular values of these attributes are useful for classification.

For each problem, thirty neural networks with 12 hidden units and 17 input units were trained starting from different initial random weights. The results of the experiments are summarized in Table II. In this table, the average accuracy rates of the networks on the training and testing data sets with and without feature selection are given. Standard deviations are given in parentheses. The average function evaluation reflects the cost of selecting the relevant features. It is the average number of times that the value and the gradient of the augmented er-

Table II
Results for the Monks Problems

Monks 1	With all features[a]	With selected features[a]
Ave. no. of features	17 (0.00)	5.07 (0.37)
Ave. acc. on training set (%)	100.00 (0.00)	100.00 (0.00)
Ave. acc. on testing set (%)	99.71 (0.67)	100.00 (0.00)
Ave. function evaluations	360.37 (114.76)	
P-value (testing set acc.)	0.09	
Monks 2	**With all features[a]**	**With selected features[a]**
Ave. no. of features	17 (0.00)	6.23 (0.43)
Ave. acc. on training set (%)	100.00 (0.00)	100.00 (0.00)
Ave. acc. on testing set (%)	98.78 (2.34)	99.54 (0.99)
Ave. function evaluations	538.63 (117.02)	
P-value (testing set acc.)	0.05	
Monks 3	**With all features[a]**	**With selected features[a]**
Ave. no. of features	17 (0.00)	3.87 (1.78)
Ave. acc. on training set (%)	100.00 (0.00)	94.23 (0.79)
Ave. acc. on testing set (%)	93.55 (1.41)	98.41 (1.66)
Ave. function evaluations	826.70 (212.86)	
P-value (testing set acc.)	$< 10^{-5}$	

[a] Standard deviation for the averages are given in parentheses.

ror function (17) are computed by the minimization algorithm SR1/BFGS. The P-value is computed to check if there is any significant increase in the accuracy of the networks with selected input features compared to the networks with the whole set of attributes as input. A smaller P-value indicates a more significant increase. Since the largest among the P-values obtained from the three sets of experiments is 0.09, we can reject at 10% level of significance the null hypothesis that there is no increase in the predictive accuracy of the networks after pruning.

The figures in Table II show that feature selection not only removes the irrelevant features, it also improves significantly the predictive accuracy of the networks. For the Monks 1 problem, all 30 networks with selected input attributes are capable of classifying all testing patterns correctly. Twenty-nine networks have the minimum five input attributes and the remaining one has seven input attributes. For the Monks 2 problem, 23 networks have the minimum six attributes and the remaining seven networks have seven attributes.

For the Monks 3 problem, most networks have either two or five input attributes. The maximum number of attributes a network has is nine. All twelve networks with five input attributes achieve 100% accuracy rate on the testing data set. All eleven networks with two input attributes have accuracy rates of 93.44% and 97.22% on the training data set and the testing data set, respectively. The 97.22% accuracy rate is the same as that reported by Thrun et al. [36]. It is worth noting that, despite the presence of six mislabeled training patterns, 14 of the 30 networks with selected attributes have a perfect 100% accuracy rate on the testing data set. None of the 30 networks with all input attributes has such accuracy. The results from running the neural network feature selection algorithm on many real-world data sets are reported in Setiono and Liu [33]. The results show that neural network classification using only selected input attributes is generally more accurate than using all the attributes in the data.

VI. DETERMINING THE NETWORK CONNECTIONS BY PRUNING

Instead of removing all the connections from an input unit to all the units in the hidden layer as described in the previous section, a finer pruning process which removes an individual weight or connection in the network may also increase the generalization capability of a neural network [37–40]. Methods for removing individual weights from a network also usually augment a penalty term to the network error function [34]. By adding a penalty term to the error function, the relevant and irrelevant network connections can be distinguished by the magnitudes of their weights or by other measures of saliency when the training process has been completed. The saliency measure of a connection gives an indication of the expected increase in the error function after that connection is eliminated from the network. In the pruning methods Optimal Brain Damage [39] and Optimal Brain Surgeon [37], the saliency of each connection is computed using a second-order approximation of the error function near a local minimum. If the saliency of a connection is below a certain threshold, then the connection is removed from the network. If the increase in the error function is larger than a predetermined acceptable error increase, the network must be retrained.

The algorithm N2P2F for neural network pruning outlined below was recently developed by Setiono [41]. Neural network pruning with penalty function (N2P2F) first trains a fully connected network by applying the SR1/BFGS method to find a set of weights that minimizes the augmented error function:

$$\hat{\theta}(w, v) = -\left(\sum_{i=1}^{k}\sum_{p=1}^{C} t_p^i \log S_p^i + \left(1 - t_p^i\right)\log\left(1 - S_p^i\right)\right) + Q(w, v), \tag{18}$$

where

$$\begin{aligned} Q(w, v) &= \epsilon_1 \sum_{j=1}^{h}\left(\sum_{\ell=1}^{n}\frac{\beta(w_\ell^j)^2}{1+\beta(w_\ell^j)^2} + \sum_{p=1}^{C}\frac{\beta(v_p^j)^2}{1+\beta(v_p^j)^2}\right) \\ &\quad + \epsilon_2 \sum_{j=1}^{h}\left(\sum_{\ell=1}^{n}\left(w_\ell^j\right)^2 + \sum_{p=1}^{C}\left(v_p^j\right)^2\right). \end{aligned} \tag{19}$$

The difference between the penalty functions (16) and (19) is that the latter also contains a penalty term for the connections between the hidden units and the output units. The reason why we also add a penalty term for each of the connections from a hidden unit and an output unit is linked to the criteria for weight removal (20) and (21) in the algorithm below.

Algorithm N2P2F: Neural network pruning with penalty function

1. Let η_1 and η_2 be positive scalars such that $\eta_1 + \eta_2 < 0.5$ (η_1 is the error tolerance, η_2 is a threshold that determines if a weight can be removed).
2. Pick a fully connected network, and train this network to minimize the error function (18) such that a prespecified accuracy level is met and the condition

$$\left|e_p^i\right| = \left|S_p^i - t_p^i\right| \le \eta_1$$

holds for all correctly classified input patterns. Let (w, v) be the weights of this network.

3. For each connection from input unit ℓ to hidden unit m, w_ℓ^m in the network, if

$$\max_p \left| v_p^m w_\ell^m \right| \leq 4\eta_2, \tag{20}$$

then remove w_ℓ^m from the network.
4. For each connection from hidden unit m to output unit p, v_p^m in the network, if

$$\left| v_p^m \right| \leq 4\eta_2, \tag{21}$$

then remove v_p^m from the network.
5. If no weight satisfies condition (20) or condition (21), then for each w_ℓ^m in the network, compute

$$\omega_\ell^m = \max_p \left| v_p^m w_\ell^m \right|.$$

Remove w_ℓ^m with the smallest ω_ℓ^m.
6. Retrain the network. If the classification rate of the network falls below the specified level, then stop and use the previous setting of network weights. Otherwise, go to Step 3.

In steps 3 and 4, N2P2F removes all the connections of the network whose magnitudes satisfy the conditions (20) or (21). In Setiono [41], we show that removal of such connections from the network does not affect the network accuracy. In step 5, we remove a network connection from an input unit to a hidden unit w_ℓ^m based on the values of its products with the weight of the connections from hidden unit m to all output units $p = 1, 2, \ldots, C$. The connection with the smallest maximum product is selected for removal. After removal of one or more connections, the network is retrained in step 6. Removed connections have their weight values fixed at 0.

The algorithm has been successfully applied to prune networks that have been trained for classification of many artificial and real-world data sets. Generally, for the problem domains tested, pruned networks have higher predictive accuracy than the fully connected networks. Among the problems tested and reported by Setiono [42] are the monks problems introduced in the previous section. For these problems, the algorithm N2P2F is able to obtain networks with fewer connections and better accuracy than those obtained by other pruning algorithms reported in the literature. Two pruned networks obtained for the Monks 1 and Monks 3 problems are shown in Figs. 5 and 6. The network in Fig. 5 correctly classifies all the patterns in the training and testing data sets of the Monks 1 problem. The network in Fig. 6 correctly identifies the six mislabeled patterns in the training data set and obtains a 100% accuracy rate on the testing patterns.

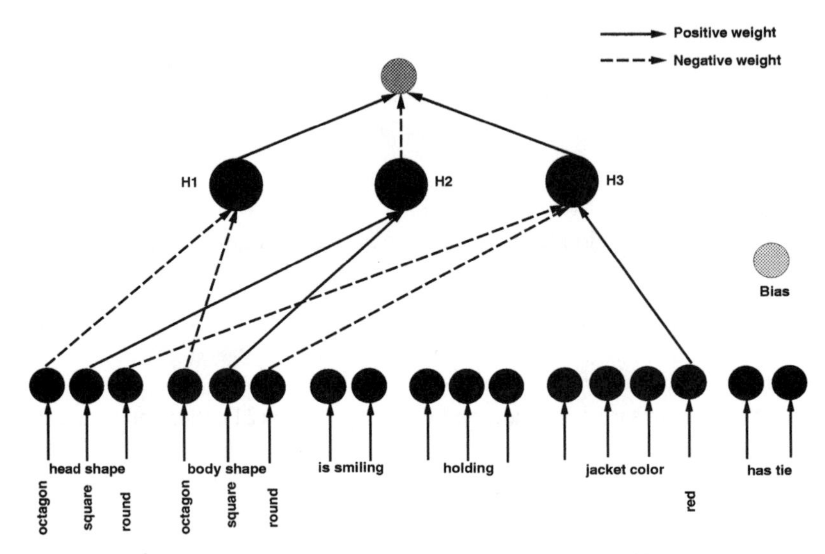

Figure 5 A pruned network that solves the Monks 1 problem.

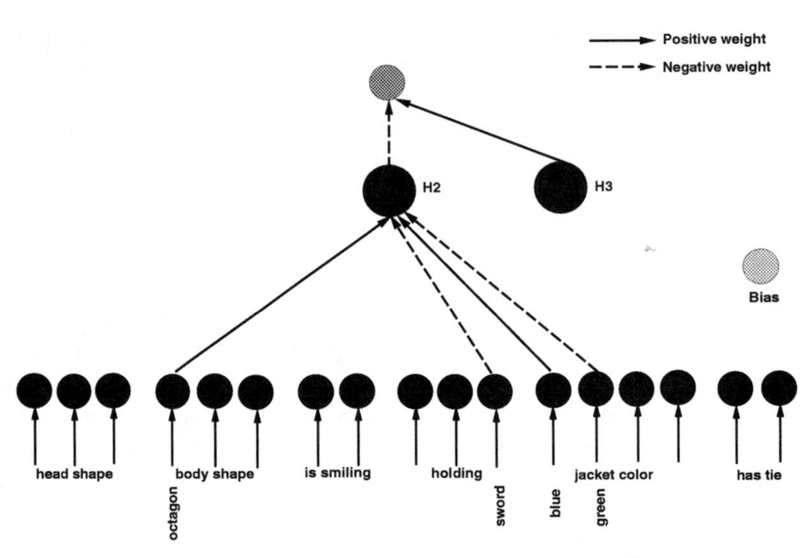

Figure 6 A pruned network that solves the Monks 3 problem.

VII. APPLICATIONS OF NEURAL NETWORKS TO DATA MINING

After pruning, the network contains only those connections that are relevant to class labels of the patterns. It is only natural for one to ask whether it is possible to express the relationship between the inputs and outputs of the network in a meaningful way. Since symbolic rules are easier to understand and verify than a collection of network weights, many attempts have been made to develop algorithms that extract symbolic rules from trained neural networks. One such algorithm is NeuroRule [43, 44]. By analyzing the activation values of the hidden units, NeuroRule generates symbolic rules that explain a network classification in terms of its input attributes.

We illustrate how NeuroRule works using the Wisconsin Breast Cancer data base [45]. The data set is available publicly via anonymous ftp from the University of California Irvine repository [46]. The data have been used as the test data for several studies on pattern classification methods using linear programming techniques [45, 47, 48] and statistical techniques [49].

Each pattern in the data set has nine attributes. The nine measurements taken from fine-needle aspirates from human breast tissues correspond to cytological characteristics of a benign or of a malignant sample. These are $\mathcal{A}_1$. clump thickness, $\mathcal{A}_2$. uniformity of cell size, $\mathcal{A}_3$. uniformity of cell shape, $\mathcal{A}_4$. marginal adhesion, $\mathcal{A}_5$. single epithelial cell size, $\mathcal{A}_6$. bare nuclei, $\mathcal{A}_7$. bland chromatin, $\mathcal{A}_8$. normal nucleoli, and $\mathcal{A}_9$. mitosis. Each of these nine attributes was graded 1 to 10 at the time of sample collection, with 1 being the closest to benign and 10 the most anaplastic. Since the attributes are integer-valued ranging from 1 to 10, we created 10 input units for each attribute. With an additional input for the bias weight at the hidden units, we have a total of 91 input units. Let us denote these inputs as $\mathcal{I}_1, \mathcal{I}_2, \ldots, \mathcal{I}_{91}$. For $i = 0, 1, \ldots, 8$, the following coding schemes for the input data are used:

$$\begin{aligned}
\mathcal{I}_{10\times i+j} &= 1 \iff \mathcal{A}_{i+1} \geq 11 - j, \;\; j = 1, 2, \ldots, 10, \\
\mathcal{I}_{10\times i+j} &= 0 \iff \mathcal{A}_{i+1} \leq 10 - j, \;\; j = 1, 2, \ldots, 9, \\
\mathcal{I}_{91} &= 1.
\end{aligned}$$

With this coding, $\mathcal{I}_{10\times j}$ is 1 for all $j = 1, 2, \ldots, 9$ for all patterns with valid attribute values in $\{1, 2, \ldots, 10\}$.

There are a total of 699 samples in the data base, of which 458 are benign samples and 241 are malignant samples. The patterns in the data set are divided randomly into a training set consisting of 350 samples and a testing set consisting of the remaining 349 samples. The target value for a benign sample is 0, while for a malignant sample the target value is 1. Figure 7 depicts a pruned network that

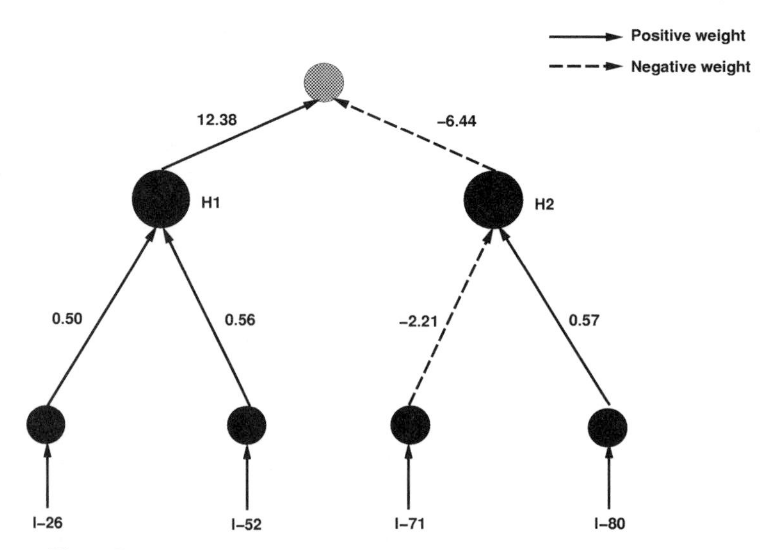

Figure 7 A pruned network for the Wisconsin Breast Cancer diagnosis problem.

has been trained to distinguish benign samples from malignant ones. Its accuracy rates on the training set and testing set are 96.57% and 93.12%, respectively.

Since the hyperbolic activation function is used at the hidden units, the activation value of a pattern can be anywhere in the interval $[-1, 1]$. In order to simplify the analysis, the continuous activation values of the patterns are discretized by clustering. A simple clustering algorithm that performs greedy clustering is given below.

A greedy clustering algorithm (GCA)

1. Find the smallest positive integer d such that if all the network hidden unit activation values are rounded to d-decimal-place, the network still retains its accuracy rate.
2. Represent each activation value α by the integer nearest to $\alpha \times 10^d$. Let $\mathcal{H}_i = \{h_{i,1}, h_{i,2}, \ldots, h_{i,k}\}$ be the set of these representations at hidden unit i for patterns $x^1, x^2, \ldots, x^k$ and let $\mathcal{H} = \{\mathcal{H}_1, \mathcal{H}_2, \ldots, \mathcal{H}_H\}$ be the set of the hidden representations of all patterns by all H hidden units.
3. Let P be an ordering sequence such that $P(i) \neq P(j)$ iff $i \neq j$ for all $i, j = 1, 2, \ldots, H$. Initialize $k = 1$.
4. Set $i = P(k)$.
5. Sort the set $\mathcal{H}$ such that the values of $\mathcal{H}_i$ are in increasing order.

6. Find a pair of distinct adjacent values $h_{i,j}$ and $h_{i,j+1}$ in the sorted set $\mathcal{H}_i$ such that if $h_{i,j+1}$ is replaced by $h_{i,j}$, no conflicting data will be generated.
7. If such a pair of values exists, replace all occurrences of $h_{i,j+1}$ by $h_{i,j}$ and repeat Step 6. Otherwise, set $k = k + 1$. If $k \leq H$, go to Step 4, else Stop.

Steps 1 and 2 of the GCA find integer representations of all hidden unit activation values. A small value for d in step 1 indicates that relatively few distinct values for the hidden unit activations are sufficient for the network to maintain its accuracy. For example, when $d = 2$, the distinct values are $-1.00, -0.99, \ldots, -0.01, 0.00, 0.01, \ldots, 0.99, 1.00$. In general, there could be up to $2 \times 10^d + 1$ distinct values. Experimental results, however, show that usually there are far fewer distinct values.

The array P contains the sequence in which the hidden units of the network is to be considered. Different ordering sequences usually result in different clusters of activation values. Once a hidden unit is selected, its discretized activation values are sorted. The values are clustered based on their distance. Step 6 of the algorithm is implemented by first finding a pair of adjacent distinct values with the shortest distance, that is, by finding a pair of distinct values $h_{i,j}$ and $h_{i,j+1}$ such that $h_{i,j+1} - h_{i,j}$ is minimum. If $h_{i,j+1}$ can be replaced by $h_{i,j}$ without causing two or more patterns from different classes to have identical discretized values, they will be merged. Otherwise, a pair with the second-shortest distance will be considered. This process is repeated until there are no more pairs of values that can be merged. The next hidden unit as determined by the array P will then be considered.

Since the network in Fig. 7 has two hidden units, the clustering can be done in two possible sequences: hidden unit 1 followed by hidden unit 2 or hidden unit 2 followed by hidden unit 1. For this network, the results of applying the two clustering sequences are the same. The range of activation values at hidden unit 1 is the interval $[0, 0.78]$, while at hidden unit 2 it is $[-0.93, 0.52]$. GCA finds two clusters each in hidden unit 1 and hidden unit 2. The clustered values at hidden unit 1 are 0 and 0.46, that is, all continuous activation values in the interval $[0, 0.46)$ can be replaced by 0, and those values in the interval $[0.46, 0.78]$ can be replaced by 0.46 without affecting the accuracy rate of the network on the training data set. At hidden unit 2, the clustered values are -0.93 and 0.52.

We only cluster the activation values of patterns in the training data set that have been correctly classified by the pruned network. There are 338 such patterns. After clustering, each of the 338 patterns is represented by one of the possible four combinations of hidden unit clusters. The classification of the samples based on the clustered activation values can be summarized as in Table III.

From Table III, we observe that a sample is classified as benign only if its activation value at hidden unit 1 is in the interval $[0, 0.46)$ and the one at hidden unit 2 is 0.52. The first hidden unit has two input connections, $\mathcal{I}_{26}$ and $\mathcal{I}_{52}$. Only the inputs $\mathcal{I}_{26} = \mathcal{I}_{52} = 0$ result in an activation value in the first interval. The

Table III

The Clustered Activation Values of the Network in Fig. 7 and Their Predicted Output

Hidden unit 1	Hidden unit 2	Predicted output	Number of samples
0.00	−0.93	Malignant	2
0.00	0.52	Benign	225
0.46	−0.93	Malignant	28
0.46	0.52	Malignant	83

second hidden unit also has two input connections, but since the value of input $\mathcal{I}_{80}$ is always 1, the activation value at the second hidden unit is practically determined by the input $\mathcal{I}_{71}$ alone. Since the weight of the connection from input $\mathcal{I}_{71}$ to hidden unit 2 is negative, the activation value of a sample will be 0.52 if only if $\mathcal{I}_{71} = 0$. The rule that can be extracted from the network is then

If $\mathcal{I}_{26} = \mathcal{I}_{52} = \mathcal{I}_{71} = 0$, then predict *benign*.
Otherwise predict *malignant*.

In terms of the original input attributes and their values, the above rule is equivalent to

If $\mathcal{A}_3 \leq 4$ and $\mathcal{A}_6 \leq 8$ and $\mathcal{A}_8 \leq 9$, then predict *benign*.
Otherwise predict *malignant*.

The accuracy rates of the rule on the training and testing data sets are the same as those of the pruned network from which they are extracted, i.e., 96.57% and 93.12%, respectively. Several other sets of rules that have been extracted from this data base are reported in [42].

VIII. SUMMARY

In this chapter, we have discussed the various aspects that are important in the construction of an effective neural network system. We presented a variant of the quasi-Newton method for fast neural network training. A fast training algorithm is crucial to the successful construction of an effective neural network system. The algorithms for finding an optimal network architecture that we have developed require retraining of the network after units are added or removed from the network.

We described how the number of hidden units required in a network can be determined by adding hidden units one at a time as they are needed to the hidden layer. We showed that adding a hidden unit to the hidden layer will decrease the cross-entropy error function. We described how the relevant network inputs and network connections can be detected during training. By having a positive penalty for each weight in the network, only the relevant network connections will have large weights. Irrelevant and redundant connections can be distinguished by their small weights and they can be removed from the network without affecting the network accuracy on the training data set. Since networks with too many weights tend to overfit the training data, removal of redundant input units and connections usually results in a higher predictive accuracy rate. The removal of unnecessary inputs and connections from a network not only increases its predictive accuracy, it also facilitates the process of rule extraction. As symbolic rules are a form of knowledge that can be verified and expanded by human experts, having symbolic rules that explain the network predictions could make the neural network system even more attractive to users.

REFERENCES

[1] E. J. Hartman, J. D. Keeler, and J. M. Kowalski. Layered neural networks with gaussian hidden units as universal function approximation. *Neural Computat.* 2:210–215, 1990.

[2] K. Hornik. Approximation capabilities of multilayer feedforward networks. *Neural Networks* 4:251–257, 1991.

[3] J. de Villiers and E. Barnard. Backpropagation neural nets with one and two hidden layers. *IEEE Trans. Neural Networks* 4:136–141, 1992.

[4] K. J. Lang and M. J. Witbrock. Learning to tell two spirals apart. In *Proceedings of the 1988 Connectionist Model Summer School* (D. Touretzky, G. Hinton, and T. Sejnowski, Eds.), pp. 52–59. Morgan Kaufmann, San Mateo, CA, 1988.

[5] O. L. Mangasarian. Mathematical programming in neural networks. *ORSA J. Comput.* 5:349–360, 1993.

[6] C. G. Broyden. The convergence of a class of double rank minimization, algorithm 2, the new algorithm. *J. Inst. Math. Appl.* 6:222–231, 1970.

[7] R. Fletcher. A new approach to variable metric algorithms. *Computer J.* 13:317–322, 1970.

[8] D. Goldfarb. A family of variable metric algorithms derived by variational means. *Math. Computat.* 24:23–26, 1970.

[9] D. F. Shanno. Conditioning of quasi-Newton methods for function minimization. *Math. Computat.* 24:647–656, 1970.

[10] M. C. Biggs. A note on minimization algorithms which make use of non-quadratic properties of the objective function. *J. Inst. Math. Appl.* 12:337–338, 1973.

[11] L. E. Scales. *Introduction to Nonlinear Optimization.* Macmillan Ltd., London, 1985.

[12] D. F. Shanno. Conjugate gradient with inexact searches. *Math. Operat. Res.* 3:224–256, 1978.

[13] J. E. Dennis, Jr. and R. B. Schnabel. *Numerical Methods for Unconstrained Optimization and Nonlinear Equations.* Prentice-Hall, Englewood Cliffs, NJ, 1983.

[14] K. H. Phua and R. Setiono. Combined quasi-Newton updates for unconstrained optimization. Technical Report TR41/92, Department of Information Systems and Computer Science, National University of Singapore, 1992.

[15] J. A. Kinsella. Comparison and evaluation of variants of the conjugate gradient method for efficient learning in feed-forward neural networks with backward error propagation. *Network* 3:27–35, 1992.

[16] G. W. Cotrell, P. Munro, and D. Zipser. Learning internal representations from gray-scale images: An example of extensional programming. In *Proceedings of the 9th Annual Conference of the Cognitive Science Society*, pp. 461–473, 1987.

[17] G. L. Sicuranza and G. Ramponi. Artificial neural network for image compression. *Electron. Lett.* 26:477–479, 1990.

[18] R. Setiono and G. Lu. A neural network construction algorithm with application to image compression. *Neural Comput. Appl.* 2:61–68, 1994.

[19] F. L. Chung and L. Lee. A node pruning algorithm for backpropagation network. *Int. J. Neural Syst.* 3:301–314, 1992.

[20] M. Hagiwara. A simple and effective method for removal of hidden units and weights. *Neurocomput.* 6:207–218, 1994.

[21] S. J. Hanson and L. Y. Pratt. Comparing biases for minimal network construction with backpropagation. In *Neural Information Processing Systems* (D. Touretzky, Ed.), pp. 177–185. Morgan Kaufmann, San Mateo, CA, 1989.

[22] M. C. Mozer and P. Smolensky. Skeletonization: A technique for trimming the fat from a network via relevance assestment. In *Neural Information Processing Systems* (D. Touretzky, Ed.), pp. 107–115. Morgan Kaufmann, San Mateo, CA, 1989.

[23] S. E. Fahlman and C. Lebiere. The cascade-correlation learning architecture. In *Neural Information Processing Systems* (D. Touretzky, Ed.), pp. 524–532. Morgan Kaufmann, San Mateo, CA, 1990.

[24] M. Mezard and J. P. Nadal. Learning in feedforward layered networks: The tiling algorithm. *J. Phys. A* 22:2191–2203, 1989.

[25] M. F. Tenorio and W. Lee. Self-organizing network for optimum supervised learning. *IEEE Trans. Neural Networks* 1:100–110, 1990.

[26] M. Frean. The upstart algorithm: A method for constructing and training feedforward neural networks. *Neural Computat.* 2:198–209, 1990.

[27] T. Ash. Dynamic node creation in backpropagation networks. *Connect. Sci.* 1:365–375, 1989.

[28] R. Setiono and L. C. K. Hui. Use of quasi-Newton method in a feedforward neural network construction algorithm. *IEEE Trans. Neural Networks* 6:273–277, 1995.

[29] R. Setiono. A neural network construction algorithm which maximizes the likelihood function. *Connect. Sci.* 7:147–166, 1995.

[30] A. van Ooyen and B. Nienhuis. Improving the convergence of the backpropagation algorithm. *Neural Networks* 5:465–471, 1992.

[31] Y. Shang and B. W. Wah. Global optimization for neural network training. *Computer* March:45–54, 1996.

[32] G. E. Robbins, M. D. Plumbey, J. C. Hughes, F. Fallside, and R. Prager. Generation and adaptation of neural networks by evolutionary techniques (GANNET). *Neural Comput. Appl.* 1:23–31, 1993.

[33] R. Setiono and H. Liu. Neural-network feature selector. *IEEE Trans. Neural Networks* 8:654–662, 1997.

[34] J. Hertz, A. Krogh, and R. G. Palmer. *Introduction to the Theory of Neural Ccomputation.* Addison-Wesley, Redwood City, CA, 1991.

[35] L. M. Belue and K. W. Bauer. Determining input features for multilayer perceptron. *Neurocomputing* 7:111–122, 1995.

[36] S. B. Thrun *et al.* The MONK's problems—A performance comparison of different learning algorithm. Preprint CMU-CS-91-197, Carnegie Mellon University, Pittsburgh, PA, 1991.

[37] B. Hassibi and D. G. Stork. Second order derivatives for network pruning: Optimal brain surgeon. In *Advances in Neural Information Processing Systems*, Vol. 5, pp. 164–171. Morgan Kaufmann, San Mateo, CA, 1993.

[38] E. D. Karnin. A simple procedure for pruning back-propagation trained neural networks. *IEEE Trans. Neural Networks* 1:239–242, 1990.

[39] Y. Le Cun, J. S. Denker, and S. A. Solla. Optimal brain damage. In *Advances in Neural Information Processing Systems*, Vol. 2, pp. 598–605. Morgan Kaufmann, San Mateo, CA, 1990.

[40] H. H. Thodberg. Improving generalization of neural networks through pruning. *Int. J. Neural Syst.* 1:317–326, 1991.

[41] R. Setiono. A penalty function approach for pruning feedforward neural networks. *Neural Computat.* 9:301–320, 1997.

[42] R. Setiono. Extracting rules from pruned neural network for breast cancer diagnosis. *Artif. Intell. Medicine* 8:37–51, 1996.

[43] R. Setiono and H. Liu. Symbolic representation of neural networks. *Computer* March:71–77, 1996.

[44] H. Lu, R. Setiono, and H. Liu. NeuroRule: A connectionist approach to data mining. In *Proceedings of 21st International Conference on Very Large Data Bases*, Zurich, Switzerland, pp. 478–489, 1995.

[45] O. L. Mangasarian, R. Setiono, and W. H. Wolberg. Pattern recognition via linear programming: Theory and application to medical diagnosis. In *Large-scale Numerical Optimization* (T. F. Coleman and Y. Li, Eds.), pp. 22–30. SIAM, Philadelphia, PA, 1990.

[46] P. M. Murphy and D. W. Aha. UCI repository of machine learning databases. Machine-readable data repository. Department of Information and Computer Science, University of California at Irvine, 1992.

[47] K. P. Bennett and O. L. Mangasarian. Neural network training via linear programming. In *Advances in Optimization and Parallel Computing* (P. M. Pardalos, Ed.), pp. 56–67. Elsevier Science Publishers B.V., Amsterdam, 1990.

[48] W. H. Wolberg and O. L. Mangasarian. Multisurface method of pattern separation for medical diagnosis applied to breast cytology. *Proc. Nat. Acad. Sci.* 87:9193–9196, 1990.

[49] W. H. Wolberg, M. A. Tanner, and W. Y. Loh. Diagnostic schemes for fine needle aspirates of breast masses, *Anal. Quantitat. Cytol. Histol.* 10:225–228, 1988.

Learning Algorithms and Applications of Principal Component Analysis

Liang-Hwa Chen
Applied Research Laboratory
Telecommunication Laboratories
Chunghwa Telecom Co., Ltd.
12, Lane 551, Min-Tsu Road, Sec. 3, Yang-Mei
Taoyuan, Taiwan, Republic of China

Shyang Chang
Department of Electrical Engineering
National Tsing Hua University
Hsin Chu, Taiwan, Republic of China

I. INTRODUCTION

Recently, inspired by the structure of human brain, artificial neural networks have been widely applied to many fields such as pattern recognition, optimization, coding, control, etc., due to their capability of solving cumbersome or intractable problems by learning directly from data. An artificial neural network usually consists of a large amount of simple processing units, i.e., neurons, via mutual interconnection. It learns to solve problems by adequately adjusting the strength of the interconnections according to input data. Moreover, it can be easily adapted to new environments by learning. At the same time, it can deal with information that is noisy, inconsistent, vague, or probabilistic. These features motivate extensive researches and developments in artificial neural networks.

The main features of artificial neural networks are their massively parallel processing architectures and the capabilities of *learning* from the presented inputs. They can be utilized to perform a specific task only by means of adequately adjusting the connection weights, i.e., by *training* them with the presented data. For each type of artificial neural network, there exists a corresponding *learning algorithm* by which we can train the network in an iterative updating manner. Gener-

Image Processing and Pattern Recognition
Copyright © 1998 by Academic Press. All rights of reproduction in any form reserved.

ally, the learning algorithms can be classified into two main categories: *supervised learning* and *unsupervised learning.*

For supervised learning, not only the input data but also the corresponding target answers are presented to the network. Learning is done in accordance with the direct comparison of the actual output of the network with known correct answers. It is also referred to as *learning with a teacher.* A special case called *reinforcement learning* is included, where the only feedback information is whether each output is right or wrong, not what the correct answer is.

On the contrary, only input data without the corresponding target answers are presented to the network for unsupervised learning. In fact, the learning goal is not defined at all in terms of specific correct examples. The available information is in the correlations of the input data. The network is expected to create categories from these correlations, and to produce output signals corresponding to the input category.

In the following, several typical artificial neural networks are briefly introduced. The first one is the Hopfield network [1]. It is a single-layer network in which neurons are mutually interconnected. The output of each neuron is fed back to all the other neurons. An energy function is associated with the network. Whenever a neuron changes state, the energy function always decreases. Training is performed by directly setting the connection weights according to the training patterns. It is essentially supervised but not iterated. The Hopfield network has been utilized as an associative memory or to solve optimization problems (e.g., [2–5]).

The back-propagation network (e.g., [6, 7]) is another typical artificial neural network. It is a multilayer network in which neurons are usually connected only to the ones belonging to the preceding and the succeeding layers. The network is trained in an *error back-propagation* manner: For each input, a signal propagates feedforward to the output layer. Then, the output is compared with the desired target and the error propagates backward to update the connection weights. This learning is obviously supervised and iterated. The back-propagation network essentially performs the *function approximation.* It has been applied to control, pattern recognition, image compression, etc. (e.g., [8–11]).

Different from the above supervised learning networks, a competitive learning network (e.g., [12–14]) is an unsupervised learning network. It consists of a single layer of competitive neurons. These neurons compete with one another for becoming the unique winner to fire. The learning is based on the *winner-take-all* rule. That is, only the weights of the winner will be updated. The aim of this type of network is essentially to cluster the input data. The weights, after learning, will be distributed, in data space, approximately in proportion to the data pattern density. Due to such characteristics, the competitive learning network can be applied to vector quantization (VQ) [15, 16] and probability density function (PDF) estimation [17] that are very useful in data compression, coding, pattern recognition, etc. (e.g., [18–23]).

The principal component analysis (PCA) learning network (e.g., [24]) is another type of unsupervised learning network. It is also a single-layer network but the neurons are linear. Figure 1 shows the schematic diagram. The learning is essentially based on the Hebb rule [25]. It is utilized to perform PCA (see, e.g., [26, 27]), i.e., to find the *principal components* embedded in the input data. PCA is one of the feature extraction methods. It extracts information by finding the directions in input space in which the inputs exhibit most variation. It then projects the inputs onto the subspace to perform a dimensionality reduction. The desired directions are in fact along those eigenvectors associated with the m largest eigenvalues of the input covariance matrix [28] if the dimension of data is expected to be reduced from n to m. In the PCA learning network, the weights will converge to the principal component vectors after learning. The PCA learning network has been applied to feature extraction, data compression, image coding, and texture segmentation (e.g., [24, 29–31]).

For the PCA learning network, a variety of neural network learning algorithms have been proposed [24, 30–36, etc.]. Most of them are based on the early work of Oja's one-unit algorithm [37]. Among these algorithms, Sanger's Generalized Hebbian algorithm (GHA) [24], which combines Gram-Schmidt orthonormalization and Oja's one-unit algorithm, is usually more useful in practical applications. This is because it can extract the principal components individually in order. In addition, it can give a reproducible result on a given data set. However, the success of the GHA is dependent on the values of the learning rate parameters. If the parameter values are too big, the learning process will diverge. On the contrary, the learning process will converge very slowly if the parameter values are too small. The appropriate values of the learning rate parameters are in fact data dependent.

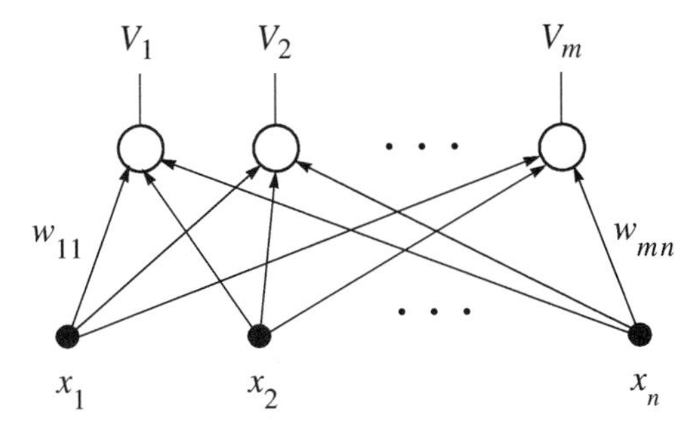

Figure 1 PCA learning network.

In order to ensure that the learning be successful when the above neural networks are applied in practical applications, we would like to develop adaptive learning algorithms that can automatically and adaptively select the appropriate parameter values according to input data. Hence, in this chapter an adaptive PCA learning algorithm is proposed. The learning rate parameters are adaptively adjusted according to the eigenvalues of the data covariance matrix that are estimated during the learning process. All weights can quickly converge to the principal component vectors in the small-eigenvalue case as well as the big-eigenvalue case. It has been applied to data compression and image coding. Excellent results have been obtained.

II. ADAPTIVE LEARNING ALGORITHM*

In this section, an adaptive learning algorithm (ALA) for PCA will be proposed (see also [38, 39]). The learning rate parameters can be selected automatically and adaptively according to the eigenvalues of the input covariance matrix that are estimated during the learning process. We will show that the m weight vectors in the network can converge quickly to the first m principal component vectors with almost the same rates. Simulation results will demonstrate that the ALA can converge quickly to the desired targets while Sanger's GHA diverges in the large-eigenvalue case and converges very slowly in the small-eigenvalue case. Finally, the ALA will be applied to image coding problems and excellent results can be obtained.

This section is organized as follows. First, some basic mathematical background and Sanger's GHA are introduced. The parameter selection problem of the GHA is then presented. Finally, our ALA is proposed and analyzed. Propositions concerning its properties are also presented in this section.

Let $\mathbf{x}$ denote the n-dimensional input data vector with probability distribution $P(\mathbf{x})$. The aim of PCA is to find a set of m orthonormal vectors in an n-dimensional data space such that they will account for as much as possible of the variance of the data. It was shown in [26] that the aforementioned orthonormal vectors were actually the m eigenvectors associated with the m largest eigenvalues of the data covariance matrix $\mathbf{T} = E\{(\mathbf{x} - \mathbf{m}_x)(\mathbf{x} - \mathbf{m}_x)^t\}$, where t denotes the transpose operator and $\mathbf{m}_x = E\{\mathbf{x}\}$. If the eigenvalues of $\mathbf{T}$ are sorted in descending order, i.e., $\lambda_1 > \lambda_2 > \cdots > \lambda_n$ with $\lambda_1 = \lambda_{\max}$, then the kth principal component direction will be along the kth eigenvector. In general, the mean values of data can be subtracted from the data. Hence, in the following, we will discuss zero-mean data exclusively. In case of zero-mean data, the covariance matrix $\mathbf{T}$ will be reduced to the correlation matrix $\mathbf{C} = E\{\mathbf{x}\mathbf{x}^t\}$.

*Portions of the following two sections are reprinted with permission from *IEEE Trans. Neural Networks* 6(5) 1255–1263, Sept. 1995 (© 1995 IEEE).

In order to find the first principal component direction vector for zero-mean data, i.e., the first eigenvector of $\mathbf{C}$, by learning directly from data, Oja [37] proposed a one-unit learning rule:

$$\Delta \mathbf{w}(t) = \eta(t) V(t)(\mathbf{x}(t) - V(t)\mathbf{w}(t)), \tag{1}$$

where $\eta(t)$ is the learning rate parameter. This rule is used to train a linear neuron whose output $V(t)$ is equal to the product of weight vector $\mathbf{w}(t)$ and input pattern $\mathbf{x}(t)$ at time t, i.e., $V(t) = \mathbf{w}^t(t)\mathbf{w}(t)$. Here, we assume that $\mathbf{x}(t)$s are independent and identically distributed with the same distribution $P(\mathbf{x})$ as before. Under the assumption that $\eta(t)$ is sufficiently small, Oja approximated Eq. (1) by a corresponding ODE

$$d\mathbf{w}/dt = \mathbf{C}\mathbf{w} - (\mathbf{w}^t\mathbf{C}\mathbf{w})\mathbf{w}$$

via the stochastic approximation theory (see, e.g., [40]). He then proved that the weight vector $\mathbf{w}(t)$ will asymptotically converge to the first normalized eigenvector of $\mathbf{C}$, i.e., $\pm\mathbf{v}_1$.

By combining Oja's rule and the Gram-Schmidt orthonormalization process, Sanger [24] proposed the so-called GHA:

$$\Delta \mathbf{w}_i(t) = \eta(t) V_i(t)\left(\mathbf{x}(t) - \sum_{j=1}^{i} V_j(t)\mathbf{w}_j(t)\right), \qquad i = 1, 2, \ldots, m, \tag{2}$$

where $V_i(t) = \mathbf{w}_i^t(t)\mathbf{x}(t)$. It is used to train a one-layer m-unit network (referring to Fig. 1) consisting of m linear neurons so as to find the first m principal components. Using the same approximation technique, the GHA was able to make $\mathbf{w}_i(t)$, $i = 1, 2, \ldots, m$, converge to the first m principal component directions, in sequential order: $\mathbf{w}_i(t) \to \pm\mathbf{v}_i$, where $\mathbf{v}_i$ is a normalized eigenvector associated with the ith largest eigenvalue λ_i of the correlation matrix $\mathbf{C}$. In fact, Eq. (2) can be rewritten as

$$\Delta \mathbf{w}_i(t) = \eta(t) V_i(t)\left[\left(\mathbf{x}(t) - \sum_{j=1}^{i-1} V_j(t)\mathbf{w}_j(t)\right) - V_i(t)\mathbf{w}_i(t)\right], \qquad i = 1, 2, \ldots, m. \tag{3}$$

Hence, Eq. (3) can be treated as Eq. (1) with the corresponding modified input $\mathbf{x}_i(t) \equiv \mathbf{x}(t) - \sum_{j=1}^{i-1} V_j(t)\mathbf{w}_j(t)$, for neuron i, where $i = 1, 2, \ldots, m$. If $\mathbf{w}_j(t)$, $j = 1, 2, \ldots, i-1$, have converged to $\mathbf{v}_j$, $j = 1, 2, \ldots, i-1$, respectively, it can be easily shown [24] that the maximal eigenvalue λ_1^i and the associated normalized eigenvector $\mathbf{v}_1^i$ of the correlation matrix of $\mathbf{x}_i$, i.e., $\mathbf{C}_i \equiv E\{\mathbf{x}_i\mathbf{x}_i^t\}$, are exactly the ith eigenvalue λ_i and the ith normalized eigenvector $\mathbf{v}_i$ of the correlation matrix of $\mathbf{x}$, i.e., $\mathbf{C}$, respectively. Hence, neuron i can find the ith normalized eigenvector of $\mathbf{C}$, i.e., $\pm\mathbf{v}_i$. In other words, the m neurons trained by

Eq. (2) can be considered to be trained by Eq. (1) with their respective modified inputs $\mathbf{x}_i$, $i = 1, 2, \ldots, m$.

In the following, we will show that the selection of $\eta(t)$ should depend on the eigenvalues λ_is of the correlation matrix $\mathbf{C}$. If $\eta(t)$ is bigger than $1/\lambda_1$, the learning process cannot converge[1] as expected. In addition, the learning rate will become very slow if the product value of $\eta(t)\lambda_1$ is very small. First, let us take the conditional expectation of Eq. (1) over the input distribution $P(\mathbf{x})$ given weight vector $\mathbf{w}(t)$, i.e.,

$$\begin{aligned} E\{\Delta\mathbf{w}(t)|\mathbf{w}(t)\} &= E\{\eta(t)V(t)(\mathbf{x}(t) - V(t)\mathbf{w}(t))|\mathbf{w}(t)\} \\ &= \eta(t)[E\{\mathbf{x}(t)\mathbf{x}^t(t)\}\mathbf{w}(t) - \mathbf{w}^t(t)E\{\mathbf{x}(t)\mathbf{x}^t(t)\}\mathbf{w}(t)\mathbf{w}(t)] \\ &= \eta(t)[\mathbf{C}\mathbf{w}(t) - \mathbf{w}^t(t)\mathbf{C}\mathbf{w}(t)\mathbf{w}(t)], \end{aligned} \tag{4}$$

where we have used the following facts for derivation: $E\{\mathbf{x}(t)\mathbf{x}^t(t)\mathbf{w}(t)|\mathbf{w}(t)\} = E\{\mathbf{x}(t)\mathbf{x}^t(t)\}\mathbf{w}(t) = \mathbf{C}\mathbf{w}(t)$, and $E\{\mathbf{w}^t(t)\mathbf{x}(t)\mathbf{x}^t(t)\mathbf{w}(t)\mathbf{w}(t)|\mathbf{w}(t)\} = \mathbf{w}^t(t) \times E\{\mathbf{x}(t)\mathbf{x}^t(t)\}\mathbf{w}(t)\mathbf{w}(t)$, where $\mathbf{x}(t)$ and $\mathbf{w}(t)$ are independent.

PROPOSITION 1. *For learning rule Eq. (1), if $\eta(t)$ selected is not smaller than $1/\lambda_1$, then weight vector $\mathbf{w}(t)$ will not converge to $\pm\mathbf{v}_1$ even if it is initially close to the target.*

For the proof, see the Appendix.

From Proposition 1, we know that $\eta(t)$ should be smaller than $1/\lambda_1$ to get the expected convergence. Under this condition, the learning rate can be estimated by the value of $\eta(t)\lambda_1$.

PROPOSITION 2. *When $\eta(t)\lambda_1 < 0.5$, the smaller the value of $\eta(t)\lambda_1$ is, the slower the convergence rate of the expectation of $\mathbf{w}(t)$ is.*

For the proof, see the Appendix.

According to these two propositions, for each neuron i in Eq. (2), the learning rate parameter $\eta(t)$ has to satisfy $\eta(t)\lambda_i < 1$ in order to converge. In the mean time, it cannot be too small in order to have a decent learning rate. However, the values of λ_is are usually unknown *a priori*. Therefore, to select properly the value of $\eta(t)$ becomes a problem when one tries to apply the GHA in practical applications. For example, if one of the eigenvalues $\lambda_i = 10^4$, the $\eta(t)$ must be smaller than 10^{-4} for $\mathbf{w}_i(t)$ to converge. However, in the GHA, the identical value setting of $\eta(t)$ for all neurons will slow down the learning rate of $\mathbf{w}_j(t)$ if for $\lambda_j \ll 10^4$ for $j > i$.

In order to overcome the aforementioned problem, an adaptive learning algorithm (ALA) for PCA will be proposed in the following. In the algorithm, the learning rate parameter for each $\mathbf{w}_i(t)$ can be selected adaptively.

[1]The convergence here is in the mean-square sense.

For n-dimensional zero-mean input pattern vector $\mathbf{x}$ with probability distribution $P(\mathbf{x})$, the ALA that will find the first m principal component vectors can be described as follows:

Step 1: Set weight vectors $\mathbf{w}_i(0) \in \mathbf{R}^n$ such that $\|\mathbf{w}_i(0)\|^2 \ll 1/2$[2] and estimate of eigenvalues $\hat{\lambda}_i(0) = \delta$ (a small positive number) > 0 for $i = 1, 2, \ldots, m$.

Step 2: Draw a new pattern $\mathbf{x}(t)$ at time t, $t \geq 1$, and present it to the network as input.

Step 3: Calculate the output V_is:

$$V_i(t) = \mathbf{w}_i^t(t)\mathbf{x}(t), \qquad i = 1, 2, \ldots, m.$$

Step 4: Estimate the eigenvalues λ_is:

$$\hat{\lambda}_i(t) = \hat{\lambda}_i(t-1) + \gamma(t)\left[\left(\mathbf{w}_i^t(t)\mathbf{x}_i(t)/\|\mathbf{w}_i(t)\|\right)^2 - \hat{\lambda}_i(t-1)\right], \qquad i = 1, 2, \ldots, m, \tag{5}$$

where $\mathbf{x}_i(t) \equiv \mathbf{x}(t) - \sum_{j=1}^{i-1} V_j(t)\mathbf{w}_j(t)$. The value of $\gamma(t)$ is set to be smaller than 1 and decreased to zero as t approaches ∞.

Step 5: Modify the weights $\mathbf{w}_i$s:

$$\mathbf{w}_i(t+1) = \mathbf{w}_i(t) + \eta_i(t)V_i(t)\left[\mathbf{x}(t) - \sum_{j=1}^{i} V_j(t)\mathbf{w}_j(t)\right], \qquad i = 1, 2, \ldots, m, \tag{6}$$

where $\eta_i(t) = \beta_i(t)/\hat{\lambda}_i(t)$. The value of $\beta_i(t)$ is set to be smaller than $2(\sqrt{2} - 1)$ and decreased to zero as t approaches ∞.

Step 6: Check the length of $\mathbf{w}_i$s:

$$\mathbf{w}_i(t+1) = \begin{cases} \sqrt{1/2}\left(\mathbf{w}_i(t+1)/\|\mathbf{w}_i(t+1)\|\right), & \text{if } \|\mathbf{w}_i(t+1)\|^2 > \dfrac{1}{\beta_i(t+1)} + \dfrac{1}{2}, \\ \mathbf{w}_i(t+1), & \text{otherwise.} \end{cases} \tag{7}$$

Step 7: Increase the time t by 1 and go back to step 2 for the next input pattern until all of the $\mathbf{w}_i$s are mutually orthonormal.

Remarks.

(i) The procedure of eigenvalue estimation in step 4 is the Grossberg learning rule [41]. When the value of $\gamma(t)$ is set smaller than 1 and decreased to zero with time, $\hat{\lambda}_i(t)$, i.e., $\hat{\lambda}_1^i(t)$, can converge to the mean of $(\mathbf{w}_i^t\mathbf{x}_i/\|\mathbf{w}_i\|)^2$ in the mean-square sense.

[2] $\|\mathbf{y}\|$ denotes in this chapter the length of a vector $\mathbf{y}$, i.e., $\|\mathbf{y}\| = (\mathbf{y}^t\mathbf{y})^{1/2}$.

(ii) Due to the fact that the estimates of λ_is may be inaccurate during the initial period of the learning process, the normalization process in step 6 is required. The details will be described in Section III.
(iii) The reason for using the upper bound of $\beta_i(t)$ in step 5 will be given in Proposition 3.

Notice that in the ALA, its learning rate parameters $\eta_i(t)$ are no longer the same for all neurons. They are adaptively selected according to the corresponding eigenvalues λ_is that are estimated by Eq. (5). We will contend in the following that $\mathbf{w}_i(t)$ can quickly converge to $\pm\mathbf{v}_i$ for all i and that they will converge with nearly the same rate.

First, let us consider the convergence of $\mathbf{w}_i(t)$ to $\pm\mathbf{v}_i$ for all i. Recall that Eq. (6) can be written as

$$\mathbf{w}_i(t+1) = \mathbf{w}_i(t) + \eta_i(t)V_i(t)\big(\mathbf{x}_i(t) - V_i(t)\mathbf{w}_i(t)\big), \tag{8}$$

where $\mathbf{x}_i(t) \equiv \mathbf{x}(t) - \sum_{j=1}^{i-1} V_j(t)\mathbf{w}_j(t)$ for all i. It can be considered just as the learning rule Eq. (1) applied to every neuron i with $\mathbf{x}_i$ as its corresponding modified input. Hence, in the following, it suffices to consider Eq. (8), where $\eta_i(t) = \beta_i(t)/\hat{\lambda}_1^i(t)$, for only *one* neuron i and show that it can converge to the first normalized eigenvector of *its* corresponding input correlation matrix $\mathbf{C}_i$.

The proof will be decomposed into several parts. It will be shown first that the mean of $\mathbf{w}_i(t)$ will approach $\mathbf{v}_1^i$. That is, its length will approach 1 and its angle from $\mathbf{v}_1^i$ will approach zero. We will then analyze the variance and show that it will decrease to zero. Notice that in the following discussion, $\mathbf{w}_i(t)$ is no longer assumed to be close $\mathbf{v}_1^i$ to initially as in Proposition 1.

PROPOSITION 3. *In the ALA, the mean of* $\mathbf{w}_i(t)$ *will approach the unit hypersphere in* $\mathbf{R}^n$ *space.*

Proof. Let $\boldsymbol{w}_i(t)$ stand for a realization of $\mathbf{w}_i(t)$ at time t. An orthonormal basis of $\mathbf{R}^n$, $\{\mathbf{u}_1^i, \mathbf{u}_2^i, \ldots, \mathbf{u}_n^i\}$, such that $\mathbf{u}_1^i$ is the unit vector along the direction of $\boldsymbol{w}_i(t)$ can be constructed. Hence, $\boldsymbol{w}_i(t)$ is represented as

$$\boldsymbol{w}_i(t) = k_{\mathrm{u}}^i(t)\mathbf{u}_1^i + \sum_{j=2}^{n} \varepsilon_{uj}^i(t)\mathbf{u}_j^i,$$

where $k_{\mathrm{u}}^i(t)$ is equal to the length of $\boldsymbol{w}_i(t)$, i.e., $\|\boldsymbol{w}_i(t)\|$, and $\varepsilon_{uj}^i(t) = 0$ for $j = 2, 3, \ldots, n$. Similar to Eq. (4), we can obtain, from Eq. (8),

$$E\{\Delta\mathbf{w}_i(t)|\boldsymbol{w}_i(t)\} = \eta_i(t)\big[\mathbf{C}_i\boldsymbol{w}_i(t) - \boldsymbol{w}_i^t(t)\mathbf{C}_i\boldsymbol{w}_i(t)\boldsymbol{w}_i(t)\big], \tag{9}$$

where $E\{\cdot|\boldsymbol{w}_i(t)\}$ stands for $E\{\cdot|\mathbf{w}_i(t) = \boldsymbol{w}_i(t)\}$. Project Eq. (9) onto, $\mathbf{u}_j^i$, $j = 1, 2, \ldots, n$; we get

$$\begin{aligned}(\mathbf{u}_1^i)^t E\{\Delta\mathbf{w}_i(t)|\boldsymbol{w}_i(t)\} \quad (\equiv \Delta k_{\mathrm{u}}^i(t))\\ = (\mathbf{u}_1^i)^t \eta_i(t)\big[\mathbf{C}_i k_{\mathrm{u}}^i(t)\mathbf{u}_1^i - k_{\mathrm{u}}^i(t)^3(\mathbf{u}_1^i)^t\mathbf{C}_i\mathbf{u}_1^i\mathbf{u}_1^i\big]\\ = \eta_i(t)k_{\mathrm{u}}^i(t)\big((\mathbf{u}_1^i)^t\mathbf{C}_i\mathbf{u}_1^i\big)\big(1 - k_{\mathrm{u}}^i(t)^2\big), \qquad \text{for } j = 1, \end{aligned} \tag{10a}$$

and

$$\begin{aligned}(\mathbf{u}_j^i)^t E\{\Delta\mathbf{w}_i(t)|\boldsymbol{w}_i(t)\} \quad (\equiv \Delta\varepsilon_{uj}^i(t))\\ = (\mathbf{u}_j^i)^t \eta_i(t)\big[\mathbf{C}_i k_{\mathrm{u}}^i(t)\mathbf{u}_1^i - k_{\mathrm{u}}^i(t)^3(\mathbf{u}_1^i)^t\mathbf{C}_i\mathbf{u}_1^i\mathbf{u}_1^i\big]\\ = \eta_i(t)k_{\mathrm{u}}^i(t)(\mathbf{u}_j^i)^t\mathbf{C}_i\mathbf{u}_1^i, \qquad \text{for } j = 2, 3, \ldots, n. \end{aligned} \tag{10b}$$

The $\Delta k_{\mathrm{u}}^i(t)$ defined in Eq. (10a) is the component of $E\{\Delta\mathbf{w}_i(t)|\boldsymbol{w}_i(t)\}$ along the direction of $\mathbf{u}_1^i$. It is referred to as the "radial weight change." The $\Delta\varepsilon_{uj}^i(t)$ defined in Eq. (10b) is the component of $E\{\Delta\mathbf{w}_i(t)|\boldsymbol{w}_i(t)\}$ along the direction of $\mathbf{u}_j^i$, $j > 1$. It is referred to as the "tangential weight change." Figure 2 is a demonstration of the case $n = 2$. It is clear that from $\boldsymbol{w}_i(t)$ to $E\{\mathbf{w}_i(t+1)|\boldsymbol{w}_i(t)\}$, the change in length is caused by $\Delta k_{\mathrm{u}}^i(t)$. On the other hand, the change in direction is caused by $\Delta\varepsilon_{uj}^i(t)$.

In order for $E\{\mathbf{w}_i(t+1)|\boldsymbol{w}_i(t)\}$ to be closer to the unit hypersphere than $\boldsymbol{w}_i(t)$, the value of $k_{\mathrm{u}}^i(t) + \Delta k_{\mathrm{u}}^i(t)$ has to be closer to 1 than $k_{\mathrm{u}}^i(t)$. From Eq. (10a), we obtain

$$k_{\mathrm{u}}^i(t) + \Delta k_{\mathrm{u}}^i(t) = \big[1 + \eta_i(t)(\mathbf{u}_1^i)^t\mathbf{C}_i\mathbf{u}_1^i\big(1 - k_{\mathrm{u}}^i(t)^2\big)\big]k_{\mathrm{u}}^i(t). \tag{11}$$

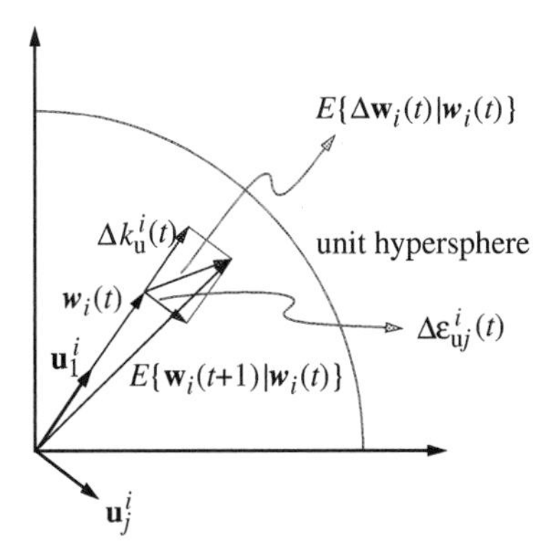

Figure 2 Illustration of radial and tangential weight changes. Reprinted with permission from L. H. Chen and S. Chang, *IEEE Trans. Neural Networks* 6:1255–1263, 1995 (©1995 IEEE).

Taking the square of both sides of Eq. (11) and letting $\lambda_u^i \equiv (\mathbf{u}_1^i)^t \mathbf{C}_i \mathbf{u}_1^i$, we get

$$\left(k_u^i(t) + \Delta k_u^i(t)\right)^2 = \left[1 + \eta_i(t)\lambda_u^i(1 - k_u^i(t)^2)\right]^2 k_u^i(t)^2. \tag{12}$$

Equation (12) takes the form of

$$z(t+1) = \left[1 + \alpha\left(1 - z(t)\right)\right]^2 z(t), \tag{13}$$

where $z(t+1) = (k_u^i(t) + \Delta k_u^i(t))^2$, $z(t) = k_u^i(t)^2$, and $\alpha = \eta_i(t)\lambda_u^i$. It can be easily checked that for Eq. (13), the relation

$$|z(t+1) - 1|/|z(t) - 1| < 1 \tag{14}$$

holds if $\alpha \in (0, 2(\sqrt{2} - 1))$ and $z(t) \in (0, 1 + 2/\alpha)$. Now, since λ_1^i is equal to $\max_{\mathbf{u}}(\mathbf{u}^t \mathbf{C}_i \mathbf{u})$ where $\mathbf{u}$ is any unit vector, thus $\eta_i(t)\lambda_u^i \leq \eta_i(t)\lambda_1^i$. In addition, the value of $\eta_i(t)$ is selected such that $\eta_i(t)\lambda_1^i = \beta_i(t) < 2(\sqrt{2} - 1)$. Hence, $\eta_i(t)\lambda_u^i < 2(\sqrt{2} - 1)$. Moreover, $k_u^i(0)^2 (= \|\boldsymbol{w}_i(0)\|^2)$ is set to be smaller than 1 in step 1 and $\boldsymbol{w}_i(t)$ will be bounded by $\|\boldsymbol{w}_i(t)\|^2 \leq 1/2 + 1/\beta_i(t) = 1/2 + 1/(\eta_i(t)\lambda_1^i) < 1 + 2/(\eta_i(t)\lambda_u^i)$ according to step 6 of the ALA. Thus, both $(k_u^i(t) + \Delta k_u^i(t))^2$ and $k_u^i(t)^2$ satisfy Eq. (14), i.e., $|(k_u^i(t) + \Delta k_u^i(t))^2 - 1|/|k_u^i(t)^2 - 1| < 1$. That is, $E\{\mathbf{w}_i(t+1)|\boldsymbol{w}_i(t)\}$ will be closer to the unit hypersphere than $\boldsymbol{w}_i(t)$. Since this is true for all realizations $\boldsymbol{w}_i(t)$, so $E\{\mathbf{w}_i(t+1)|\mathbf{w}_i(t)\}$ will be closer to the unit hypersphere than $\mathbf{w}_i(t)$. Hence, $E\{\mathbf{w}_i(t+1)\} = E\{E\{\mathbf{w}_i(t+1)|\mathbf{w}_i(t)\}\}$ will be closer to the unit hypersphere than $E\{\mathbf{w}_i(t)\}$. We conclude that the length of the mean of $\mathbf{w}_i(t)$ will approach 1 as t goes to ∞. ■

Next, we will show that the direction of $\mathbf{w}_i(t)$ will approach that of $\mathbf{v}_1^i$.

PROPOSITION 4. *The angle between the mean of $\mathbf{w}_i(t)$ and $\mathbf{v}_1^i$ in the ALA will approach zero.*

Proof. First, express $\boldsymbol{w}_i(t)$ in the following form:

$$\boldsymbol{w}_i(t) = k^i(t)\mathbf{v}_1^i(t) + \mathbf{s}_i(t) = k^i(t)\mathbf{v}_1^i + \sum_{j=2}^{n} \varepsilon_j^i(t)\mathbf{v}_j^i, \tag{15}$$

where $k^i(t)$ is the magnitude of $\boldsymbol{w}_i(t)$ along eigenvector $\mathbf{v}_1^i$ and $\mathbf{s}_i(t) = \sum_{j=2}^n \varepsilon_j^i(t)\mathbf{v}_j^i$ the component of $\boldsymbol{w}_i(t)$ perpendicular to $\mathbf{v}_1^i$. Notice that $\varepsilon_j^i(t)$ is the magnitude of $\boldsymbol{w}_i(t)$ along $\mathbf{v}_j^i$. For a given $\boldsymbol{w}_i(t)$, its average new location after one iteration of learning will be $E\{\mathbf{w}_i(t+1)|\boldsymbol{w}_i(t)\} = \boldsymbol{w}_i(t) + E\{\Delta\mathbf{w}_i(t)|\boldsymbol{w}_i(t)\}$. Thus,

$$\left(\mathbf{v}_1^i\right)^t E\left\{\mathbf{w}_i(t+1)|\boldsymbol{w}_i(t)\right\} \quad \left(\equiv k^i(t+1)\right) = k^i(t) + (\mathbf{v}_1^i)^t E\left\{\Delta\mathbf{w}_i(t)|\boldsymbol{w}_i(t)\right\}$$

and

$$\left(\mathbf{v}_1^i\right)^t E\left\{\mathbf{w}_i(t+1)|\boldsymbol{w}_i(t)\right\} \quad \left(\equiv \varepsilon_j^i(t+1)\right)$$
$$= \varepsilon_j^i(t) + (\mathbf{v}_j^i)^t E\{\Delta\mathbf{w}_i(t)|\boldsymbol{w}_i(t)\}, \qquad \text{for } j = 2, 3, \ldots, n.$$

According to Eq. (9) and Eq. (15), $(\mathbf{v}_j^i)^t E\{\Delta\mathbf{w}_i(t)|\boldsymbol{w}_i(t)\}$, $j = 1, 2, \ldots, n$, can be written as

$$\left(\mathbf{v}_1^i\right)^t E\left\{\Delta\mathbf{w}_i(t)|\boldsymbol{w}_i(t)\right\} = \eta_i(t)\left(\lambda_1^i - \sigma_\lambda^i(t)\right)k^i(t) \tag{16a}$$

and

$$(\mathbf{v}_j^i)^t E\{\Delta\mathbf{w}_i(t)|\boldsymbol{w}_i(t)\} = \eta_i(t)(\lambda_j^i - \sigma_\lambda^i(t))\varepsilon_j^i(t), \qquad j = 2, 3, \ldots, n, \tag{16b}$$

where $\sigma_\lambda^i(t) = \boldsymbol{w}_i^t(t)\mathbf{C}_i\boldsymbol{w}_i(t) = \lambda_1^i k^i(t)^2 + \sum_{j=2}^n \lambda_j^i \varepsilon_j^i(t)^2$. Then, we get

$$k^i(t+1) = [1 + \eta_i(t)(\lambda_1^i - \sigma_\lambda^i(t))]k^i(t)$$

and

$$\varepsilon_j^i(t+1) = [1 + \eta_i(t)(\lambda_j^i - \sigma_\lambda^i(t))]\varepsilon_j^i(t).$$

Let us denote the angle between $\mathbf{w}_i$ and $\mathbf{v}_1^i$ by $\text{Ang}(\mathbf{w}_i)$. Then,

$$\tan^2(\text{Ang}(\boldsymbol{w}_i)) = \frac{\|\mathbf{s}_i\|^2}{(k^i)^2} = \frac{\sum_{j=2}^n (\varepsilon_j^i)^2}{(k^i)^2}.$$

To prove that $\text{Ang}(E\{\mathbf{w}_i(t+1)|\boldsymbol{w}_i(t)\})$ will be smaller than $\text{Ang}(\boldsymbol{w}_i(t))$, it suffices to show that $\tan^2(\text{Ang}(E\{\mathbf{w}_i(t+1)|\boldsymbol{w}_i(t)\}))$ will be smaller than $\tan^2(\text{Ang}(\boldsymbol{w}_i(t)))$. That is,

$$\frac{\varepsilon_j^i(t+1)^2}{k^i(t+1)^2} = \frac{[1 + \eta_i(t)(\lambda_j^i - \sigma_\lambda^i(t))]^2}{[1 + \eta_i(t)(\lambda_1^i - \sigma_\lambda^i(t))]^2}\frac{\varepsilon_j^i(t)^2}{k^i(t)^2} < \frac{\varepsilon_j^i(t)^2}{k^i(t)^2}, \qquad \text{for } j = 2, 3, \ldots, n. \tag{17}$$

It can be easily checked that if

$$\left\|\boldsymbol{w}_i(t)\right\|^2 \leq \frac{1}{\eta_i(t)\lambda_1^i} + \frac{1}{2}, \tag{18}$$

then

$$\sigma_\lambda^i(t) \leq \lambda_1^i\left\|\boldsymbol{w}_i(t)\right\|^2 \leq \frac{2 + \eta_i(t)\lambda_1^i}{2\eta_i(t)},$$

and then

$$\frac{[1+\eta_i(t)(\lambda_j^i - \sigma_\lambda^i(t))]^2}{[1+\eta_i(t)(\lambda_1^i - \sigma_\lambda^i(t))]^2} < 1, \quad \text{for } j = 2, 3, \ldots, n.$$

That is, Eq. (17) can hold. Since $\boldsymbol{w}_i(t)$ is, by Eq. (7), bounded such that $\|\boldsymbol{w}_i(t)\|^2 \le 1/2 + 1/\beta_i(t) = 1/2 + 1/(\eta_i(t)\lambda_1^i)$, the condition Eq. (18) will be satisfied. Thus, $\mathrm{Ang}(E\{\mathbf{w}_i(t+1)|\boldsymbol{w}_i(t)\})$ will be smaller than $\mathrm{Ang}(\boldsymbol{w}_i(t))$. Since this is true for all realizations $\boldsymbol{w}_i(t)$, so $E\{\mathbf{w}_i(t+1)|\mathbf{w}_i(t)\}$ will have a smaller angle from $\mathbf{v}_1^i$ than $\mathbf{w}_i(t)$. Hence, $E\{\mathbf{w}_i(t+1)\} = E\{E\{\mathbf{w}_i(t+1)|\mathbf{w}_i(t)\}\}$ will have a smaller angle from $\mathbf{v}_1^i$ than $E\{\mathbf{w}_i(t)\}$. That is, the angle between the mean of $\mathbf{w}_i(t)$ and $\mathbf{v}_1^i$ will approach zero. ■

In the following, we will analyze the variance of $\Delta\mathbf{w}_i(t)$, i.e., $\mathrm{Var}(\Delta\mathbf{w}_i(t))$.

PROPOSITION 5. *Given that* $\mathbf{x}$ *is normally distributed and* $\|\mathbf{w}_i(t)\| \le 1$, *then* $\mathrm{Var}(\Delta\mathbf{w}_i(t))$ *is bounded above by* $3(n-1)(\eta_i(t)\lambda_1^i)^2$, *where* n *is the dimension of the input pattern.*

Proof. Recall that

$$\mathrm{Var}(\Delta\mathbf{w}_i(t)) = E\{\|\Delta\mathbf{w}_i(t)\|^2\} - \|E\{\Delta\mathbf{w}_i(t)\}\|^2 \le E\{\|\Delta\mathbf{w}_i(t)\|^2\}.$$

According to Eq. (8), we get

$$\begin{aligned} E\{\|\Delta\mathbf{w}_i(t)\|^2|\boldsymbol{w}_i(t)\} &= \eta_i^2(t)E\{V_i^2(t)\|\mathbf{x}_i(t) - V_i(t)\boldsymbol{w}_i(t)\|^2|\boldsymbol{w}_i(t)\} \\ &= \eta_i^2(t)E\{V_i^2(t)(\|\mathbf{x}_i(t)\|^2 - 2V_i^2(t) \\ &\quad + V_i^2(t)\|\boldsymbol{w}_i(t)\|^2)|\boldsymbol{w}_i(t)\}. \end{aligned}$$

If $\|\boldsymbol{w}_i(t)\| \le 1$, then

$$\begin{aligned} E\{\|\Delta\mathbf{w}_i(t)\|^2|\boldsymbol{w}_i(t)\} &\le \eta_i^2(t)E\{V_i^2(t)(\|\mathbf{x}_i(t)\|^2 - V_i^2(t))|\boldsymbol{w}_i(t)\} \\ &\le \sum_{j=2}^{n} \eta_i^2(t)E\{((\mathbf{u}_1^i)^t\mathbf{x}_i(t))^2((\mathbf{u}_j^i)^t\mathbf{x}_i(t))^2\}, \end{aligned}$$

where $\mathbf{u}_j^i$, $j = 1, 2, \ldots, n$, have been defined in Proposition 3. Notice that

$$E\{((\mathbf{u}_1^i)^t\mathbf{x}_i(t))^2((\mathbf{u}_j^i)^t\mathbf{x}_i(t))^2\} \le 3E\{((\mathbf{u}_1^i)^t\mathbf{x}_i(t))^2\}E\{((\mathbf{u}_j^i)^t\mathbf{x}_i(t))^2\},$$

if $\mathbf{x}_i$ is normally distributed. In addition, $E\{(\mathbf{u}^t\mathbf{x}_i)^2\} = \mathbf{u}^t E\{\mathbf{x}_i\mathbf{x}_i^t\}\mathbf{u} = \mathbf{u}^t\mathbf{C}_i\mathbf{u} \leq (\mathbf{v}_1^i)^t\mathbf{C}_i\mathbf{v}_1^i = \lambda_1^i$, where $\mathbf{u}$ is any unit vector. Thus,

$$\begin{aligned} E\{\|\Delta\mathbf{w}_i(t)\|^2|\boldsymbol{w}_i(t)\} &\leq \sum_{j=2}^{n} 3\eta_i(t)^2\big((\mathbf{v}_1^i)^t\mathbf{C}_i\mathbf{v}_1^i\big)^2 \\ &= 3(n-1)(\eta_i(t)\lambda_1^i)^2, \quad \text{for } \|\boldsymbol{w}_i(t)\| \leq 1. \end{aligned}$$

Since it is true for all realizations $\boldsymbol{w}_i(t)$ with $\|\boldsymbol{w}_i(t)\| \leq 1$, so

$$\begin{aligned} E\big\{E\{\|\Delta\mathbf{w}_i(t)\|^2|\mathbf{w}_i(t)\}|\|\mathbf{w}_i(t)\| \leq 1\big\} &= E\big\{\|\Delta\mathbf{w}_i(t)\|^2|\|\mathbf{w}_i(t)\| \leq 1\big\} \\ &\leq 3(n-1)\big(\eta_i(t)\lambda_1^i\big)^2. \end{aligned}$$

Thus, $\mathrm{Var}(\Delta\mathbf{w}_i(t))$ for $\|\mathbf{w}_i(t)\| \leq 1$ will be bounded above by

$$3(n-1)(\eta_i(t)\lambda_1^i)^2. \qquad \blacksquare \qquad (19)$$

The variance $\mathrm{Var}(\Delta\mathbf{w}_i(t))$ can be decomposed into the sum of the variances along $\mathbf{u}_j^i$s, i.e., $\mathrm{Var}(\Delta\mathbf{w}_i(t)) = \sum_{j=1}^{n}\mathrm{Var}((\mathbf{u}_j^i)^t\Delta\mathbf{w}_i(t))$. Here, the radial variance of $\Delta\mathbf{w}_i(t)$, i.e., the variance along the direction of $\boldsymbol{w}_i(t)$, is

$$\mathrm{Var}\big((\mathbf{u}_1^i)^t\Delta\mathbf{w}_i(t)|\boldsymbol{w}_i(t)\big) = E\big\{\big((\mathbf{u}_1^i)^t\Delta\mathbf{w}_i(t)\big)^2\big|\boldsymbol{w}_i(t)\big\} - \big(\Delta k_{\mathrm{u}}^i(t)\big)^2, \qquad (20\mathrm{a})$$

and the tangential variances of $\Delta\mathbf{w}_i(t)$, i.e., the variances along the directions perpendicular to $\boldsymbol{w}_i(t)$, are

$$\mathrm{Var}\big((\mathbf{u}_j^i)^t\Delta\mathbf{w}_i(t)|\boldsymbol{w}_i(t)\big) = E\big\{\big((\mathbf{u}_j^i)^t\Delta\mathbf{w}_i(t)\big)^2\big|\boldsymbol{w}_i(t)\big\} - \big(\Delta\varepsilon_{uj}^i(t)\big)^2, \quad j = 2, 3, \ldots, n. \qquad (20\mathrm{b})$$

According to Eq. (8), we get

$$\begin{aligned} &E\big\{\big((\mathbf{u}_1^i)^t\Delta\mathbf{w}_i(t)\big)^2|\boldsymbol{w}_i(t)\big\} \\ &\quad = \eta_i(t)^2k_{\mathrm{u}}^i(t)^2\big(1 - k_{\mathrm{u}}^i(t)^2\big)^2E\big\{\big((\mathbf{u}_1^i)^t\mathbf{x}_i(t)\mathbf{x}_i^t(t)\mathbf{u}_1^i\big)^2\big\}, \qquad (21\mathrm{a}) \\ &E\big\{\big((\mathbf{u}_j^i)^t\Delta\mathbf{w}_i(t)\big)^2|\boldsymbol{w}_i(t)\big\} \\ &\quad = \eta_i(t)^2k_{\mathrm{u}}^i(t)^2E\big\{\big((\mathbf{u}_j^i)^t\mathbf{x}_i(t)\mathbf{x}_i^t(t)\mathbf{u}_1^i\big)^2\big\}, \qquad j = 2, 3, \ldots, n. \qquad (21\mathrm{b}) \end{aligned}$$

Substituting the square of Eq. (10) and Eq. (21) into Eq. (20), we get

$$\begin{aligned} \mathrm{Var}\big((\mathbf{u}_1^i)^t\Delta\mathbf{w}_i(t)|\boldsymbol{w}_i(t)\big) &= \eta_i(t)^2k_{\mathrm{u}}^i(t)^2(1 - k_{\mathrm{u}}^i(t)^2)^2\big[E\big\{\big((\mathbf{u}_1^i)^t\mathbf{x}_i(t)\mathbf{x}_i^t(t)\mathbf{u}_1^i\big)^2\big\} \\ &\quad - \big((\mathbf{u}_1^i)^t\mathbf{C}_i\mathbf{u}_1^i\big)^2\big], \qquad (22\mathrm{a}) \\ \mathrm{Var}\big((\mathbf{u}_j^i)^t\Delta\mathbf{w}_i(t)|\boldsymbol{w}_i(t)\big) &= \eta_i(t)^2k_{\mathrm{u}}^i(t)^2\big[E\big\{\big((\mathbf{u}_j^i)^t\mathbf{x}_i(t)\mathbf{x}_i^t(t)\mathbf{u}_1^i\big)^2\big\} \\ &\quad - \big((\mathbf{u}_j^i)^t\mathbf{C}_i\mathbf{u}_1^i\big)^2\big], \qquad j = 2, 3, \ldots, n. \qquad (22\mathrm{b}) \end{aligned}$$

It is obvious that the tangential variances $\mathrm{Var}((\mathbf{u}_j^i)^t \Delta\mathbf{w}_i(t)|\boldsymbol{w}_i(t))$ increase as the length of $\boldsymbol{w}_i(t)$, i.e., $\|\boldsymbol{w}_i(t)\|$ or $k_{\mathrm{u}}^i(t)$, increases. On the other hand, the radial variance $\mathrm{Var}((\mathbf{u}_1^i)^t \Delta\mathbf{w}_i(t)|\boldsymbol{w}_i(t))$ vanishes as the length of $\boldsymbol{w}_i(t)$ approaches 1. It leads to the following propositions.

PROPOSITION 6. *When $\mathbf{w}_i(t)$ reaches the stable fixed point $\mathbf{v}_1^i$, it will fluctuate around $\mathbf{v}_1^i$ only in direction but not in length. In addition, the range θ_f can be estimated by $\theta_f \le \tan^{-1}(\sqrt{3(n-1)}\eta_i(t)\lambda_1^i)$.*

Proof. Take $\mathbf{w}_i(t) = \mathbf{v}_1^i$. Then, $k_{\mathrm{u}}^i(t) = 1$ and $\mathbf{u}_1^i = \mathbf{v}_1^i$. As a result, the radial variance $\mathrm{Var}((\mathbf{u}_1^i)^t \Delta\mathbf{w}_i(t)|\mathbf{v}_1^i)$ [referring to Eq. (22a)] vanishes because $k_{\mathrm{u}}^i(t) = 1$. That is, there is no radial fluctuation of $\mathbf{w}_i(t)$ to influence its length. On the other hand, the tangential variance $\mathrm{Var}((\mathbf{u}_j^i)^t \Delta\mathbf{w}_i(t)|\mathbf{v}_1^i)$ [referring to Eq. (22b)] still exists. That is, there exists tangential fluctuation of $\mathbf{w}_i(t)$. Hence, $\mathbf{w}_i(t)$ fluctuates only in direction not in length. According to Proposition 5, the range θ_f of such fluctuation in direction can be estimated as follows:

$$\tan^{-1} \left.\frac{\sqrt{\mathrm{Var}(\Delta\mathbf{w}_i(t))}}{\|\mathbf{w}_i(t)\|}\right|_{\mathbf{w}_i(t)=\mathbf{v}_1^i} \le \tan^{-1} \frac{\sqrt{3(n-1)}\eta_i(t)\lambda_1^i}{1}$$
$$= \tan^{-1}(\sqrt{3(n-1)}\eta_i(t)\lambda_1^i). \quad \blacksquare \quad (23)$$

In accordance with the above propositions, it can be seen that the learning rate of $\mathbf{w}_i(t)$ and its variance can be estimated by the value of $\eta_i(t)\lambda_1^i$. For instance, from Eq. (19) and Eq. (22), one can see that the size of the variance $\mathrm{Var}(\Delta\mathbf{w}_i(t))$ can be estimated by the value of $(\eta_i(t)\lambda_1^i)^2$. It decreases to zero as $(\eta_i(t)\lambda_1^i)^2$ decreases to zero. Since the value of $\beta_i(t)$ $[= \eta_i(t)\lambda_1^i]$ is monotone decreasing, $\mathbf{w}_i(t)$ will then converge to $\mathbf{v}_1^i$ in the *mean-square sense* due to the decreasing variance. On the other hand, the learning rate of the length of $\mathbf{w}_i(t)$ depends, from Eq. (12), on the value of $\eta_i(t)\lambda_{\mathrm{u}}^i$. It increases as $\eta_i(t)\lambda_{\mathrm{u}}^i$ increases. Since $\lambda_{\mathrm{u}}^i < \lambda_1^i$, we can then use $\eta_i(t)\lambda_1^i$ to estimate the rate. Similarly, from Eq. (17), the learning rate of the direction of $\mathbf{w}_i(t)$ depends on the values of $\eta_i(t)\lambda_j^i$s. It increases as $\eta_i(t)\lambda_j^i$s increase for a given data set. Since λ_1^i is the largest eigenvalue, it is then reasonable to estimate the rate by using the value of $\eta_i(t)\lambda_1^i$. Therefore, for different neurons corresponding to different eigenvalues, i.e., λ_1^i, the same level of learning rate can be obtained by choosing $\eta_i(t)$ such that the values of $\eta_i(t)\lambda_1^i$s are the same. Hence, the following proposition can be obtained.

PROPOSITION 7. *The learning rates of all $\mathbf{w}_i(t)$ in the ALA are nearly the same if $\beta_i(t)$ is the same for all i.*

Proof. Since $\beta_i(t)$ is the same for all i, the value of $\eta_i(t)\lambda_1^i$ will be the same for all i. Hence, the learning rates of all $\mathbf{w}_i(t)$ in the ALA will be nearly the same due to previous discussion. $\blacksquare$

According to Proposition 7, the learning rate of $\mathbf{w}_i(t)$ will not decrease as that of the GHA when λ_i decreases. Hence, the learning of the ALA can be faster than that of the GHA. Simulation results in the next section will confirm the effectiveness of the ALA.

III. SIMULATION RESULTS

First, let us demonstrate that the ALA can converge quickly to the desired target in the small-eigenvalue case as well as the large-eigenvalue case while the GHA fails to do so. Two sets of three-dimensional randomly generated zero-mean data are adopted as input. The maximal eigenvalue λ_1 of the covariance matrices of the two data sets is 0.0086 and 25.40, respectively. One neuron, i.e., $m = 1$, is considered for the moment. The weight vector $\mathbf{w}_1(t)$ is initially set to be nearly perpendicular to its target $\mathbf{v}_1$ with length about 0.2. Such initial setting is adopted in all of the following experiments. The GHA (it is reduced to Oja's rule here since $m = 1$) is first used to train the network. The value of $\eta(t)$ is set to be exponentially decreased with time from 0.1 to the final value 0.008. Its time function is set as $\max(0.1(0.01/0.1)^{t/1000}, 0.008)$. Figure 3a and b shows the time histories of learning corresponding to the two data sets, respectively. From Fig. 3a, one can see that the convergence of the learning process is very slow. The angle $\theta_1(t)$ between $\mathbf{w}_1(t)$ and $\mathbf{v}_1$ is still near 90° and the length of $\mathbf{w}_1(t)$, i.e., $\|\mathbf{w}_1(t)\|$ is still much smaller than 1. The reason is that the value of λ_1 (0.0086) is so small such that $\eta(t))\lambda_1$ is even smaller and the convergence rate for the learning process is very slow. On the other hand, if the same value of $\eta(t)$ is used for the other data set with $\lambda_1 = 25.40$, the learning process will fail because it is too big for this set of data. As shown in Fig. 3b, $\|\mathbf{w}_1(t)\|$ grows to infinity and $\theta_1(t)$ cannot decrease to zero.

On the contrary, the ALA can succeed in both cases with the parameters $\beta_1(t)$ set as the $\eta(t)$ of the GHA mentioned above and $\gamma(t)$ set as a constant value 0.01. First, let us discuss the procedure of eigenvalue estimation. In our experiments, $\mathbf{w}_1(t)$ is initially set far from $\mathbf{v}_1$ purposely; then the $\hat{\lambda}_1$ estimated by Eq. (5) is much smaller than its true value during the initial period of the learning process. As a result, $\eta_1(t)$ $[= \beta_1(t)/\hat{\lambda}_1(t)]$ becomes much bigger than the desired value $\beta_1(t)/\lambda_1$. That is, $\eta_1(t)\lambda_1$ becomes much bigger than $\beta_1(t)$, the value we set, or even the upper bound $2(\sqrt{2} - 1)$. According to Eq. (11), the length of $\mathbf{w}_1(t)$ will diverge. However, this minor problem can be remedied by the normalization procedure in step 6 of the ALA. In step 6, once $\|\mathbf{w}_1(t)\| > \sqrt{1/\beta_1(t) + 1/2}$, it is normalized to $\sqrt{0.5}$. Otherwise, no normalization is required. From Eq. (18), one can see that the directional convergence of $\mathbf{w}_1(t)$ will then hold. Moreover, the convergence rate will, from Eq. (17), be faster with the bigger value of $\eta_1(t)\lambda_1$. Hence, $\mathbf{w}_1(t)$ will be close to $\mathbf{v}_1$ in direction and the mean of $(\mathbf{w}_1^t\mathbf{x}/\|\mathbf{w}_1\|)^2$ will

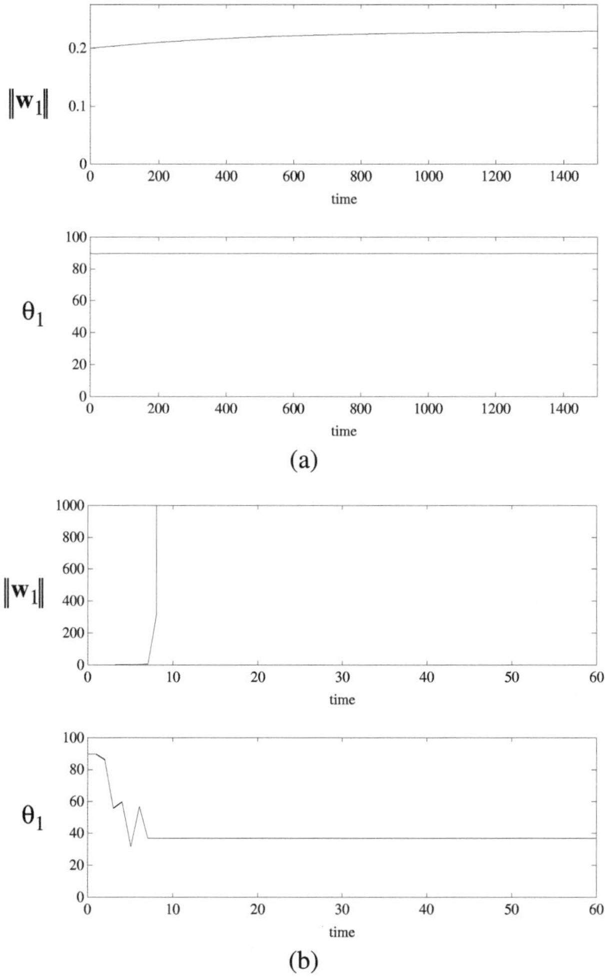

Figure 3 Simulation results of the GHA for (a) $\lambda_1 = 0.0086$, (b) $\lambda_1 = 25.40$. Reprinted with permission from L. H. Chen and S. Chang, *IEEE Trans. Neural Networks* 6:1255–1263, 1995 (©1995 IEEE).

then approach the desired value λ_1. As a result, $\hat{\lambda}_1(t)$ will converge very quickly. Figures 4a and 5a clearly illustrate this point for these two data sets. With the accurate estimate of eigenvalue, $\mathbf{w}_1(t)$ will converge to $\mathbf{v}_1$ in the mean-square sense as indicated by Propositions 3–6. It is illustrated in Figs. 4b and 5b. It can be seen from the figures that the length of $\mathbf{w}_1(t)$ and angle $\theta_1(t)$ between $\mathbf{w}_1(t)$ and $\mathbf{v}_1$ converge quickly to 1 and 0, respectively, for both data sets. From the $k^1(t)$

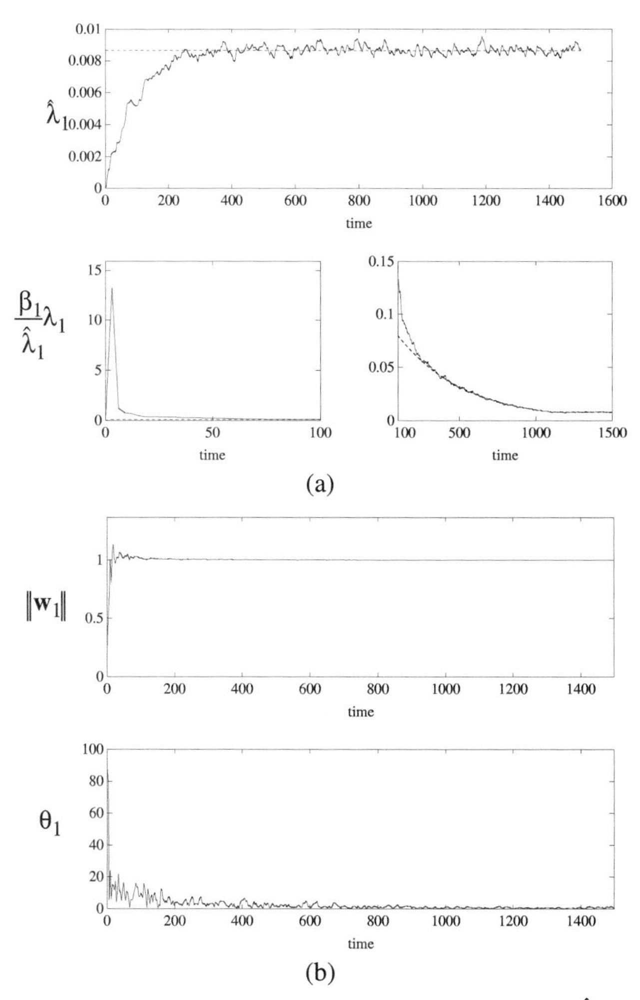

Figure 4 Simulation results of the ALA for $\lambda_1 = 0.0086$. (a) Time histories of $\hat{\lambda}_1$ (top figure) and $(\beta_1/\hat{\lambda}_1)\lambda_1$ (bottom figure for two different time scales). The dashed line and curve denote the values of λ_1 and $\beta_1(t)$, respectively. (b) Time histories of $\|\mathbf{w}_1\|$ and θ_1 as functions of iterations. (c) Trajectory of $\mathbf{w}_1$ on the k^1 versus $\|\mathbf{s}_1\|$ plane. Parts (a) and (b) reprinted with permission from L. H. Chen and S. Chang, *IEEE Trans. Neural Networks* 6:1255–1263, 1995 (©1995 IEEE).

versus $\|\mathbf{s}_1(t)\|$ plots in Figs. 4c and 5c, one can see that each $\mathbf{w}_1(t)$ approaches the unit hypersphere (the dotted curve) quickly and then reaches its target $\mathbf{v}_1$ which is located at the coordinate $(1, 0)$ of the plot. To sum up, the learning process of ALA is successful for both data sets.

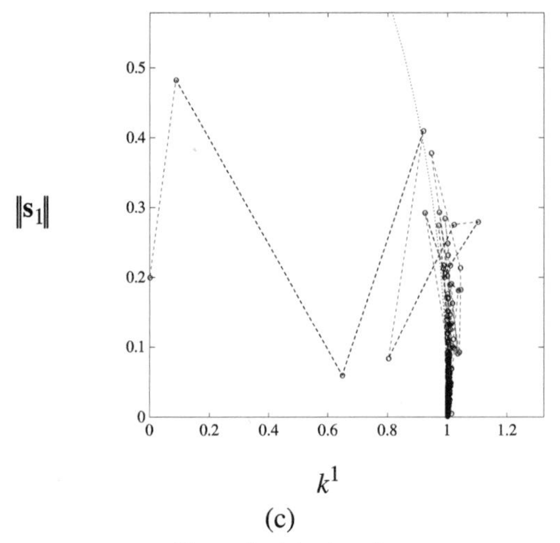

(c)

Figure 4 (*Continued*)

In the following, simulations will be used to demonstrate that the ALA can make all of the m weight vectors $\mathbf{w}_i(t)$, $i = 1, 2, \ldots, m$, converge quickly to the desired targets independent of the eigenvalues and eigenvalue spread. Three data sets are used here: Sandpapers, IRIS [42], and XO8. The set of Sandpapers contains 96 four-dimensional patterns describing the texture measurements of the images of four kinds of sand. The IRIS data contains 150 four-dimensional patterns of three classes of flowers. Finally, the set of XO8 contains 45 eight-dimensional patterns describing the characters "X," "O," and "8." The ALA can still handle such non-zero-mean data, well as zero-mean data by estimating the data mean and subtracting it from the patterns. We estimate the data mean with the equation $\mathbf{m}_x(t+1) = \mathbf{m}_x(t) + (\mathbf{x}(t) - \mathbf{m}_x(t))/(t+1)$. Notice that the patterns in the data set are drawn randomly and repeatedly as the inputs presented to the network. In addition, the number of output, m, is set to n now. The parameters $\beta_i(t)$, for $i = 1, 2, \ldots, m$, are all set to the same value as $\beta_1(t)$ in the previous experiments except time delay and the final value denoted by β_f. For the time delay, the learning time of neuron i starts later than that of neuron $i - 1$ with a time delay t_p which is set to 500. The goal is to make all $\mathbf{w}_j(t)$ come close to $\mathbf{v}_j$ for $j < i$ when neuron i begins to learn. Moreover, in order to make the final angle error between $\mathbf{w}_i(t)$ and $\mathbf{v}_i$ be smaller than $1.5°$, the final value β_f of $\beta_i(t)$ is set, according to Eq. (23), to 0.008 for Sandpapers and IRIS, and 0.005 for higher-dimensional XO8, respectively. The results for Sandpapers and IRIS are shown in Figs. 6 and 7, respectively. It is clear from the time histories of $\|\mathbf{w}_i(t)\|$s and $\theta_i(t)$s that all of $\mathbf{w}_i(t)$s can converge quickly to their corresponding $\mathbf{v}_i$s re-

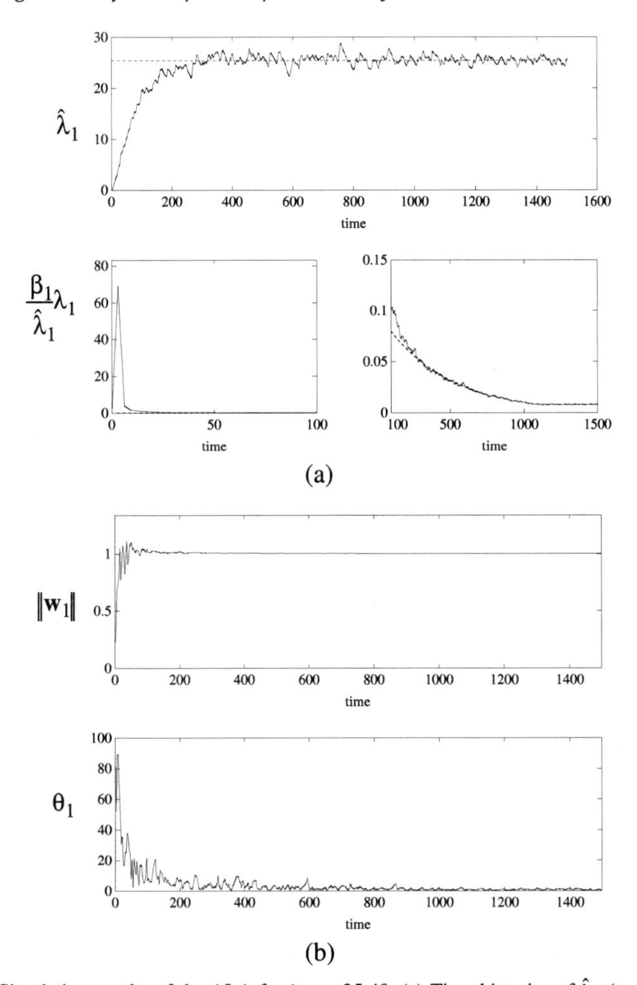

Figure 5 Simulation results of the ALA for $\lambda_1 = 25.40$. (a) Time histories of $\hat{\lambda}_1$ (top figure) and $(\beta_1/\hat{\lambda}_1)\lambda_1$ (bottom figure for two different time scales). The dashed line and curve denote the values of λ_1 and $\beta_1(t)$, respectively. (b) Time histories of $\|\mathbf{w}_1\|$ and θ_1 as functions of iterations. (c) Trajectory of $\mathbf{w}_1$ on the k^1 versus $\|\mathbf{s}_1\|$ plane. Parts (a) and (b) reprinted with permission from L. H. Chen and S. Chang, *IEEE Trans. Neural Networks* 6:1255–1263, 1995 (©1995 IEEE).

spectively even in the Sandpapers case where the second and third eigenvalues are very close. Moreover, one should notice that, although the differences among the eigenvalues are great, the learning rates of neurons are all nearly the same after they start learning. The eigenvalue spread λ_1/λ_n reaches about 200. However, the learning rate of neuron i in the ALA will not be slowed down as i increases.

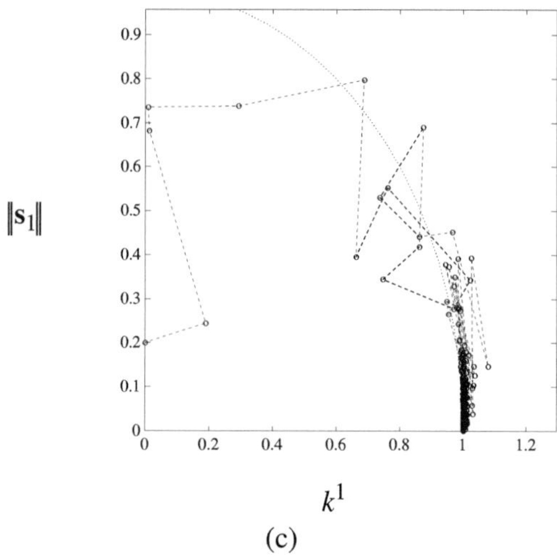

(c)

Figure 5 (*Continued*)

Table I

Final Values for Sandpapers Using ALA

Neuron i	θ_i	$\|\mathbf{w}_i\|$	$\hat{\lambda}_1$	λ_1
1	0.5831	1.0002	0.0266	0.0265
2	0.8085	0.9985	0.0011	0.0011
3	0.4997	0.9931	0.0009	0.0008
4	0.5086	0.9921	0.0001	0.0001

Reprinted with permission from L. H. Chen and S. Chang, *IEEE Trans. Neural Networks* 6:1255–1263, 1995 (©1995 IEEE).

Table II

Final Values for IRIS Using ALA

Neuron i	θ_i	$\|\mathbf{w}_i\|$	$\hat{\lambda}_1$	λ_1
1	0.3112	1.0003	4.1045	4.2001
2	0.8727	0.9982	0.2358	0.2411
3	0.9051	0.9951	0.0729	0.0777
4	0.8935	0.9886	0.0233	0.0237

Reprinted with permission from L. H. Chen and S. Chang, *IEEE Trans. Neural Networks* 6:1255–1263, 1995 (©1995 IEEE).

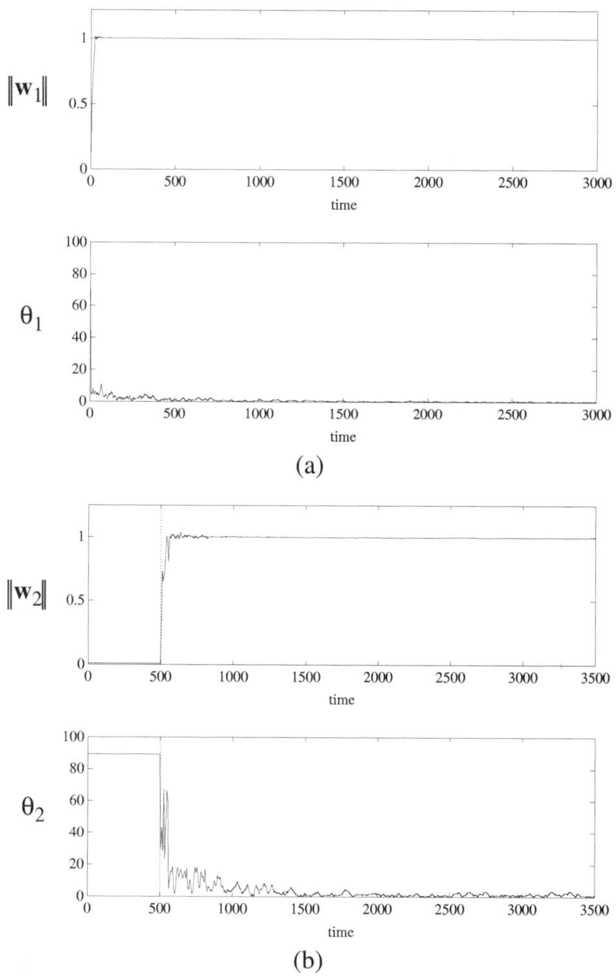

Figure 6 Simulation results of the ALA for Sandpapers data. (a)–(d) Learning times histories of $\mathbf{w}_1$, $\mathbf{w}_2$, $\mathbf{w}_3$, and $\mathbf{w}_4$, respectively. The vertical dotted lines denote the starting time of learning. Reprinted with permission from L. H. Chen and S. Chang, *IEEE Trans. Neural Networks* 6:1255–1263, 1995 (©1995 IEEE).

Tables I and II list the final values of $\|\mathbf{w}_i(t)\|$s and $\theta_i(t)$s as well as the eigenvalue estimates $\hat{\lambda}_i(t)$s at the end of the learning process. It is obvious that the results are quite accurate compared with the theoretical values. Table III is the simulation result for the data set XO8. It demonstrates that the ALA also works well for higher-dimensional data.

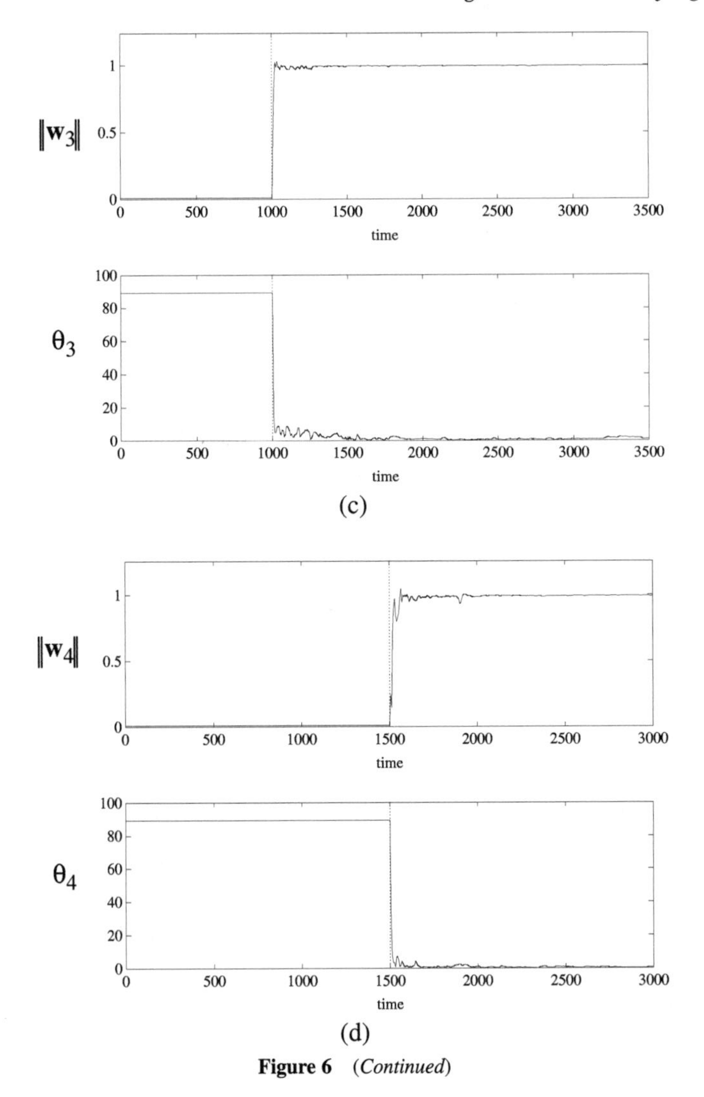

Figure 6 (*Continued*)

From these experiments, it is clear that the ALA can properly and automatically select the learning rate parameters such that $\mathbf{w}_i(t)$ can converge to $\mathbf{v}_i$ with almost the same rate for each i no matter what values the eigenvalues λ_is are. Hence, the ALA is a very effective way to execute PCA.

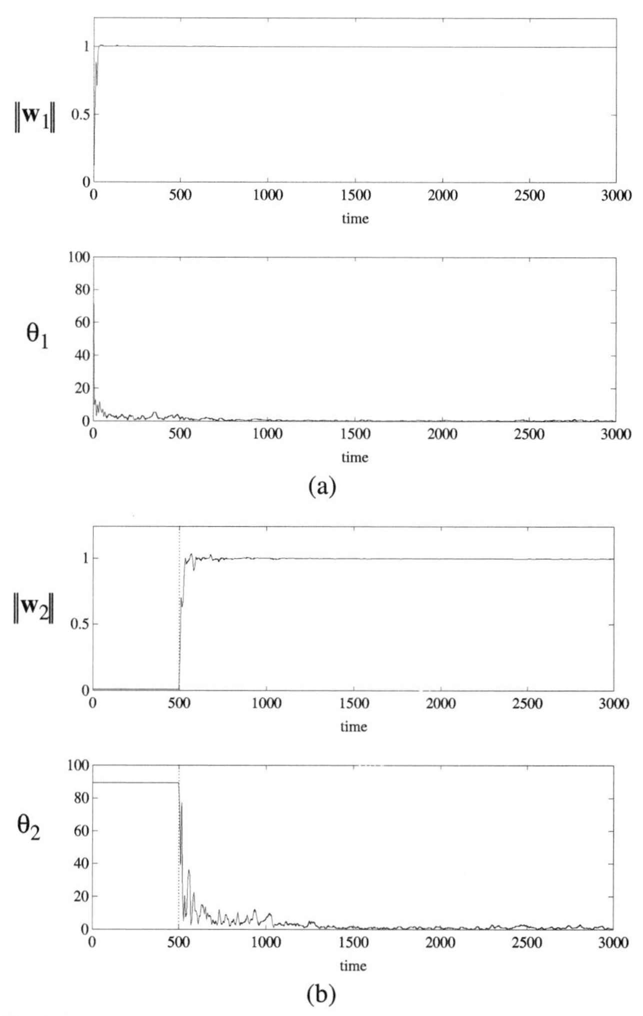

Figure 7 Simulation results of the ALA for IRIS data. (a)–(d) Learning times histories of $\mathbf{w}_1$, $\mathbf{w}_2$, $\mathbf{w}_3$, and $\mathbf{w}_4$, respectively. The vertical dotted lines denote the starting time of learning.

IV. APPLICATIONS

Due to the aforementioned effective computing capability for PCA, the ALA can then be applied to data compression. By not removing the data mean, the PCA is equivalent to the Karhunen–Loève transform (KLT) [43, 44], by which higher-dimensional data can be transformed to lower-dimensional data. From

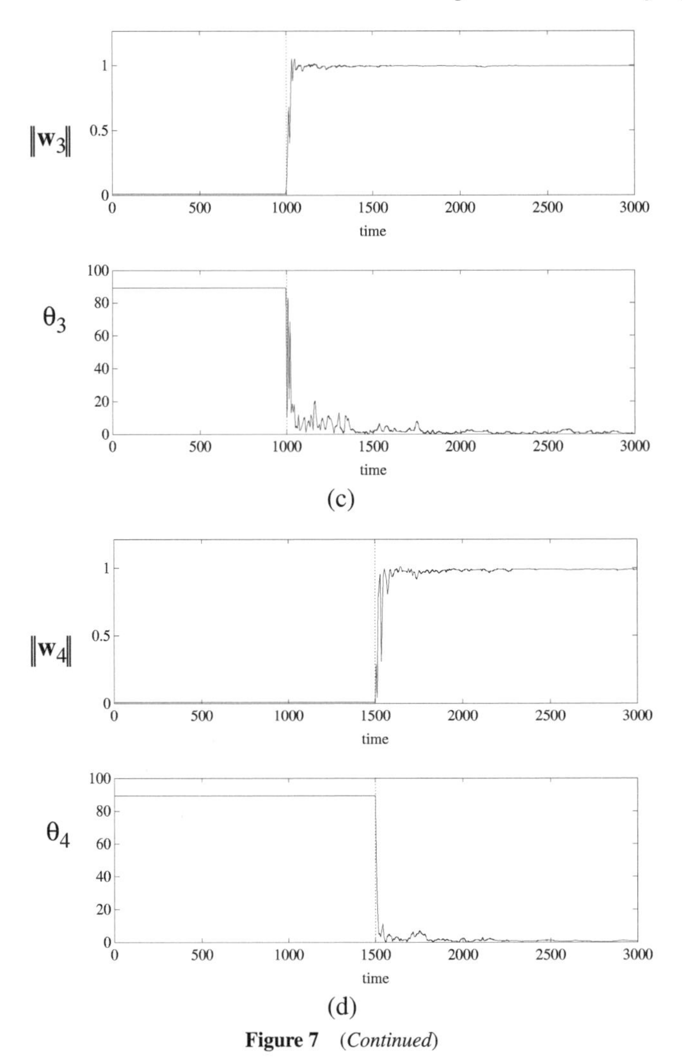

Figure 7 (*Continued*)

these lower-dimensional data, we can reconstruct the higher-dimensional data with minimal mean-square error. That is, data are compressed via dimensionality reduction. The ALA can exactly be utilized to quickly perform such a task.

The ALA can also be applied to image coding, texture segmentation, and development of receptive fields due to its effective executing capability of the KLT. These applications can be found in [24]. The following are two examples of image coding. Figure 8a is the original "pepper" image with 256 × 256 pixels and 256 graylevels. A PCA learning network with 64 inputs and 8 outputs is adopted

Table III

Final Values for XO8 Using ALA

Neuron i	θ_i	$\|\mathbf{w}_i\|$	$\hat{\lambda}_1$	λ_1
1	1.2992	1.0029	15.822	15.842
2	1.5005	0.9996	9.9920	10.145
3	1.1511	0.9994	6.9473	7.0847
4	1.6035	1.0001	4.8364	5.0343
5	1.7126	0.9966	3.7637	3.9537
6	1.5847	0.9909	2.2885	2.3659
7	0.6891	0.9935	1.2062	1.1999
8	0.6647	0.9884	0.6284	0.6376

here. That is, the number of neurons, m, is equal to 8 now. The image is first divided into nonoverlapped 8×8 blocks. These 8×8 blocks are then presented in a random order to the network as training samples. The parameter γ is set here as 0.005 while β_i is set to be exponentially decreased with time from 0.1 to 0.002. It decays to 0.01 as t comes to 500. The final value of β_i, i.e., 0.002, is still set according to Eq. (23) to make the final angle error between the 64-dimensional weight vector $\mathbf{w}_i$ and its target $\mathbf{v}_i$ to be smaller than 1.5°. In addition, the time delay t_p is set to 50.

The eight outputs are then used to represent the input 8×8 image block. This is the image coding. To reconstruct the image, each 8×8 block is approximated by adding together the weight vectors multiplied by the corresponding outputs. The performance of coding is evaluated by calculating the normalized mean-square error (NMSE) [11]:

$$\mathrm{NMSE} = \frac{E[(I_{n,m} - \hat{I}_{n,m})^2]}{E[I_{n,m}^2]}, \tag{24}$$

where $I_{n,m}$ is the pixel value at position n, m of the original image and $\hat{I}_{n,m}$ is the corresponding pixel value in the reconstructed image.

Figure 8b shows the reconstructed pepper image, and in the top part of Fig. 8c are the learned eight weight vectors, each of which is arranged as an 8×8 mask. The bottom part of Fig. 8c is the plot of the time history of the NMSE. It decays very quickly. At $t = 798$, the NMSE has been decreased to lower than 0.01! The final value reaches only 0.006. On the contrary, the similar experiment by the GHA can only obtain a NMSE higher than 0.02 even if the iterations have been over 2000 [24]. In addition, observing from the estimated eigenvalues listed in Table IV, the first eigenvalue extends 10^6 and is almost 900 times the eighth one. It needs the learning rate parameters smaller than 10^{-6} in order to make the learning process converge! Moreover, these parameters should be increased

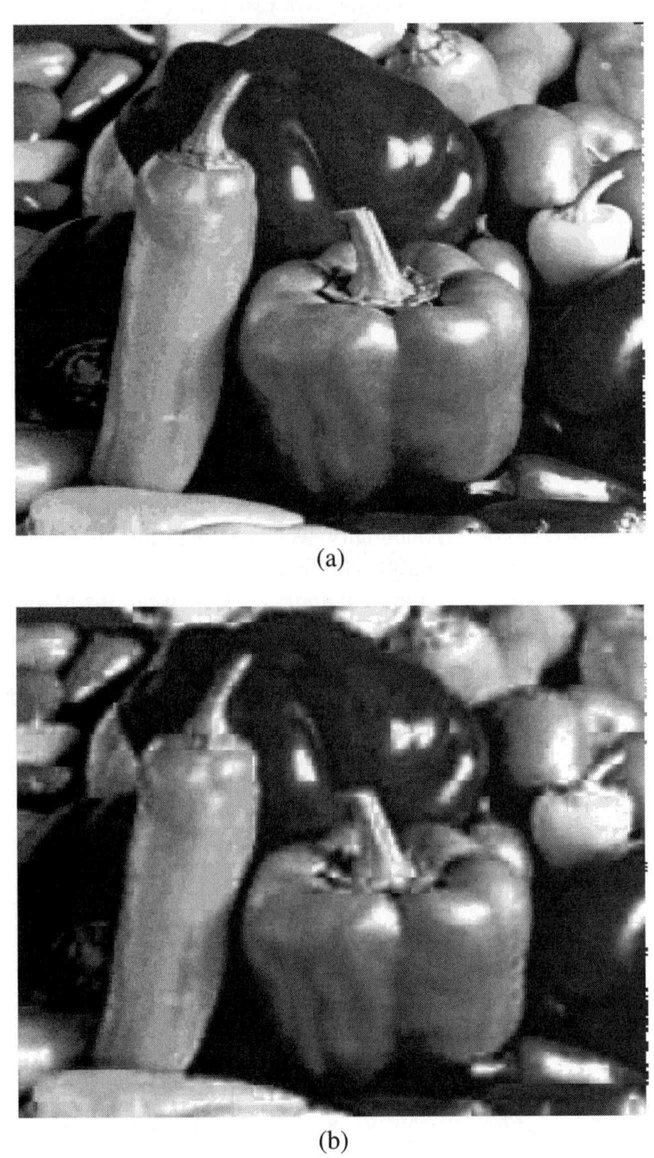

(a)

(b)

Figure 8 Experimental results of the ALA for image coding. (a) Original "pepper" image. (b) Reconstructed image. (c) Learned eight weight vectors and the NMSE learning curve.

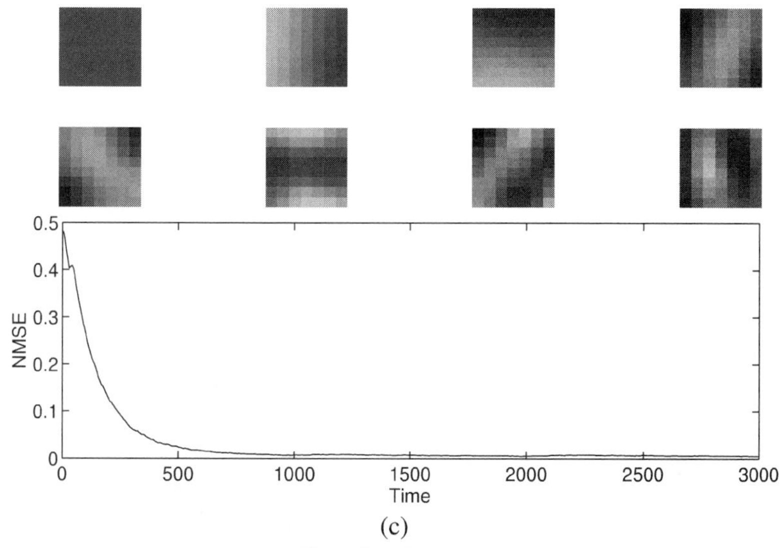

(c)

Figure 8 (*Continued*)

as the index of neuron increases such that the last one becomes 900 times the first one in order to keep the learning speed from decaying as the index increases. All of these requirements are automatically and adaptively achieved by our ALA. The learning process then not only converges but also converges very quickly as indicated by the NMSE learning curve. Figure 9 and Table V are results for another similar experiment for the "Wha" image. All of these experimental results obviously demonstrate again the power of the ALA.

Table IV

Estimated Eigenvalues of the "Pepper" Image

Neuron i	$\hat{\lambda}_i$
1	1.2 e6
2	2.1 e4
3	1.5 e4
4	5.7 e3
5	4.0 e3
6	2.3 e3
7	1.6 e3
8	1.4 e3

(a)

(b)

Figure 9 Experimental results of the ALA for image coding. (a) Original "Wha" image. (b) Reconstructed image. (c) Learned eight weight vectors and the NMSE learning curve.

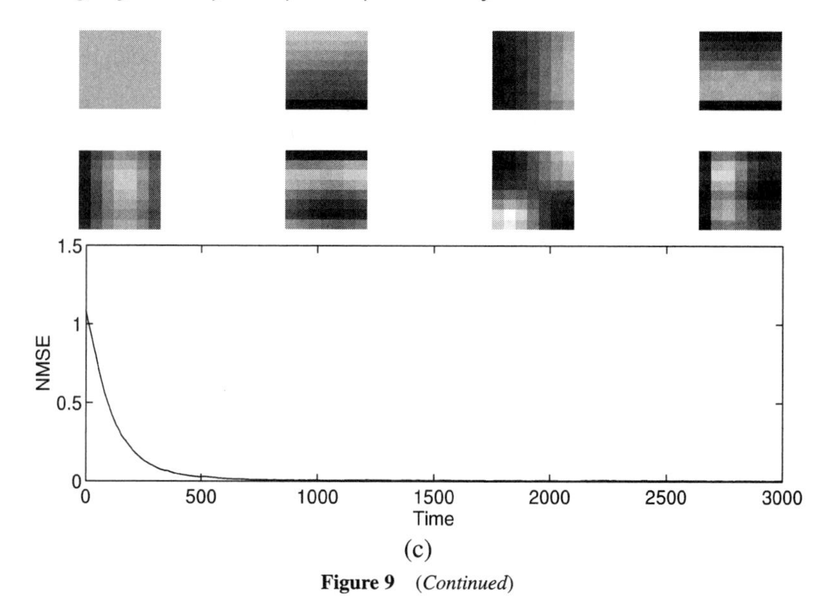

(c)

Figure 9 (*Continued*)

V. CONCLUSION

For PCA learning networks, we have proposed an adaptive learning algorithm (ALA). By adaptively selecting the learning rate parameters according to the eigenvalues of the input covariance matrix, it has been shown that the ALA can

Table V
Estimated Eigenvalues of the "Wha" Image

Neuron i	$\hat{\lambda}_i$
1	1.2 e6
2	1.7 e4
3	1.3 e4
4	6.7 e3
5	3.4 e3
6	2.6 e3
7	1.9 e3
8	1.0 e3

make the m weight vectors in the network converge quickly to the first m principal component vectors with almost the same learning rates. From the simulation results, it has been confirmed that the ALA can converge very quickly to the desired target in the large-eigenvalue case as well as in the small-eigenvalue case. On the other hand, the conventional GHA diverges in the former case and converges very slowly in the latter case. In addition, from the simulation results of the three real data sets—Sandpapers, IRIS, and XO8—one can see that the ALA has been able to find quickly all principal component vectors even if the eigenvalue spread is quite big. The ALA is thus a very effective way to execute the PCA. Based on such capability, the ALA has been applied to data compression and image coding. Excellent experimental results have been obtained.

In the future, it is expected to implement the above adaptive learning algorithms in VLSI architectures in order to facilitate practical applications.

VI. APPENDIX

Proof of Proposition 1. Take $\mathbf{w}(t)$ to be close to $\mathbf{v}_1$, i.e., $\mathbf{w}(t) = \mathbf{v}_1 + \mathbf{e}(t)$, where $\mathbf{C}\mathbf{v}_1 = \lambda_1\mathbf{v}_1$, $\|\mathbf{v}_1\| = 1$, and $\|\mathbf{e}(t)\| \ll 1$. Thus, $E\{\Delta\mathbf{w}(t)|\mathbf{w}(t)\} = E\{\Delta\mathbf{e}(t)|\mathbf{e}(t)\}$, and we get by Eq. (4) $E\{\Delta\mathbf{e}(t)|\mathbf{e}(t)\} = \eta(t)\{\mathbf{C}(\mathbf{v}_1+\mathbf{e}(t)) - [(\mathbf{v}_1^t + \mathbf{e}^t(t))\mathbf{C}(\mathbf{v}_1+\mathbf{e}(t))](\mathbf{v}_1+\mathbf{e}(t))\} = \eta(t)[\mathbf{C}\mathbf{e}(t) - 2\lambda_1\cdot\mathbf{e}^t(t)\mathbf{v}_1\mathbf{v}_1 - \lambda_1\mathbf{e}(t) + O(\mathrm{e}^2)]$, where $O(\mathrm{e}^2)$ denotes the higher-order terms of $\mathbf{e}(t)$. Ignoring the $O(\mathrm{e}^2)$ terms, it becomes $E\{\Delta\mathbf{e}(t)|\mathbf{e}(t)\} \cong \eta(t)[\mathbf{C}\mathbf{e}(t) - 2\lambda_1\mathbf{e}^t(t)\mathbf{v}_1\mathbf{v}_1 - \lambda_1\mathbf{e}_1(t)]$. Recall that the normalized eigenvectors associated with distinct eigenvalues of symmetric matrix $\mathbf{C}$ are orthonormal. They can form a basis spanning the $\mathbf{R}^n$ space. As a result, we can represent $\mathbf{w}(t)$, $\mathbf{e}(t)$, $E\{\Delta\mathbf{e}(t)\}$, etc., by their components along the directions of the normalized eigenvectors of $\mathbf{C}$. Thus, along the direction of $\mathbf{v}_j$, $j = 1, 2, \ldots, n$, the component of $E\{\Delta\mathbf{e}(t)|\mathbf{e}(t)\}$ is $\mathbf{v}_j^t E\{\Delta\mathbf{e}(t)|\mathbf{e}(t)\} \cong -2\eta(t)\lambda_1\mathbf{v}_1^t\mathbf{e}(t)$, if $j = 1$; $-\eta(t)(\lambda_1 - \lambda_j)\mathbf{v}_j^t\mathbf{e}(t)$, if $j \neq 1$. Therefore,

$$\mathbf{v}_j^t E\{\mathbf{e}(t+1)|e(t)\} = \mathbf{v}_j^t e(t) + \mathbf{v}_j^t E\{\Delta\mathbf{e}(t)|e(t)\}$$
$$\cong \begin{cases} (1 - 2\eta(t)\lambda_1)\mathbf{v}_1^t e(t), & \text{if } j = 1, \\ [1 - \eta(t)(\lambda_1 - \lambda_j)]\mathbf{v}_j^t e(t), & \text{if } j \neq 1, \end{cases} \tag{25}$$

where $e(t)$ stands for the realization of $\mathbf{e}(t)$ at time t. It can be seen that if $\eta(t) < 1/\lambda_1$, then $|1 - 2\eta(t)\lambda_1| < 1$ and $|1 - \eta(t)(\lambda_1 - \lambda_j)| < 1$ for $j = 1, 2, \ldots, n$. As a result, from Eq. (25), $|\mathbf{v}_j^t E\{\mathbf{e}(t+1)|e(t)\}|$ will be smaller than $|\mathbf{v}_j^t e(t)|$ along all directions $\mathbf{v}_j$, $j = 1, 2, \ldots, n$. Since this is true for all realizations $e(t)$, the expectation of error will thus decrease. It implies that the expectation of $\mathbf{w}(t)$ will approach $\mathbf{v}_1$. Hence, if $\eta(t) > 1/\lambda_1$, then the expectation of $\mathbf{w}(t)$ cannot approach $\mathbf{v}_1$ and therefore $\mathbf{w}(t)$ cannot converge to $\mathbf{v}_1$. This completes the proof. ■

Proof of Proposition 2. When $\eta(t)\lambda_1 < 0.5$, the values of $|1 - 2\eta(t)\lambda_1|$ and $|1 - \eta(t)(\lambda_1 - \lambda_j)|$, $j = 2, 3, \ldots, n$, will be closer to 1 if the value of $\eta(t)\lambda_1$ is closer to zero. As a result, from Eq. (25), the expectation of error will decrease much more slowly. ■

REFERENCES

[1] J. J. Hopfield. Neural networks and physical systems with emergent collective computational abilities. *Proc. Nat. Acad. Sci. USA* 79:2554–2558, 1982.

[2] W. C. Wu, R. M. Chen, and S. Chang. An analog architecture on parameter estimation of ARMA models. *IEEE Trans. Signal Process.* 41:2846–2953, 1993.

[3] A. D. Culhane, M. C. Peckerar, and C. R. K. Marrian. A neural net approach to discrete Hartly and Fourier transforms. *IEEE Trans. Circuits Syst.* 36:695–703, 1989.

[4] A. Tabatabai and T. P. Troudet. A neural net based architecture for the segmentation of mixed gray-level and binary pictures. *IEEE Trans. Circuits Syst.* 38:66–77, 1991.

[5] R. J. McEliece, E. C. Posner, E. R. Rodemich, and S. S. Venkatesh. The capacity of the Hopfield associative memory. *IEEE Trans. Inform. Theory* 33:461–482, 1987.

[6] D. E. Rumelhart, G. E. Hinton, and R. J. Williams. Learning representations by back-propagating errors. *Nature* 323:533–536, 1986.

[7] D. E. Rumelhart, J. L. McClelland, and the PDP Research Group. *Parallel Distributed Processing: Explorations in the Microstructure of Cognition*, 2 vols. MIT Press, Cambridge, MA, 1986.

[8] D. A. Pomerleau. ALVINN: An autonomous land vehicle in a neural network. In *Advances in Neural Information Processing Systems*, (D. S. Touretzky, Ed.), pp. 305–313. Morgan Kaufmann, San Mateo, CA, 1989.

[9] R. P. Gorman and T. J. Sejnowski. Learned classification of sonar targets using a massively-parallel network. *IEEE Trans. Acoust. Speech Signal Process.* 36:1135–1140, 1988.

[10] Y. Le Cun, B. Boser, J. S. Denker, D. Henderson, R. E. Howard, W. Hubbard, and L. D. Jackel. Backpropagation applied to handwritten Zip code recognition. *Neural Computat.* 1:541–551, 1989.

[11] G. W. Cottrell, P. Munro, and D. Zipser. Learning internal representations from gray-scale images: An example of extensional programming. In *Ninth Annual Conference of the Cognitive Science Society*, pp. 462–473. Seattle, WA, 1987.

[12] F. Rosenblatt. *Principles of Neurodynamics*. Spartan, New York, 1962.

[13] Ch. von der Malsburg. Self-organization of orientation sensitive cells in the striate cortex. *Kybernetika* 14:85–100, 1973.

[14] D. E. Rumelhart and D. Zipser. Feature discovery by competetive learning. *Cognitive Sci.* 9:75–112, 1985.

[15] J. Hertz, A. Krogh, and R. G. Palmer. *Introduction to the Theory of Neural Computation*. Addison-Wesley, Redwood City, CA, 1991.

[16] B. Kosko. Stochastic competitive learning. *IEEE Trans. Neural Networks* 2:522–529, 1991.

[17] R. Hecht-Nielsen. *Neuroncomputing*, Addison-Wesley, Reading, MA, 1989.

[18] J. Makhoul, S. Roucos, and H. Gish. Vector quantization in speech coding. *Proc. IEEE* 73:1551–1588, 1985.

[19] R. Pieraccini and R. Billi. Experimental comparison among data compression techniques in isolated word recognition. In *Proceedings of the IEEE International Conference on Acoustics, Speech, and Signal Processing*, pp. 1025–1028. Boston, MA, 1983.

[20] M. O. Dunham and R. M. Gray. An algorithm for design of labeled-transition finite-state vector quantizers. *IEEE Trans. Commun.* COM-33:83–89, 1985.

[21] N. M. Nasrabadi and R. A. King. Image coding using vector quantization: A review. *IEEE Trans. Commun.* 36:957–971, 1988.
[22] N. R. Dixon and T. B. Martin, Eds. *Automatic Speech and Speaker Recognition*. IEEE Press, New York, 1979.
[23] B. W. Silverman. *Density Estimation for Statistics and Data Analysis*. Chapman and Hall, New York, 1986.
[24] T. D. Sanger. Optimal unsupervised learning in a single-layer linear feedforward neural network. *Neural Networks* 2:459–473, 1989.
[25] D. Hebb. *The Organization of Behavior*. Wiley, New York, 1949.
[26] I. T. Jolliffe. *Principal Component Analysis*. Springer-Verlag, New York, 1986.
[27] P. Common and G. H. Golub. Tracking a few extreme singular values and vectors in signal processing. *Proc. IEEE* 78:1327–1343, 1990.
[28] K. Hornik and C.-M. Kuan. Convergence analysis of local feature extraction algorithms. *Neural Networks* 5:229–240, 1992.
[29] R. Linsker. Self-organization in a perceptual network. *Computer* March:105–117, 1988.
[30] R. H. White. Competitive Hebbian learning: Algorithm and demonstrations. *Neural Networks* 5:261–275, 1992.
[31] A. L. Yuille, D. M. Kammen, and D. S. Cohen. Quadrature and the development of orientation selective cortical cells by Hebb rules. *Biolog. Cybernet.* 61:183–194, 1989.
[32] P. Baldi and K. Hornik. Neural networks and principal component analysis: Learning from examples without local minima. *Neural Networks* 2:53–58, 1989.
[33] E. Oja. Neural networks, principal components, and subspaces. *Internat. J. Neural Syst.* 1:61–68, 1989.
[34] P. Földiák. Adaptive network for optimal linear feature extraction. In *Proceedings of the International Joint Conference on Neural Networks*, Vol. I, pp. 401–405. San Diego, CA, 1989.
[35] J. Rubner and P. Tavan. A self-organizing network for principal component analysis. *Europhys. Lett.* 10: 693–698, 1989.
[36] R. Lenz and M. Österberg. Computing the Karhunen–Loève expansion with a parallel, unsupervised filter system. *Neural Computat.* 4:382–392, 1992.
[37] E. Oja. A simplified neuron model as a principal component analyzer. *J. Math. Biol.* 15:267–273, 1982.
[38] L. H. Chen and S. Chang. An improved learning algorithm for principal component analysis. In *International Conference on Signal Processing Applications & Technology*, ICSPAT '93, Vol. II, pp. 1049–1057. Santa Clara, California, 1993.
[39] L. H. Chen and S. Chang. An adaptive learning algorithm for principal component analysis. *IEEE Trans. Neural Networks* 6:1255–1263, 1995.
[40] H. J. Kushner and D. S. Clark. *Stochastic Approximation Methods for Constrained and Unconstrained Systems*. Springer-Verlag, New York, 1978.
[41] D. M. Clark and K. Ravishankar. A convergence theorem for Grossberg learning. *Neural Networks* 3:87–92, 1990.
[42] R. A. Fisher. The use of multiple measurements in taxonomic problems. *Ann. Eugen.* 7:179–188, 1936.
[43] H. Karhunen. Über lineare Methoden in der Wahrscheinlichkeitsrechnung. *Ann. Acad. Sci. Fenn*, Ser. A. I., Vol. 37, Helsinki, 1947.
[44] M. Loève. Fonctions aleatoires de seconde ordre. In *Processus Stochastiques et Mouvement Brownien* (P. Levy, Ed.). Hermann, Paris, 1948.
[45] E. Oja and J. Karhunen. On stochastic approximation of the eigenvectors and eigenvalues of the expectation of a random matrix. *J. Math. Anal. Appl.* 106:69–84, 1985.

Learning Evaluation and Pruning Techniques

Leda Villalobos
Engineering School
University of Texas at El Paso
El Paso, Texas 79968-0521

Francis L. Merat
Electrical Engineering Department
Case Western Reserve University
Cleveland, Ohio 44106-7221

I. INTRODUCTION

The fundamental goal of supervised learning is to synthesize functions which capture the *underlying relationships* between input patterns and outputs of some particular task of interest. For learning to be truly satisfactory, these functions must provide good estimates of the outputs corresponding to input patterns not used during training. This ability is normally referred to as generalization.

Clearly, the architecture of a neural network—its layers, connectivity, and especially hidden neurons—defines the number of adjustable parameters available to synthesize the functions. Large networks have more parameters than small ones and, therefore, do better at fitting the training patterns. Too small networks may not even be able to bring the training error below some desired minimum value. However, if the number of parameters is far larger than needed, the network will actually learn the idiosyncrasies of the data, an effect known as *tuning to the noise* or *overfitting*, and will exhibit rather poor generalization.

It is widely accepted that good generalization results when the number of hidden neurons is close to the minimum required to learn representative training patterns with a small quadratic error. Hence, it is desirable to assess learning ability with respect to the training samples, and find a reduced architecture which ensures proper generalization.

Several strategies have been proposed to estimate upper and/or lower bounds on the number of hidden neurons required to learn specific tasks within the desired generalization accuracy. Many of these approaches are based on the seminal learning theory paper by Valiant [1], or apply the theory of large deviations in

Image Processing and Pattern Recognition
Copyright © 1998 by Academic Press. All rights of reproduction in any form reserved.

its uniform version by Vapnik and Chervonenkis. Not surprisingly, it has been usually found that a high generalization error is expected when the number of samples is small compared with the system's complexity.

Blumer *et al.* [2], for instance, relate the size of the learning system to the number of training samples. Along the same line of thought, Baum and Haussler [3] give theoretical upper and lower bounds to relate the number of training patterns to weights so that valid generalization can be expected. After making a few simplifying assumptions—such as considering neurons with linear threshold activation functions or Boolean valued—their derivations suggest the number of training samples should be roughly the number of weights divided by some accuracy parameter, ε. Hence, if $\varepsilon = 0.1$ (for a 90% generalization accuracy), the number of weights should be approximately 0.1 times the number of patterns. A similar rule of thumb had been previously suggested and proven to be effective [4].

Other methods are tailored to specific architectures, usually with only one hidden layer [5], or with particular conditions on the input data and the activation functions [6]. Igelnik and Pao [7] derived a lower bound for the size of the single hidden layer in the random functional-link net, a higher-order network [8]. This bound guarantees the training error will be smaller than some prescribed level. Sometimes, the practical usefulness of these methods is rather marginal. Arai [9], for example, found that a binary-valued network with a single hidden layer needs a maximum of $(P - 1)$ neurons to learn P patterns.

In any case, upper and lower bounds can only serve as initial guidelines in the construction of an effective network. Arriving at an optimal or near-optimal architecture usually requires sound heuristics and an iterative building process during training.

A. Simplifying Architecture Complexity

Techniques which iteratively build networks with good generalization properties and reduced architectures fall under two categories: *network growing* and *network pruning*.

In network growing, we start with an architecture which is smaller than needed to learn the task at hand. This architecture is then progressively increased during training, adding more neurons until the learning error falls below a specified threshold. Hirose *et al.* [10] start with a network that has only one hidden neuron. Then back-propagation (BP) is applied until the learning error does not decrease significantly any longer. At this point, a new hidden neuron is added to the architecture, and the cycle repeated. Eventually, there will be enough neurons to synthesize appropriate classification rules so that the error falls below a desired value. Subsequently, a reduction phase is initiated which consists in pruning one neuron at a time. After pruning a neuron, the network is retrained until it converges. If it does not, the algorithm stops and the last converging network is used. Other variations on the same theme include Zhang's SEL and SELF architectures [11], Fahlman and Lebiere's cascade correlation [12], Lee *et al.*'s [13] structure-level adaptation, and Lee *et al.*'s [14] separating hyperplanes incremental model.

In network pruning, we start with an architecture which is larger than the minimum needed for learning. Such architecture is then progressively reduced by pruning or weakening neurons and synaptic weights. Network pruning techniques and their application constitute the main focus of this chapter.

B. Applications

Architecture reduction paradigms have multiple applications. As will become apparent in future sections, some paradigms are better suited for certain applications than others. Consequently, algorithm selection should be the result of evaluating the problem at hand, as well as any hardware or computational limitations. In general, the following list comprises the most important applications of network pruning.

- *Improve generalization.* This is the primary aim of *all* pruning techniques and, as previously indicated, is an essential characteristic of any successfully prepared network.
- *Speed up on-line operation.* Obviously, smaller networks with few connections and neurons need less time to generate an output value after the input pattern is presented. This application is particularly important when the trained network becomes part of a real-time estimation or control system.
- *Reduce hardware requirements.* In commercial applications where the networks might have to be realized in hardware, product costs can be cut down by identifying reduced architectures.
- *Understand behavior in terms of rules.* In networks with reduced connectivity, it is easier to identify those features exerting the most effect on the output functions. Generally, those features propagate their effects to the output neurons through large weights. Consequently, it is rather easy to derive gross rules relating process features and outputs.
- *Evaluate and improve feature space.* Several pruning strategies, particularly constraint optimization and iterative pruning, can be readily applied to the assessment of feature spaces. This way, features deemed irrelevant or carrying limited information are eliminated, while features with higher information content are added. Improving feature space quality has an unmeasurable value: a pattern recognition problem cannot be solved without good feature representation.
- *Speed up learning.* At first glance, talking about learning time could appear to be irrelevant once a reduced architecture has been found. However, small net-

works usually show a strong dependency with the initial weights values during training. For this reason, sometimes it is necessary to retrain a reduced network. Clearly, smaller networks train faster than larger ones.

C. Outline

Many network pruning schemes have been proposed since the late 1980s. Based on the principles and heuristics they exploit, we have broadly grouped them into the following six categories.

- *Complexity Regularization.* The first type of pruning algorithms reported, complexity regularization, attempts to embed architecture simplification within the gradient descent training rule. The primary assumption is that complexity and weight magnitude are equivalent. Hence, the training rules are modified so that large weights are penalized. After training, connections with very small values can be pruned altogether, rendering a smaller network.
- *Sensitivity Estimation.* Magnitude is not always a good indicator of a weight's importance. Frequently, small weights provide highly desirable resistance to static noise by smoothing output functions. For this reason, their prevention could actually deteriorate performance. A more effective pruning approach consists in eliminating weights that show little effect on the synthesized outputs. Through heuristics and other simplifying assumptions, sensitivity methods estimate the relevance of a weight or neuron on the network's output, and deactivate elements with low relevance.
- *Optimization Through Constraint Satisfaction.* There are similarities between a supervised learning task and a resource allocation problem. The architecture's complexity can be treated as a limited resource, while the learning requirements set forth by the training patterns can be treated as constraints. Within this framework, performance depends on the network's ability to satisfy the constraints with its available resources. Constraint satisfaction techniques set up the resource allocation problem in ways which are appropriate for optimal pruning.
- *Bottlenecks.* It has been observed that generalization improves when a bottleneck—a hidden layer with significantly fewer neurons than previous layers—is imposed in the architecture. In these paradigms, bottlenecks are created during gradient descent training by compressing the dimensionality of the spaces spanned by the hidden layer's weights.
- *Interactive Pruning.* In trained networks, it is sometimes possible to identify neurons whose behavior duplicates other neurons, or which have limited discrimination ability. Interactive pruning techniques identify these neurons through heuristics, and eliminate them.
- *Other Methods.*

In the remaining sections, we discuss the fundamental principles behind each one of these categories. Similarly, we describe relevant algorithms appearing in the open literature. Our interest is not to give an exhaustive review of all available techniques, but rather to present in sufficient detail the most important ones under each category, as well as comment on their potential and limitations.

II. COMPLEXITY REGULARIZATION

As mentioned before, to improve generalization it is important to reduce the size or complexity of a network, so that the synthesized output function captures the essence of the training data rather than its idiosyncrasies. Since the purpose of the training session is to construct a nonlinear model of the phenomena originating the data, it is convenient and simple to include *as part of the training criteria* some measure of complexity. This way, pattern learning and complexity reduction can be accomplished simultaneously as part of the same process.

Complexity regularization was one of the first proposed paradigms aimed at reducing size. In its general form, it consists of adding to the usual quadratic error function, E_o, an extra penalty term E_1 that measures complexity. The result is a total error function E, which penalizes both training data misfit and complexity. Hence, the learning rule is obtained by applying gradient descent to

$$E(\underline{\mathbf{w}}) = E_o(\underline{\mathbf{w}}) + \lambda E_1(\underline{\mathbf{w}}), \tag{1}$$

where λ is the regularization parameter dictating the relative importance of E_o with respect to E_1. A large λ favors solutions with reduced complexity, while a small one gives more importance to accurate training pattern learning. It should be mentioned that λ can be modified as needed during training.

The procedures described in this section differ between each other by their penalty term(s), the selection or modification of the regularization parameter, and the training stopping criteria. However, all of them share a—sometimes unintended—association of complexity with weight magnitudes. As a consequence, they tend to reduce the weights' absolute values rather than the actual number of connections in the network's architecture.

A. WEIGHT DECAY

Weight decay (WD) [15, 16], the simplest complexity regularization method, directly equates network complexity to weight magnitude. The idea is to reduce connections by adding to the standard BP quadratic error function a term which penalizes large weights, so that

$$E = \sum_{p \in T} (t_p - o_p)^2 + \lambda \sum_{i \in C} w_i^2, \tag{2}$$

where t_p and o_p denote the desired and actual output for training pattern p, T and C represent the sets of all training patterns and all weights, and w_i is the ith weight.

Since the derivative of the complexity term is proportional to w_i, each weight will have a tendency to decay toward zero at a rate proportional to its magnitude. Such decay is only prevented by the reinforcement introduced through the gradient of the quadratic error term. Thus, the learning rule not only causes pattern learning but also favors weight magnitude reduction. Once training has finished, the weights can be grouped into two sets: one with relatively large values and influence on the network's performance, and another with small values. This latter set contains the so-called *excess weights*, which can be removed to avoid overfitting and improve generalization [15].

Krogh and Hertz [17] provide analytical arguments to explain why WD improves generalization. They conclude that improvement occurs for two reasons: (1) WD chooses the smallest vector which learns the training patterns, and (2) if properly selected, it can reduce some of the effects of static noise. However, appropriate dynamic modification of λ during training is of critical importance. It has been found that a poorly selected or constant λ could preclude learning and generalization altogether [18], while an adaptive one renders better results [19]. Also, it has been argued that a $|w|$ regularizer is more appropriate than w^2 [20].

B. WEIGHT ELIMINATION

A disadvantage of the penalty term included in Eq. (2) is that all weights are subjected to the same decaying criteria. Better performance can be obtained if biases are designed so that only weights within particular range values are affected [21]. Weight elimination (WE), proposed by Weigend *et al.* [22, 23], is a procedure which selectively biases weights. In WE the modified error function is given by

$$E = \sum_{p \in T} (t_p - o_p)^2 + \lambda \sum_{i \in C} \frac{w_i^2}{w_i^2 + w_o^2}, \tag{3}$$

where w_o is a preassigned weight-decay parameter. Thus, the importance of a weight depends on the magnitude of its value relative to w_o. As shown in Fig. 1, the complexity term approaches zero when $|w_i| \ll w_o$, and unity when $|w_i| \gg w_o$. Hence, the BP training will promote the appearance of small weights,

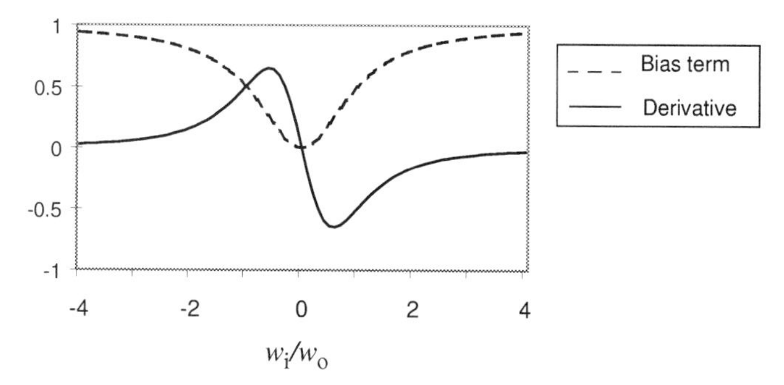

Figure 1 Bias complexity term in weight elimination and its derivative.

and penalize large ones. Note that WD is then a special case of WE: for large w_o, Eq. (3) reduces to Eq. (2), except for a scaling factor.

Just as with WD, here too performance is particularly sensitive to the selection of λ. A small λ lets the network exploit all of its weights and pay more attention to learning the training patterns; a large λ assigns more importance to the reduction of weight magnitude in an attempt to improve generalization. For this reason, a few heuristics have been derived to dynamically adjust λ after every epoch depending on the current value of the error over the training set, E_o. These adjustments are of three types: small increments, small decrements, and cut down.

Suppose E_n denotes the error after the nth epoch. Initially, training starts with $\lambda = 0$. To modify λ after every epoch, E_n is compared against three quantities: (1) the quadratic error in the previous epoch, E_{n-1}; (2) an exponentially weighted value of the error,

$$A_n = \gamma A_{n-1} + (1 - \gamma)E_n, \qquad \gamma \approx 1; \tag{4}$$

and (3) a desired minimum error, D.

If $E_n < D$ and/or $E_n < E_{n-1}$, it can be inferred that training is going well. Consequently, we proceed to increment λ by a small amount $\Delta\lambda$, on the order 10^{-6}. If $E_n \geq E_{n-1} \wedge E_n < A_n \wedge E_n \geq D$, the error is increasing but still improving with respect to the long-term average. Hence, we decrement λ by $\Delta\lambda$. If the new λ is negative, then $\lambda = 0$. Finally, if $E_n \geq E_{n-1} \wedge E_n \geq A_n \wedge E_n \geq D$, the error is definitely deteriorating; in an attempt to prevent weight elimination from permanently damaging the net, λ is now set to 0.9λ.

Weigend *et al.* [23] have used WE to predict yearly sunspot averages, and currency exchange rates. For sunspot series prediction the procedure rendered not only a smaller network (just three hidden neurons), but also one with half the out-of-sample error of the benchmark model by Tong [24]. Similarly, out-of-sample exchange rate predictions were much better than chance. In both cases, the resulting networks successfully ignored irrelevant information.

C. Smoothness Constraint Generalization

A broad class of ill-posed inverse problems describing physical phenomena—including early vision—have been regularized through *smoothness constraints*. In those cases, generalization is reduced to finding a solution function that smoothly interpolates between the training patterns. In a neural network, the output function is smooth when the weights are relatively small, but it can exhibit abrupt transitions if the weights are large. This is so due to the dependency between a sigmoidal neuron's response and its input weights.

Inspired by the good performance of smoothness constraints, Ji *et al.* [25] propose a complexity reduction algorithm which consists of modifying the quadratic error function with two heuristic terms, one to reduce the number of hidden neurons and the other to minimize the weights' magnitudes. The first term eliminates spurious local extrema in the output function, while the second one avoids unnecessary transitions.

The procedure is introduced in the context of a network with one linear input, one hidden layer of N sigmoidal neurons, and one linear output. Due to the architecture's simplicity, the input and output weights of the ith hidden neuron are denoted as u_i and v_i, respectively. A hidden neuron is assumed to be significant only when connected to the input and output through weights of large magnitude. Hence, the significance of the ith neuron could be quantified as

$$S_i = \sigma(u_i) \cdot \sigma(v_i), \tag{5}$$

where $\sigma(w_{ijk}) = w_{ijk}^2/(1 + w_{ijk}^2)$.

To procure solutions with few significant neurons, the term

$$E_1(\underline{\mathbf{w}}) = \lambda \sum_{i=1}^{N} \sum_{j=1}^{i-1} S_i S_j \tag{6}$$

is added to the quadratic learning error, E_o. Note how this modification only affects connection weights—it does *not* affect biases. After applying the gradient decent algorithm, the learning rule for *weights* then becomes

$$w_{ijk} = w_{ijk} - \eta \frac{\partial E_o}{\partial w_{ijk}}(\underline{\mathbf{w}}, \underline{\mathbf{b}}) - \lambda \frac{\partial E_1}{\partial w_{ijk}}(\underline{\mathbf{w}}), \tag{7}$$

with $\underline{\mathbf{w}}$, $\underline{\mathbf{b}}$ denoting the vectors of weights and biases, respectively.

Both gradients in Eq. (7) could be in conflict and thus originate spurious equilibria. To avoid this undesirable situation, it is convenient to have a dynamic λ which increases in value as the error E_o decreases. Ji *et al.* [25] use $\lambda = \lambda_o \exp(-\kappa E_o)$, where κ specifies the E_o value below which the neuron elimination term kicks in, and λ_o is on the order of 10^{-2}.

To reduce weight magnitudes, a small amount is subtracted from *both* weights and thresholds in every epoch, thus resembling weight decay. The amount subtracted is $\mu \tanh(w_{ijk})$, with, $\mu = \mu_o|\Delta E_o|$, μ_o on the order of 10^{-3}. This term seems to reduce the larger weights more effectively than other methods [25], and its effects diminish as convergence occurs.

Neuron pruning works as follows. Once an acceptable learning error E_o is reached, weights with small magnitudes are periodically eliminated as training progresses. After all weights connected to a particular neuron have been eliminated, the neuron itself is removed.

Simulation results showed this algorithm produces smoother response functions than standard BP, and architectures with fewer hidden neurons. This behavior is nevertheless obtained at the expense of a slower convergence rate.

D. Constrained Back-Propagation

The regularization methods described so far operate by expanding the error function with terms which directly penalize weight magnitude. On a more ambitious path, Chauvin [26–28] has proposed and tested a variety of penalty terms to reduce as well the magnitude of the hidden neurons' outputs over the training set. The underlying assumption is that a neuron's "energy"—how much its output changes across the training set—indicates relevance. Naturally, neurons carrying significant information have large energy values, while less important ones have little internal energy.

Constrained back-propagation (CBP), perhaps the most elaborate method explored by Chauvin [27], adds two terms to the usual error function. The first term reduces large weights just as WE, while the second term reduces the outputs of the hidden layers across the training set. The combined error function becomes

$$E = \eta \sum_{p \in T} (t_p - o_p)^2 + \lambda_w \sum_{i \in C} \frac{w_i^2}{w_i^2 + 1} + \lambda_o \sum_k \sum_p \frac{o_{kp}^2}{o_{kp}^2 + 1}. \tag{8}$$

Obviously, the second and third terms effectively introduce a *selective* parameter-decay force into the learning rule. For example, the gradient of the weight-dependent term approximates $2\lambda_w w_i$ for small weights. Hence, when a parameter's magnitude—either weight or output—is much smaller than unity, the learning dynamics will tend to decrease that parameter even more. The final result is not only a network with smaller weights but also one with possibly several inactive hidden neurons. These neurons can be identified and pruned as training progresses.

Chauvin extensively tested CBP in the difficult task of phonemic classification from spectrograms [28]. His results indicated that:

- Overfitting depends on both the size of the network and the number of training cycles. CBP basically eliminates overtraining despite long training times. Regardless of the original network size, the generalization performance remained approximately constant during the entire training session.
- With CBP the hidden neurons' energy rapidly decreases to a low level at the start of training. On the other hand, with BP the energy continues increasing, though slightly, throughout training.
- The learning error decreases more slowly in CBP than in the regular BP.

III. SENSITIVITY CALCULATION

According to several researchers including Mozer and Smolensky [29], and Hanson and Pratt [21], the penalty parameters in regularization methods are difficult to adjust, and it is often impossible to avoid local minima. Historically, this drawback motivated work into alternative pruning algorithms based on sensitivity analyses. The idea is that a neuron or a weight to which the output of a *trained* network is insensitive can be eliminated without much detriment to generalization performance. On the other hand, if the output's sensitivity is high, then this is an indication that the weight has captured important information contained in the training patterns. As a result, the weight or neuron should remain as part of the core—the *skeleton*—of the network.

In principle, calculating the sensitivity with respect to a weight is simple: just make that weight equal to zero, and then find the resulting increment in the error function E. If the increment is small, the weight can be pruned, and the architecture's complexity reduced. However, the problem with this brute force approach is the computational time required. In serial computers, a forward propagation of an input pattern takes $O(W)$ time, where W is the number of weights. Hence, assuming we have P training patterns and one single output, the time needed to make one pruning pass is $O(PW^2)$. Furthermore, since a weight's elimination affects other sensitivities, a more conservative algorithm would prune only one weight after each pruning pass. This would increase the time to $O(PW^3)$. Clearly, such an exhaustive approach is infeasible for all but the smaller networks.

The pruning methods presented in this section try to *approximate* sensitivity through more efficient means. They differ among each other in the way the approximation is formulated. Nevertheless, most of them share the following characteristics:

- They attempt to prune weights or neurons, rather than reducing their magnitudes. This represents a significant departure from the objective of regularization.
- Sensitivity is approximated from information which is available as training progresses.
- Actual pruning only takes place after some training, usually—but not always—until convergence. Thus, the error function is at a near local minimum with respect to the weights of the trained architecture.
- Some retraining or other weight modification is needed after pruning.
- Pruning can be repeated several times, until further architecture reduction starts deteriorating learning performance.

A. Neuron Relevance

The idea of pruning *neurons* rather than individual connections was first introduced by Mozer and Smolensky [29]. The underlying idea in their strategy is rather simple: iteratively train the network to a certain performance criterion, compute some meaningful functionality or *relevance* metric to quantify how important each neuron is, and then eliminate those neurons which are less relevant. The process can be repeated after a number of epochs, so the net is trimmed little by little.

Suppose we measure performance by calculating the quadratic error E over the training set. Conceptually the relevance ρ_{ik} of the ith neuron in the kth layer is the increment in error experienced as a result of eliminating that neuron. This is,

$$\rho_{ik} = E_{\text{without neuron}} - E_{\text{with neuron}}. \tag{9}$$

Since calculating E requires a complete pass on the training set, the cost of computing all relevances will be $O(NP)$, where N and P represent the number of neurons and training patterns, respectively. A more efficient solution can be obtained by finding an estimate $\hat{\rho}_{ik}$. To this end, the gating coefficient α_{ik} was introduced and a neuron's output expressed as

$$o_{j(k+1)} = f\left(\sum_i w_{ijk}\alpha_{ik}o_{ik}\right), \tag{10}$$

where f denotes the sigmoidal squashing function. By taking the derivative of the error function with respect to α_{ik}, and through some rather crude approximations, it can be shown that [29]

$$\hat{\rho}_{ik} \approx -\left.\frac{\partial E}{\partial \alpha_{ik}}\right|_{\alpha_{ik}=1}. \tag{11}$$

Thus, when $\hat{\rho}_{ik}$ falls below a certain small threshold, its corresponding neuron can be pruned. Since the error derivative fluctuates significantly in time, the exponentially decaying average

$$\hat{\rho}_{ik}(t+1) = 0.8\hat{\rho}_{ik}(t) + 0.2\frac{\partial}{\partial \alpha_{ik}}E(t) \tag{12}$$

is used instead. Also, even though the typical sum of squared errors is applied during training, the error function that measures relevance is the sum of absolute values of the errors. Hence, two separate error functions must be computed; this could be considered a disadvantage.

Segee and Carter [30] tested the fault tolerance of pruned networks trained to produce the sine value of inputs on the interval $[-\pi, \pi]$. Pruning consisted of computing relevances every 500 epochs, and eliminating the neuron with the lowest relevance. Training was stopped when the error reached a specified threshold. After training, fault tolerance was measured by calculating the increment in the RMS error over the training set which resulted from zeroing weights and neurons one at a time.

Three revealing results were found from this study. First, the algorithm basically eliminates neurons with small weights; the pruned networks did not have weights with small values. Second, not surprisingly the larger the magnitude of a *weight*, the higher its relevance. There is also a strong correlation between the relevance of a neuron and the magnitude of its largest weight. Finally, it was concluded that the pruned networks were not less fault tolerant than the unpruned ones.

B. Weight Sensitivity

As mentioned before, it is a disadvantage to have network training and relevance evaluation as separate processes. To eliminate this drawback, Karnin [31] proposed an improved version for pruning *weights* which does not require computation of two error functions. This way, both training and sensitivity (relevance) estimation take place simultaneously without interfering with one another. The algorithm is derived as follows.

Suppose that *after training*, the weight w_{ijk} was eliminated. The sensitivity of the error function to this pruning can be expressed as

$$S_{ijk} = -\frac{E(\mathbf{w}^f) - E(0)}{w_{ijk}^f - 0} w_{ijk}^f, \tag{13}$$

where $\mathbf{w}^f$ represents the collection of all synapses after training.

When training starts, synapses are initialized to some random, usually small value. Suppose the initial value of w_{ijk} is fairly small and given by w_{ijk}^i. Then the sensitivity can be approximated by

$$S_{ijk} \cong -\frac{E(\mathbf{w}^f) - E(\mathbf{w}^i)}{w_{ijk}^f - w_{ijk}^i} w_{ijk}^f, \tag{14}$$

in which $\mathbf{w}^i$ represents the weights after training, but with $w_{ijk} = w_{ijk}^i$. This approximation is advantageous because the difference in the numerator corresponds to the variation the error function experiences during *training* as a result of updating w_{ijk}, assuming all other synapses remain fixed at their final values. Consequently, the difference can be expressed as

$$E(\mathbf{w}^f) - E(\mathbf{w}^i) = \int_I^F \frac{\partial E(\mathbf{w})}{\partial w_{ijk}} dw_{ijk}, \tag{15}$$

where **I** and **F** are the initial and final points in weight space. This integral can, in turn, be approximated by a summation along the learning trajectory in weight space throughout the total number of epochs, R. Hence,

$$\hat{S}_{ijk} = -\sum_{r=0}^{R-1} \frac{\partial E(r)}{\partial w_{ijk}} \Delta w_{ijk}(r) \frac{w_{ijk}^f}{w_{ijk}^f - w_{ijk}^i}. \tag{16}$$

For implementation purposes, a general expression for the partial derivative of the error function with respect to the synapse should be found in terms of the training parameters (such as the gain factor η, or momentum β), and the synapse modifications. For example, if training takes place using the basic BP without momentum, the sensitivities would then be calculated with

$$\hat{S}_{ijk} = \sum_{r=0}^{R-1} [\Delta w_{ijk}(n)]^2 \frac{w_{ijk}^f}{\eta(w_{ijk}^f - w_{ijk}^i)}. \tag{17}$$

Note how all the data needed to compute $\hat{S}_{ijk}$ according to Eq. (16) would be available during training.

C. Optimal Brain Damage

Since the previous two pruning algorithms are based on the simplistic approximation of Eq. (11), they favor pruning small weights (or neurons with small weights.) As has been found by several researchers, this elimination criterion sometimes actually leads to sensible increments in the error function. As pointed out by Le Cun *et al.* [32], a more effective approach is to construct a local model of E to analytically predict the effects of eliminating *weights*.

Applying Taylor series, it is easy to show that a small perturbation $\delta\underline{\mathbf{w}}$ in the weights will produce a variation δE in the error function, with

$$\delta E = \left(\frac{\partial E}{\partial \underline{\mathbf{w}}}\right)^T \delta\underline{\mathbf{w}} + \frac{1}{2}\delta\underline{\mathbf{w}}^T \cdot \mathbf{H} \cdot \delta\underline{\mathbf{w}} + O(\|\delta\underline{\mathbf{w}}\|^3), \tag{18}$$

where $\mathbf{H} \equiv \partial^2 E/\partial\underline{\mathbf{w}}^2$ is the Hessian matrix with all second-order derivatives.

If the network has been trained to some local minimum, then the first term in Eq. (18) vanishes. By ignoring the third- and higher-order terms in the expansion, we get the simplified expression

$$\delta E = \tfrac{1}{2}\delta\underline{\mathbf{w}}^T \cdot \mathbf{H} \cdot \delta\underline{\mathbf{w}}. \tag{19}$$

To reduce the computational cost, Le Cun *et al.* [32] approximate the Hessian by its diagonal [33]. This approximation assumes the δE produced by changing several weights is equal to the sum of the δEs produced by changing each weight individually. Hence, the saliency of weight w_{ijk} can be computed as

$$\delta E_{ijk} = \hat{S}_{ijk} = \frac{1}{2} w_{ijk}^2 \frac{\partial^2 E}{\partial w_{ijk}^2}. \tag{20}$$

The resulting iterative pruning algorithm, normally referred to as Optimal Brain Damage (OBD), is as follows:

Step 1. Select a network with a reasonable architecture.
Step 2. Train the network to a local minimum or a satisfactory solution.
Step 3. Compute the saliencies according to Eq. (20).
Step 4. If weights with low saliencies are identified, prune them. Otherwise, stop.
Step 5. Iterate to step 2.

OBD has been used successfully in real-world applications not only to reduce network size but also to interactively find better architectures [32, 34].

D. OPTIMAL BRAIN SURGEON

OBD presents two drawbacks. First, after deleting a few weights, the network has to be retrained, increasing training time significantly. Second, and perhaps most importantly, using the Hessian's diagonal rather than the Hessian seems to cause incorrect pruning. Hassibi *et al.* [35, 36] have reported better generalization and size reduction with a variation of OBD called Optimal Brain Surgeon (OBS).

In OBD, retraining is required after pruning because the error E is no longer at a local minimum. OBS takes care of this inconvenience by providing a method to analytically determine the modifications $\delta\underline{\mathbf{w}}$ needed to bring E back to a minimum. Suppose a single weight in a trained network is to be selected for elimination. The *objective* of this selection will be to find that weight whose pruning minimizes the increment δE in Eq. (19). If this weight is represented by w_{ijk}, then pruning (setting it to zero) can be expressed as the *constraint*

$$\underline{\mathbf{e}}_{ijk} \cdot \delta\underline{\mathbf{w}} + w_{ijk} = 0, \tag{21}$$

where $\underline{\mathbf{e}}_{ijk}$ is the unit vector in weight space which corresponds to w_{ijk}. Hence, we have a constraint optimization problem, solvable with the Lagrangian

$$S = \tfrac{1}{2}\delta\underline{\mathbf{w}}^T \cdot \mathbf{H} \cdot \delta\underline{\mathbf{w}} + \lambda(\underline{\mathbf{e}}_{ijk} \cdot \delta\underline{\mathbf{w}} + w_{ijk}), \tag{22}$$

where λ is the Lagrangian multiplier.

After taking the derivative of S with respect to $\delta\underline{\mathbf{w}}$, applying Eq. (21), and some algebra, we find the optimum change in the weight vector is

$$\delta\underline{\mathbf{w}} = -\frac{w_{ijk}}{[\mathbf{H}^{-1}]_{ijk,ijk}}\left(\mathbf{H}^{-1} \cdot \underline{\mathbf{e}}_{ijk}\right), \tag{23}$$

while the saliency of w_{ijk} becomes

$$S_{ijk} = \frac{w_{ijk}^2}{2[\mathbf{H}^{-1}]_{ijk,ijk}}. \tag{24}$$

In general, the Hessian produced by BP is always nonsingular but almost rank-deficient. Nevertheless, Hassibi *et al.* [35] also present an elegant way to compute the inverse for a fully trained network, independently of training method. However, it should be pointed out that the computation requirements in OBS are more significant than in OBD.

Summarizing, the sequence of steps in the OBS pruning algorithm is as follows:

Step 1. Select a network with a reasonably large architecture.
Step 2. Train the network to a local minimum.
Step 3. Compute $\mathbf{H}^{-1}$.

Step 4. Find and delete the weight with smallest saliency by using Eq. (24). Otherwise, stop.
Step 5. Update all weights using Eq. (23).
Step 6. Iterate to step 3.

IV. OPTIMIZATION THROUGH CONSTRAINT SATISFACTION

There are similarities between a supervised learning task and a resource allocation problem. A network's architecture—number of layers, neurons, and activation function characteristics—can be treated as interrelated, limited resources which must satisfy the constraints set forth by the training patterns. Within this framework, performance depends on the network's ability to satisfy the constraints, and it can be readily measured through mathematical programming. Work on this area was first introduced during the 1960s, with the idea of deciding whether the pattern classes a perceptron had to learn were linearly separable [37, 38]. More recently, constraint optimization has been exploited to solve more challenging problems, such as network pruning and feature space optimization [39, 40].

Learning performance is usually assessed with the quadratic error E: training is normally considered successful if E falls below a small, *nonzero* value. This implicit discrepancy tolerance acts as an inequality constraint, and can be exploited to find optimal architectures and feature spaces.

To show the effect the tolerance has on the training process, consider a particular training pattern with K features $(f_i;\ i = 1, 2, \ldots, K)$, and one desired output t_p. Assume the allowed discrepancy tolerance is specified by an upper bound δ_+ and a lower bound δ_-. In such a case, the network would have learned the pattern if its actual output falls in the range $[t_p - \delta_-, t_p + \delta_+]$. Hence, the pattern is learned if the constraints $O \geq (t_p - \delta_-)$ and $O \leq (t_p + \delta_+)$ are satisfied. Since it is possible to specify tolerance levels for every pattern, this procedure can be extended to include all patterns used to train the network. Consequently, learning can be posed as an inequality constraint satisfaction problem.

A. CONSTRAINTS IN HIGHER-ORDER NETWORKS

For simplicity, we concentrate our work on the class of higher-order networks shown in Fig. 2, which are universal approximators [41] that have been used in pattern recognition [42], character recognition [43], and system identification and control [44] applications. Although focused on higher-order networks, our analysis can be extended to other architectures.

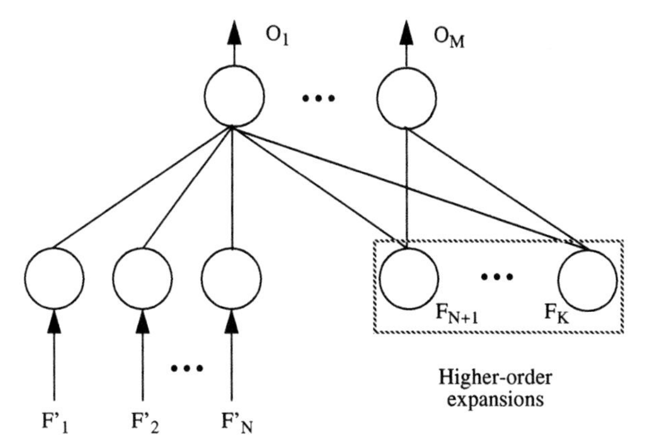

Figure 2 Architecture of a higher-order network. The original feature vector has N components. Through nonlinear transformations, higher-order terms are added to the original feature vector, effectively expanding it.

The mapping learned by the architecture is given by

$$\begin{aligned} O_m &= g(\text{net}_m), \\ \text{net}_m &= \sum_{j=1}^{N} W_{jm} F'_j + \sum_{i=1}^{K-N} W_{im} h_i(\underline{\mathbf{F}}') + \theta_m, \end{aligned} \tag{25}$$

where O_m is the mth output, W_{im} is the weight from the ith feature to the mth output neuron, $g(\cdot)$ is the output neuron's activation function, θ is the output neuron's bias, F'_j is the jth element of the original $N - D$ feature vector $\underline{\mathbf{F}}'$, and $h_i(\cdot)$ is the ith higher-order feature expansion function.

To simplify the analysis, we concatenate the original feature vector $\underline{\mathbf{F}}'$ with the higher-order feature expansions and create the $K - D$ vector $\underline{\mathbf{F}}$. Thus, the mth output can be expressed as

$$O_m = g\left(\sum_{j=1}^{K} W_{jm} F_j + \theta_m\right). \tag{26}$$

Suppose there are P training patterns of K features (including the higher-order expansions), and one output. Their information can be captured in a $P \times K$ matrix (**F**) with the feature vectors, and a $P - D$ vector $\underline{\mathbf{T}}$ with the desired outputs. These data can be learned with the desired accuracy if there exists at least one $K - D$

vector $\underline{\mathbf{W}}$, and a scalar θ which satisfy

$$\begin{bmatrix} \mathbf{F} & 1 \\ \mathbf{F} & 1 \end{bmatrix} \begin{bmatrix} \underline{\mathbf{W}} \\ \theta \end{bmatrix} = \begin{bmatrix} \leq g^{-1}(\underline{\mathbf{T}} + \underline{\delta}_{+}) \\ \geq g^{-1}(\underline{\mathbf{T}} - \underline{\delta}_{-}) \end{bmatrix}, \tag{27}$$

where $\underline{\delta}_{+}$ and $\underline{\delta}_{-}$ are $P - D$ vectors with the upper and lower bound tolerances, and $g^{-1}(\cdot)$ is the inverse of the activation function. According to Eq. (27), successful learning occurs when the set of solutions of the linear inequality constraints is nonempty.

B. Linear Programming Formulations

Linear inequality constraints like those of Eq. (27) are solved with linear programming (LP) algorithms such as the Simplex method [45]. Simplex operates in two stages. First, it finds out whether the constraints have a nonempty set of feasible solutions; then, if there is at least one feasible solution, the algorithm searches the space of feasible solutions guided by an objective function.

LP algorithms require the variables in the inequalities to take only nonnegative values. Hence, a variable without sign restrictions must be expressed as the difference of two nonnegative variables [46]. For this reason, the formulation of Eq. (27) has to be rewritten as

$$\begin{bmatrix} \mathbf{F} & 1 \\ \mathbf{F} & 1 \end{bmatrix} \begin{bmatrix} \underline{\mathbf{W}}_A - \underline{\mathbf{W}}_B \\ \theta_A - \theta_B \end{bmatrix} = \begin{bmatrix} \leq g^{-1}(\underline{\mathbf{T}} + \underline{\delta}_{+}) \\ \geq g^{-1}(\underline{\mathbf{T}} - \underline{\delta}_{-}) \end{bmatrix}, \tag{28}$$

where $\underline{\mathbf{W}}_A$ and $\underline{\mathbf{W}}_B$ are $K - D$ vectors of nonnegative variables and $\underline{\mathbf{W}}_A - \underline{\mathbf{W}}_B = \underline{\mathbf{W}}$. Similarly, θ_A and θ_B are nonnegative variables such that $\theta_A - \theta_B = \theta$.

The solution of Eq. (28) indicates whether the patterns can be learned with the desired accuracy. Should learning be possible, feasible solutions will be identifiable which correspond to connectivities satisfying all the accuracy constraints. Otherwise, the set of feasible solutions will be empty, prompting Simplex to stop after its first phase.

If not all patterns can be learned with the desired accuracy, the formulation would not be appropriate for identifying any of the nonlearnable patterns. To address this problem, we modify the constraints such that they have a default feasible solution. With this modification, we can define an objective function such that the optimum feasible solution indicates which patterns are nonlearnable.

Introducing a default feasible solution We guarantee the existence of a default feasible solution by introducing "pad variables," one for every pattern, into the LP formulation.

Let S_i be the pad variable associated with the ith pattern. S_i is allowed to take any real number and so we express it as the difference of the nonnegative variables S_{Ai} and S_{Bi}. We introduce the pad variables in the formulation as shown in Eq. (29):

$$\begin{bmatrix} \mathbf{F} & 1 & \mathbf{I} \\ \mathbf{F} & 1 & \mathbf{I} \end{bmatrix} \begin{bmatrix} \underline{\mathbf{W}}_A - \underline{\mathbf{W}}_B \\ \theta_A - \theta_B \\ \underline{\mathbf{S}}_A - \underline{\mathbf{S}}_B \end{bmatrix} = \begin{bmatrix} \leq g^{-1}(\underline{\mathbf{T}} + \underline{\delta}_+) \\ \geq g^{-1}(\underline{\mathbf{T}} - \underline{\delta}_-) \end{bmatrix}, \tag{29}$$

with $\mathbf{I}$ the $P \times P$ identity matrix. Clearly, the default feasible solution consists of making $\underline{\mathbf{W}}$ and θ equal to zero, and assigning each pad variable a value which falls in the range of satisfactory learning. Should one or more of the feasible solutions have all pad variables equal to zero, the learning task would be feasible. If none of the feasible solutions has all pad variables equal to zero, satisfactory learning would not be possible.

Specifying an optimization criterion Among all the feasible solutions for Eq. (29), one is particularly informative: the feasible solution with the largest number of zeroed pad variables. Let this solution be $C^* = [\underline{\mathbf{W}}^* \theta^* \underline{\mathbf{S}}^*]$. Then, the patterns whose associated pad variables appear zeroed in C^* form the largest set of patterns the structure can learn with the desired accuracy. For this statement to hold true, every learnable pattern must have its pad variable zeroed in C^*, *and* every nonlearnable pattern must have its pad variable different from zero. From here, it follows that our optimization criterion should be the minimization of the sum of nonzero pad variables. After some work [40], the complete LP formulation can be expressed as

$$\text{Objective Function} = \text{Min} \sum_{i=1}^{P} H_i,$$

subject to

$$\begin{bmatrix} \mathbf{F} & 1 & \mathbf{I} & \mathbf{0} & \mathbf{0} \\ \mathbf{F} & 1 & \mathbf{I} & \mathbf{0} & \mathbf{0} \\ \mathbf{0} & 0 & \mathbf{0} & \mathbf{I} & -\mathbf{L} \end{bmatrix} \begin{bmatrix} \underline{\mathbf{W}}_A - \underline{\mathbf{W}}_B \\ \theta_A - \theta_B \\ \underline{\mathbf{S}}_A - \underline{\mathbf{S}}_B \\ \underline{\mathbf{S}}_A + \underline{\mathbf{S}}_B \\ \underline{\mathbf{H}} \end{bmatrix} = \begin{bmatrix} \leq g^{-1}(\underline{\mathbf{T}} + \underline{\delta}_+) \\ \geq g^{-1}(\underline{\mathbf{T}} - \underline{\delta}_-) \\ \leq 0 \end{bmatrix}, \tag{30}$$

where L_i is a sufficiently large upper bound for $|S_i|$, and H_i is an integer $0/1$ variable used to test whether S_i is different from zero.

Remarks. The solution of Eq. (30) provides us with the following information:

- It indicates whether the network can effectively learn the training patterns with the desired level of accuracy.

- If the structure is appropriate for learning the training patterns, the solution gives a connectivity which corresponds to satisfactory learning.
- If the network is not capable of learning all the patterns, one or more of the integer variables will remain nonzero. The patterns whose associated pad variables are nonzero form the smallest set of nonlearnable patterns.

C. Optimizing Feature Space

If we have confirmed that a particular structure is appropriate for learning the information contained in the training patterns, the next step would be to identify those features, if any, which can be eliminated from the feature space without diminishing performance.

To explain our feature space pruning technique, let us assume the jth feature in the feature space can be eliminated. This implies the jth connection weight W_j can be made equal to zero in a feasible solution, which means that

$$|W_j| = W_{Aj} + W_{Bj} = 0. \tag{31}$$

It is possible to test whether W_j can be made equal to zero following a procedure similar to the one used to test pad variables [40]. The only difference is that the objective function should now be the minimization of the number of connection weights different from zero in the optimum solution. If the integer variable Q_j is used for testing W_j, our LP formulation becomes

$$\text{Objective Function} = \text{Min} \sum_{j=1}^{K} Q_j,$$

subject to

$$\begin{bmatrix} \mathbf{F} & 1 & \mathbf{0} & \mathbf{0} \\ \mathbf{F} & 1 & \mathbf{0} & \mathbf{0} \\ \mathbf{0} & 0 & \mathbf{I} & -\mathbf{L} \end{bmatrix} \begin{bmatrix} \underline{\mathbf{W}}_A - \underline{\mathbf{W}}_B \\ \theta_A - \theta_B \\ \underline{\mathbf{W}}_A + \underline{\mathbf{W}}_B \\ \underline{\mathbf{Q}} \end{bmatrix} = \begin{bmatrix} \leq g^{-1}(\underline{\mathbf{T}} + \underline{\delta}_+) \\ \geq g^{-1}(\underline{\mathbf{T}} - \underline{\delta}_-) \\ \leq 0 \end{bmatrix}, \tag{32}$$

where $\underline{\mathbf{Q}}$ and $\underline{\mathbf{L}}$ are $K - D$ vectors of integer variables and constant values, respectively.

V. LOCAL AND DISTRIBUTED BOTTLENECKS

Researchers report generalization improvement when a *bottleneck*—a hidden layer with significantly fewer neurons than previous layers—is imposed on the architecture of a network. Some methods actually introduce localized bottlenecks, either through weight or neuron deactivation. Kruschke [47, 48] has argued that this hardware minimization presents some disadvantages in terms of noise and damage resistance. As an alternative, he proposes a duplication of the bottleneck's *functional properties*—particularly complexity and weight-space dimension compression—without the actual hardware reduction.

Consider the two consecutive layers $k-1$ and k, with B and A neurons, respectively. The weights connecting layer $k-1$ to layer k can be arranged in the $B \times A$ matrix $\mathbf{W}^s$, of rank R. If $R < A$, layer k forms a bottleneck. Additionally, if $B = R$, the bottleneck is *local*, while if $B > R$, it is *distributed*. Hence, to improve generalization we want to decrease the functional dimensionality of R, and decrease the number of neurons B in layer k. This corresponds to compressing the weight space, and clustering the weights within that space. Shepard [49] described an algorithm to accomplish both objectives. It is based on increasing the variance of the distances between weights by further stretching large distances and reducing small ones.

Suppose $\mathbf{w}_{jk} = [w_{1jk} w_{2jk} \ldots w_{Ajk}]$ is the vector with the weights connecting layer $k-1$ to node j in layer k. The Euclidean distance between vectors $\mathbf{w}_{jk}$ and $\mathbf{w}_{ik}$ is $d_{ijk} = \|w_{jk} - w_{ik}\|$, which has a mean value d_k. We can define a cost function proportional to the variance of the distances, say

$$D = -\tfrac{1}{4} \sum_{i=1}^{B} \sum_{j=1}^{B} (d_{ijk} - \bar{d}_k)^2, \tag{33}$$

which produces the gradient descent

$$\Delta \mathbf{w}_{ik} = -\frac{\partial D}{\partial \mathbf{w}_{ik}} = \lambda \sum_{j=1}^{B} (d_{ijk} - \bar{d}_k) \cdot \frac{(\mathbf{w}_{ik} - \mathbf{w}_{jk})}{d_{ijk}}. \tag{34}$$

Thus, *after* every standard BP epoch, the weights have to be modified according to Eq. (34).

Although promising, this procedure requires nonlocal computations. An easy way to improve it consists in redefining the distance so that now

$$d_{ijk} = -\sum_{p \in T} \mathrm{net}_{ikp} \cdot \mathrm{net}_{jkp}, \tag{35}$$

where net_{ikp} represents the net input to neuron i for pattern p [50]. Recall that $\mathrm{net}_{ik} = \langle \mathbf{w}_{ik}, \mathbf{o}_{k-1} \rangle$. Using this distance in Eq. (33), and applying gradient decent, we get

$$\Delta \mathbf{w}_{ik} = -\lambda \sum_{j=1}^{B} (d_{ijk} - \bar{d}_k) \cdot \sum_{p \in T} \mathrm{net}_{jkp} \cdot \mathbf{o}_{(k-1)p}. \tag{36}$$

Substantial simplification occurs when we make $d_k = 0$, for example by including the magnitude of the weight vectors as part of the error function in BP [50]. As a side effect, this also prevents the variance of the weights from growing too large. Just as in some previously discussed procedures [22, 25], selection of the parameter λ is critical. With a very large λ, all weight vectors will collapse into two antiparallel directions, effectively acting like one neuron. A solution is to dynamically modify λ. When learning is going well, λ can be increased; otherwise, it is decreased and even reversed in sign.

VI. INTERACTIVE PRUNING

Interactive pruning strategies work by training a somewhat oversized network up to a local minimum, and then heuristically identifying and pruning *redundant* hidden neurons. Once pruning takes place, the skeletonized network is trained again to a local minimum.

A. NEURON REDUNDANCY

Sietsma and Dow [51, 52] have proposed an *interactive* off-line pruning procedure to eliminate hidden *neurons* in trained networks. It is based on heuristics carried out in two steps. The first phase identifies and prunes redundant neurons whose outputs remain nearly constant across the training set, or mimic the outputs of other neurons. To some extent, this resembles one of the objectives of bottlenecks [48], namely, to group together neurons whose weights are parallel or antiparallel. The main difference resides in that grouping here takes place interactively and after training.

Suppose the hidden output $o_{i(k-1)}$ falls within the range $(o \pm \delta o)$ across the training set, where δo is fairly small. Then, its respective neuron can be eliminated, and its effects compensated for, by modifying the biases of the neurons on the kth layer according to

$$b_{jk} = b_{jk} + w_{ijk}o. \tag{37}$$

Similarly, if the hidden output $o_{i(k-1)}$ is approximately the same as $o_{m(k-1)}$ across the training patterns, then one of the two neurons can be eliminated. When the pruned output is $o_{m(k-1)}$, then all weights w_{ijk} originating from $o_{i(k-1)}$ have to be modified so that

$$w_{ijk} = w_{ijk} + w_{mjk}. \tag{38}$$

It is also possible to find $o_{i(k-1)} \approx 1 - o_{m(k-1)}$. In this case, the elimination of $o_{m(k-1)}$ is done by modifying the biases and weights of all neurons fed by $o_{m(k-1)}$

and $o_{i(k-1)}$:

$$b_{jk} = b_{jk} + w_{mjk}, \tag{39}$$

$$w_{ijk} = w_{ijk} - w_{mjk}. \tag{40}$$

The second pruning phase is aimed at identifying and removing neurons which, at the level of their respective hidden layer, do not contribute to the separation of pattern classes. These neurons are considered as transmitting unnecessary information to the next layer [52]. Such a form of pruning could lead to the outputs of the trimmed layer being linearly inseparable with respect to the classes of the *following* layer. To deal with this problem, a technique for adding more layers has been proposed. It consists in training a small network to receive the outputs of the trimmed layer as inputs and produce the outputs of the following layer, and then inserting it into the original network. As a result, the pruned networks are narrow and have many layers.

Sietsma and Dow conducted several tests to evaluate the hypothesis that narrow, many-layered networks generalize better than broad, shallow ones [53]. Networks were first trained with patterns contaminated with different levels of noise, and then pruned. These tests showed that [51]:

- Better generalization and more hidden neuron utilization occur when the training patterns are noisy. This happens because the noise smears the basins of attraction, making overfitting more difficult.
- Generalization deteriorates when the networks are trimmed to the smallest possible size during the second pruning phase. This observation indicates there are circumstances when minimum size is not a guarantee of better performance. However, it is important to keep in mind that there was no extra training after the rather crude neuron elimination. Consequently, this result cannot be extrapolated to other pruning procedures.
- The long and narrow networks performed poorly. However, there is no reason to believe this result will apply to other algorithms.

B. Information Measure

Information measure (IM), an indicator of how well a feature discriminates between members of different classes [54], has been used in several decision tree induction schemes. Basically, it measures the entropy reduction attained by knowing the value of a given feature attached to the classes. In a particular pattern recognition problem, the idea is to select features with high IMs, because they define a good discriminant. Consider for instance the case illustrated in Fig. 3, where two features (f_1 and f_2) help define two linear discriminant functions. The function obtained with f_2 separates the classes very well, far better than the func-

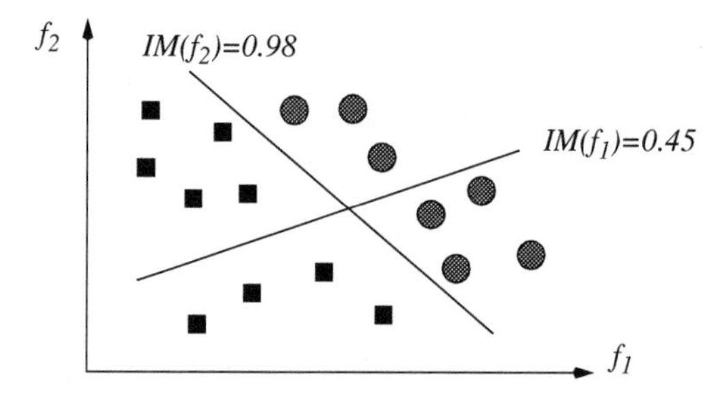

Figure 3 Feature discriminating power and information measure (IM). A feature's IM is related to the quality of the class discriminants it generates; good discriminants receive high IMs, while poor ones receive low IMs. Because of the little information they convey, features with low IMs can be eliminated.

tion obtained with f_1. Consequently, $IM(f_2)$ takes a large value, while $IM(f_1)$ takes a fairly small one. Of course, f_1 could be eliminated from the feature set used to describe the classes.

Ramachandran and Pratt [55] have proposed a technique to prune already trained networks by estimating the hidden neurons' IMs. These estimates are calculated by first thresholding the output of each neuron for every training pattern. If the actual output is above 0.5, the neuron is assumed to have a 1.0 output. Otherwise, the output is assumed to be 0.0. This thresholding makes it possible to easily estimate IMs [54]. It is also possible to make a multivalue thresholding, thus producing a more accurate estimate of a neuron's true significance.

After thresholding, the hidden neurons are treated as discrete-valued features, and the idea is to figure out whether one or more of them are either redundant or have limited discrimination ability. Neurons with little discrimination power have small IMs and can be eliminated without inflicting significant damage to the architecture's classification potential. On the other hand, important neurons have large IMs and should remain as part of the skeleton. Of course, after pruning, the resulting network has to be retrained.

VII. OTHER PRUNING METHODS

Even though the paradigms described so far are probably the most important ones, many others have been reported in the open literature. Some of them use, for example, Boltzmann methods [56], sequential function estimation [57], switching theory [58], and class entropy [59]. In this section, we consider techniques grounded on genetic algorithms and evolutionary programming.

A. Genetic Algorithms

The main idea in this case consists in defining a *parent* network whose complexity is sufficient to learn the task of interest, and then applying genetic algorithms so as to generate smaller *offspring* which can still learn the required information. It should be pointed out that the goal is not necessarily to obtain a *particular* pruned network—called a phenotype, but rather an architecture prototype—the genotype [60]. This is important because the actual performance of a phenotype is initial-weight dependent, while the performance of a genotype is not.

Miller *et al.* [61] present a very simple approach in which an untrained network functions as parent, and the genetic operators swap functional substructures during recombination. The crossover operator swaps all the links leading into some node. The offspring is then trained for a fixed number of cycles, and its genetic quality is measured by the final training error. The offspring with the lowest error would be the final pruned network.

Obviously, two drawbacks plague this method. First, although it performs very well with small problems, the computational time in more complex ones becomes truly significant. Lacking a mechanism to favor the generation of some networks over others, the genetic algorithm has to evaluate all possible offspring. With a parent that has 50 connections, for example, retraining could be needed for perhaps 2000 different networks, or more [62]. Second, there is no effective reward assigned to smaller networks; each offspring is trained the same number of cycles. The result is that larger networks have more opportunity to get lower training errors, which reduces pruning potential.

Witley and Bogart [62] propose a more refined approach which takes care of these two issues. In their method, the parent is an already trained network and its weights are assigned to each offspring. Consequently, offspring retraining is much faster. Also, the number of allowed training cycles increases linearly with the number of pruned weights. This operates as a reward given to the smaller, leaner offspring. Additionally, instead of generating all possible pruned versions of a parent, more reproductive trials are assigned to the networks which got smaller training errors. This increases the probability of generating very good offspring early in the process.

B. Evolutionary Programming

Evolutionary programming is a global optimization paradigm through systematically stochastic search. Applied to neural architecture, the search can be applied for various purposes: to reduce the number of weights and/or neurons, find appropriate weight values, or guide architecture enhancement by adding extra neurons during network growing [12, 13].

Within the scope of pruning, McDonnell and Waagen [63] present three strategies in which stochastic search simultaneously finds the weights and the number of hidden neurons. Weights are stochastically modified through Gaussian mutations proportional to the learning error, while the architecture is modified using the standard deviation of the neurons' activation over all the training patterns. Their results suggest that smaller networks can be obtained by artificially constraining the search.

VIII. CONCLUDING REMARKS

By eliminating the chance of pattern overfitting, pruning techniques are of fundamental importance to improving the generalization capabilities of neural networks. In this chapter, we have described the principles underlying a variety of pruning paradigms such as complexity regularization, sensitivity analysis, constraint optimization, iterative pruning, and others.

As we have explained, no one paradigm or algorithm gives optimal results for all learning tasks and applications. For example, if the main concern is to obtain efficient compact networks for hardware implementation or real-time estimation, then architecture optimizing algorithms are probably most appropriate despite their stronger training computation requirements. On the other hand, if derivation of rules relating features and outputs is more important, then complexity regularization methods could be better suited. Similarly, if the goal is to identify irrelevant input features to improve feature space, then optimization or iterative pruning are more effective. In summary, algorithm selection must be tailored to the application being considered.

REFERENCES

[1] L. G. Valiant. A theory of the learnable. *Commun. ACM* 27:1134–1142, 1984.

[2] A. Blumer, A. Ehrenfeucht, D. Haussler, and M. K. Warmuth. Learnability and the Vapnik-Chervonenkis dimension. *J. Assoc. Comput. Machinery* 36:929–965, 1989.

[3] E. B. Baum and D. Haussler. What size net gives valid generalization?. *Neural Computat.* 1:151–160, 1989.

[4] B. Widrow. ADALINE and MADALINE—1963. In *Proceedings of the IEEE First International Conference on Neural Networks*, Vol. I, pp. 143–158. San Diego, CA, 1987.

[5] T. Onoda. Neural network information criterion for the optimal number of hidden units. In *IEEE International Conference on Neural Networks*, Vol. 1, pp. 275–280. Perth, Australia, 1995.

[6] B. Amirikian and H. Nishimura. What size network is good for generalization of a specific task of interest?. *Neural Networks* 7:321–329, 1994.

[7] B. Igelnik and Y.-H. Pao. Estimation of size of hidden layer on basis of bound of generalization error. In *IEEE International Conference on Neural Networks*, Vol. 4, pp. 1923–1927. Perth, Australia, 1995.

[8] Y.-H. Pao. *Adaptive Pattern Recognition and Neural Networks*. Addison-Wesley, Reading, MA, 1989.

[9] M. Arai. Bounds on the number of hidden units in binary-valued three-layer neural networks. *Neural Networks* 6:855–860, 1993.

[10] Y. Hirose, K. Yamashita, and S. Hijiya. Back-propagation algorithm which varies the number of hidden units. *Neural Networks* 4:61–66, 1991.

[11] B.-T. Zhang. An incremental learning algorithm that optimizes network size and sample size in one trial. In *Proceedigns of the IEEE International Conference on Neural Networks*, Vol. I, pp. 215–220. Orlando, FL, 1994.

[12] S. E. Fahlman and C. Lebiere. The cascade-correlation learning architecture. In *Advances in Neural Information Processing II* (D. S. Touretzky, Ed.), pp. 524–532. Morgan Kaufmann, San Mateo, CA, 1990.

[13] T.-C. Lee, A. M. Peterson, and J.-C. Tsai. A multi-layer feed-forward neural network with dynamically adjustable structures. In *IEEE International Conference on System, Man, and Cybernetics*, pp. 367–369. Los Angeles, 1990.

[14] J. C. Lee, Y. H. Kim, W. D. Lee, and S. H. Lee. A method to find the structure and weights of layered neural networks. In *Proceedings of the International World Conference on Neural Networks*, Vol. 3, pp. 552–555. Portland, OR, 1993.

[15] D. C. Plaut, S. J. Nowlan, and G. E. Hinton. Experiments on learning by back propagation. Technical Report CMU-CS-86–126, Carnegie-Mellon University, 1986.

[16] G. E. Hinton. Connectionist learning procedures. *Artif. Intell.* 40:185–234, 1989.

[17] A. Krogh and J. A. Hertz. A simple weight decay can improve generalization. In *Advances in Neural Information Processing IV* (J. Moody, S. J. Hanson, and R. P. Lippmann, Eds.), pp. 951–957. Morgan Kaufmann, San Mateo, CA, 1992.

[18] F. Hergert, W. Finnoff, and H. G. Zimmermann. A comparison of weight elimination methods for reducing complexity in neural networks. In *Proceedings of the International Joint Conference on Neural Networks*, Vol. 3, pp. 980–987. San Diego, CA, 1992.

[19] L. K. Hansen and C. E. Rasmussen. Pruning from adaptive regularization. *Neural Computat.* 6:1223–1232, 1994.

[20] P. Williams. Bayesian regularization and pruning using a laplace prior. *Neural Computat.* 7:117–143, 1995.

[21] S. J. Hanson and L. Y. Pratt. Comparing biases for minimal networks construction with back-propagation. In *Advances in Neural Information Processing I* (D. S. Touretzky, Ed.), pp. 177–185. Morgan Kaufmann, San Mateo, CA, 1989.

[22] A. S. Weigend, D. E. Rumelhart, and B. A. Huberman. Generalization by weight-elimination applied to currency exchange rate prediction. In *Proceedings of the International Joint Conference on Neural Networks*, Vol. I, pp. 837–841. Seattle, WA, 1991.

[23] A. S. Weigend, D. E. Rumelhart, and B. A. Huberman. Generalization by weight-elimination with application to forecasting. In *Advances in Neural Information Processing III* (R. P. Lippmann, J. Moody, and D. S. Touretzky, Eds.), pp. 875–882. Morgan Kaufmann, San Mateo, CA, 1991.

[24] H. Tong. *Non-linear Time Series: A Dynamical System Approach*. Oxford University Press, New York/London, 1990.

[25] C. Ji, R. R. Snapp, and D. Psaltis. Generalization smoothness constraints from discrete samples. *Neural Computat.* 2:188–197, 1990.

[26] Y. Chauvin. A back-propagation algorithm with optimal use of hidden units. In *Advances in Neural Information Processing I* (D. S. Touretzky, Ed.), pp. 519–526. Morgan Kaufmann, San Mateo, CA, 1989.

[27] Y. Chauvin. Dynamic behavior of constrained back-propagation networks. In *Advances in Neural Information Processing II* (D. S. Touretzky, Ed.), pp. 642–649. Morgan Kaufmann, San Mateo, CA, 1990.

[28] Y. Chauvin. Generalization performance of overtrained back-propagation networks. In *Neural Networks, Proceedings of the EUROSIP Workshop* (L. B. Almeida and C. J. Wellekens, Eds.), pp. 46–55. Springer-Verlag, Berlin/New York, 1990.

[29] M. C. Mozer and P. Smolensky. Skeletonization: A technique for trimming the fat from a network via relevance assessment. In *Advances in Neural Information Processing I* (D. S. Touretzky, Ed.), pp. 107–115. Morgan Kaufmann, San Mateo, CA, 1989.

[30] B. E. Segee and M. J. Carter. Fault tolerance of pruned multilayer networks. In *Proceedings of the International Joint Conference on Neural Networks*, Vol. II, pp. 447–452. Seattle, WA, 1991.

[31] E. D. Karnin. A simple procedure for pruning back-propagation trained neural networks. *IEEE Trans. Neural Networks* 1:239–242, 1990.

[32] Y. Le Cun, J. Denker, and S. A. Solla. Optimal Brain Damage. In *Advances in Neural Information Processing II* (D. S. Touretzky, Ed.), pp. 598–605. Morgan Kaufmann, San Mateo, CA, 1990.

[33] Y. Le Cun. Generalization and network design strategies. In *Connectionism in Perspective* (R. Pfeifer, Z. Schreter, F. Fogelman, and L. Steels, Eds.). Elsevier, Zurich, 1989.

[34] Y. Le Cun, B. Boser, J. Denker, D. Henderson, R. E. Howard, W. Hubbard, and L. D. Jackel. Back-propagation applied to handwritten zip code recognition. *Neural Computat.* 1:541–551, 1989.

[35] B. Hassibi and D. G. Stork. Second order derivatives for network pruning: Optimal Brain Surgeon. In *Advances in Neural Information Processing V* (S. J. Hanson, J. D. Cowan, and C. L. Giles, Eds.), pp. 164–171. Morgan Kaufmann, San Mateo, CA, 1993.

[36] B. Hassibi, D. G. Stork, and G. Wolff. Optimal Brain Surgeon: Extensions and performance comparisons. In *Advances in Neural Information Processing VI* (D. S. Touretzky, Ed.), pp. 263–270. Morgan Kaufmann, San Mateo, CA, 1994.

[37] F. W. Smith. Pattern classifier design by linear programming. *IEEE Trans. Computers* C-17:367–372, 1968.

[38] O. L. Mangasarian. Linear and nonlinear separation of patterns by linear programming. *Oper. Res.* 444–452, 1965.

[39] P. Rujan. A fast method for calculating the perceptron with maximal stability. *J. Phys. I* 3:277–290, 1993.

[40] L. Villalobos and F. L. Merat. Learning capability assessment and feature space optimization for higher-order neural networks. *IEEE Trans. Neural Networks* 6:267–272, 1995.

[41] K. Hornik, M. Stinchcombe, and H. White. Multilayer feedforward networks are universal approximators. *Neural Networks* 2:359–366, 1989.

[42] Y. H. Pao and D. J. Sobajic. Combined use of unsupervised and supervised learning for dynamic security assessment. *IEEE Trans. Power Syst.* 7:878, 1992.

[43] S. J. Perantonis and P. J. G. Lisboa. Translation, rotation, and scale invariant pattern recognition by high-order neural networks and moment classifiers. *IEEE Trans. Neural Networks* 3:241–251, 1992.

[44] D. J. Sobajic, Y. H. Pao, and D. T. Lee. Autonomous adaptive synchronous machine control. *Internat. J. Elec. Power Energy* 14:166, 1992.

[45] S. I. Gass. *Linear Programming: Methods and Applications*. McGraw-Hill, New York, 1985.

[46] K. G. Murty. *Linear Programming*. John Wiley, New York, 1983.

[47] J. K. Kruschke. Improving generalization in back-propagation networks with distributed bottlenecks. In *Proceedings of the Joint International Conference on Neural Networks*, Vol. 1, pp. 443–447. Washington, DC, 1989.

[48] J. K. Kruschke. Creating local and distributed bottlenecks in hidden layers of back-propagation networks. In *Proceedings of the 1988 Connectionist Models Summer School* (D. Touretzky, G. Hinton, and T. Sejnowski, Eds.), pp. 120–126, 1988.

[49] R. N. Shepard. The analysis of proximities: Multidimensional scaling with an unknown distance function, I and II. *Psychometrika* 27:125–140, 219–246, 1962.

[50] J. K. Kruschke. Improving generalization in back-propagation networks with distributed bottlenecks. In *Proceedings of the International Joint Conference on Neural Networks*, Vol. II, pp. 163–168. Seattle, WA, 1991.

[51] J. Sietsma and R. J. F. Dow. Creating artificial neural networks that generalize. *Neural Networks* 4:67–79, 1991.

[52] J. Sietsma and R. J. F. Dow. Neural net pruning—Why and how?. In *Proceedings of the IEEE International Conference on Neural Networks*, Vol. 1, pp. 325–332. San Diego, CA, 1988.

[53] D. E. Rumelhart. Parallel distributed processing. Plenary Session, *IEEE International Conference on Neural Networks*, San Diego, CA, 1988.

[54] J. R. Quinlan. Induction of decision trees. *Mach. Learn.* 1:81–106, 1986.

[55] S. Ramachandran and L. Y. Pratt. Information measure based skeletonization. In *Advances on Neural Information Processing IV* (J. Moody, S. J. Hanson, and R. P. Lippmann, Eds.), pp. 1080–1087. Morgan Kaufmann, San Mateo, CA, 1992.

[56] O. M. Omidvar and C. L. Wilson. Optimization of neural network topology and information content using Boltzmann methods. In *Proceedings of the International Joint Conference on Neural Networks*, Vol. IV, pp. 594–599. Baltimore, MD, 1992.

[57] C. Molina and M. Niranjan. Pruning with replacement on limited resource allocating networks by F-projections. *Neural Computat.* 8:855–868, 1996.

[58] K.-H. Lee, H.-Y. Hwang, and D.-S. Cho. Determining the optimal number of hidden nodes and their corresponding input and output patterns in threshold logic network. In *Proceedings of the World Conference on Neural Networks*, Vol. 3, pp. 484–487. Portland, OR, 1993.

[59] S. Ridella, G. Speroni, P. Trebino, and R. Zunino. Pruning and rule extraction using class entropy. In *Proceedings of the IEEE International Conference on Neural Networks*, Vol. 1, pp. 250–256. San Francisco, CA, 1993.

[60] N. Dodd. Optimisation of network structure using genetic techniques. In *Proceedings of the International Joint Conference on Neural Networks*, Vol. 3, pp. 965–970. San Diego, CA, 1990.

[61] G. Miller, P. Todd, and S. Hedge. Designing neural networks using genetic algorithms. In *Proceedings of the Third International Conference on Genetic Algorithms*, Morgan Kaufmann, San Mateo, CA, 1989.

[62] D. Witley and C. Bogart. The evolution of connectivity: Pruning neural networks using genetic algorithms. In *Proceedings of the International Joint Conference on Neural Networks*, Vol. 1, pp. 134–137. Washington, DC, 1990.

[63] J. R. McDonnell and D. Waagen. Determining neural network hidden layer size using evolutionary programming. In *Proceedings of the World Conference on Neural Networks*, Vol. 3, pp. 564–567. Portland, OR, 1993.

Index

N

O

P

Q

R

S

T

U

W